NANO-GLASS CERAMICS

NANO-GLASS CERAMICS

Processing, Properties and Applications

VAHAK MARGHUSSIAN
School of Metallurgy and Materials Engineering
Iran University of Science and Technology (IUST)
Tehran, Iran

ELSEVIER

AMSTERDAM • BOSTON • HEIDELBERG • LONDON • NEW YORK • OXFORD
PARIS • SAN DIEGO • SAN FRANCISCO • SINGAPORE • SYDNEY • TOKYO
William Andrew is an imprint of Elsevier

William Andrew is an imprint of Elsevier
The Boulevard, Langford Lane, Kidlington, Oxford, OX5 1GB, UK
225 Wyman Street, Waltham, MA 02451, USA

ISBN: 978-0-323-35386-1

British Library Cataloguing-in-Publication Data
A catalogue record for this book is available from the British Library

Library of Congress Cataloging-in-Publication Data
A catalog record for this book is available from the Library of Congress

For Information on all William Andrew publications
visit our website at http://store.elsevier.com/

Typeset by MPS Limited, Chennai, India
www.adi-mps.com

CONTENTS

PREFACE

The nano-structured materials today are the cutting edge of modern advanced materials development. Among these materials, nano-structured glass ceramics, or **nano-glass ceramics**, possessing uniformly dispersed crystals <100 nm in size, have become established in various technical and domestic applications and also offer great promise for many potential new applications. These applications are widely varied, including thermally resistant cookware and radiant stove tops, telescope mirror blanks, radoms, and biomedical materials. They are also very promising candidates for many optical, optoelectronic, and electromagnetic applications, such as lasers, optical amplifiers, solar cells, information storage systems, and ferrofluid technology.

It seems that despite the great potential of the nano-glass ceramics, the importance of these materials has not appreciated enough. No book has been published yet, dealing with these materials.

Hence, this book has been written to fill this gap. It provides detailed coverage of the synthesis and processing methods of nano-glass ceramic materials, their selected properties, and the present and future applications.

In writing this book, while both the theoretical and practical aspects of the subject have been considered, the emphasis has been placed upon the experimental and practical aspects of the subject. The methods of investigation adopted in various stages of synthesis and processing of glass ceramics by researchers and the current state of art in developing nano-structured glass ceramics have been discussed in detail.

Throughout the discussions, wherever possible, the material scientist's point of view regarding the relationship between properties and the parameters of synthesis as well as the microstructure and properties of the resulting glass ceramics has been adopted.

In the discussion of the application of nano-glass ceramics, considerations were given to both the fields of application in which the materials are firmly established as well as the fields where great promise exists for their future exploitation.

While it has been tried to avoid deep scientific discussions, the basic principles of glass science, especially the subject of nucleation and crystallization in glass, have been utilized in many discussions. Hence, prior knowledge of the glass science, at an introductory level, would help better understanding the subjects discussed here. For the readers with little or no prior knowledge in glass science, the existing textbooks in glass science may be useful in providing the necessary background of the subject.

This book appeals to the graduate students of materials science and engineering, ceramics, and nano-science and nanotechnology courses. The application-oriented perspective of the book also makes it suitable for engineers and especially younger researchers who are interested in nano-structured materials. The newcomers of the research field of nano-glass ceramics may find the numerous practical examples and the details of experimental methods represented in the book useful in entering the new field.

INTRODUCTION

By definition, glass ceramics are engineering materials that are first prepared as glass and then are subjected to a carefully controlled heat-treatment procedure in order to convert them to fully or partially crystalline materials. With selecting appropriate glass compositions and controlling crystal nucleation and growth processes in the glass, glass ceramics with tailor-made properties can be fabricated. The development of practical glass ceramics is relatively recent. As far as we know, Re'aumur, a French chemist, was the first person to produce practical glass ceramics. He showed that if glass bottles were heat-treated in a bed of gypsum and sand mixture for several days, they would be converted to a porcelain-like opaque object. Although Re'aumur succeeded in converting glass to a "glass ceramic" material, unfortunately the low strength and deformation of the new product was frustrating for him. Some 200 years after Re'aumur's experiment, S.D. Stookey of Corning Glass Works (Corning, NY) in the United States made an important discovery when one of his photosensitive glass specimens, that was to be opacified by irradiation, was accidentally subjected to a higher than usual temperature during heat treatment; it was converted to an opaque ceramic material, the first true glass ceramic, exhibiting much higher mechanical strength than the original glass, with no distortion. This experiment opened the way for the development of other types of nucleation agents that did not require irradiation of the glass and eventually led to the commercialization of the first glass ceramic product "pyroceram" by Corning Glass Works.

Tremendous developments have occurred in the field of glass-ceramic research and technology since Stookey's pioneering work five decades ago. Today glass ceramics have become established in a wide variety of technical and domestic applications ranging from thermally resistant cookware and radiant stove tops to telescope mirror blanks and radoms and from tiles and artificial stones to biomedical materials and, more recently, as quite interesting potential candidates in some electromagnetic and optoelectronic applications. This can be explained by the nature of the glass ceramic production route that is based on the easier, faster, and less expensive technology of glass manufacturing method. In addition to the ease of fabrication process and higher automation, glass ceramics also possess a uniformity of microstructure and subsequent reproducibility of properties that result from the homogeneity of the starting glass. Usually the conversion of glass into glass ceramic involves only minor overall volume changes with zero or little porosity which is essential for some applications. Furthermore, glass ceramics provide a wide range of physical and chemical properties that often are not readily obtained in glasses or ceramics. The method of glass ceramic fabrication has its own limitations. It is limited to the materials that relatively easily and at reasonable temperatures can be melted and formed as glass and in subsequent heat treatment in a controllable manner converted into glass ceramics.

Among various types of glass ceramics, **nano-glass ceramics**, containing tiny crystals of <100 nm, are currently utilized in many products and offer great promise for many potential new applications.

Some nano-structured transparent glass ceramics possessing high thermal stability and thermal shock resistance are currently used in high optical applications, such as telescope mirror blanks, cookware, radiant stove tops, and fire doors. Other types of transparent glass ceramics, based on mullite, are considered as host media for luminescent transition metal ions, especially Cr^{3+}, for potential uses in certain fields, such as solar concentrators and tunable lasers. Applications have also been anticipated for nano-structured spinel glass ceramics as optical amplifiers and passive Q-switchers in lasers.

Oxyfluoride glass ceramics have potential applications in the field of up-conversion lasers, fiber amplifiers, and the optical devices for three-dimensional displays.

Nano-structured, transparent glass ceramics containing alkali niobate ferroelectric crystalline phases, because of their large electro-optic effect and strong second harmonic generating ability have great promise in many applications in the fabrication of optical waveguides, modulators, switches, and optical circuits.

Nano-crystalline glass ceramics with high elastic modulus and moderate strength and toughness, based on spinel and enstatite phase assemblages, have been developed for potential use as magnetic disk substrates in computer hard drives. The preparation of nano-crystalline glass ceramics for biomedical application has also attracted considerable attention in recent years as coatings, orthopedic devices, dentistry, as well as magnetic nano-glass ceramics for therapeutic (e.g., hyperthermia and drug-targeting) and diagnostic (e.g., magnetic resonance imaging) applications.

This book covers the details of processing, properties, and application of the most important types of nano-glass ceramics, as known at the present time.

1

GLASS CRYSTALLIZATION

CHAPTER OUTLINE

Glass ceramics are materials composed of one or more glass and crystal phases. By preparing base glasses with appropriate compositions and by controlling crystal nucleation and growth in these glasses, glass ceramics with tailor-made properties can be fabricated. The key to preparing these types of materials, especially nano-glass ceramics, is the control of the nucleation process. Nucleation is followed by one or more higher-temperature treatments, usually called growth process, to promote crystallization of

V.K. Marghussian, Nano-Glass Ceramics, ISBN: 978-0-323-35386-1.

DOI: http://dx.doi.org/10.1016/B978-0-323-35386-1.00001-3

the primary phase or phases and development of the desired microstructure. As stated above, the first step in the glass crystallization process which is critical in preparing nano-glass ceramics is the nucleation process.

1.1 Nucleation in Glass

Although it is not the intention in the present book to cover comprehensively the details of all aspects of crystal nucleation in glass systems; however considering the importance of the nucleation process in nano-glass ceramics fabrication and properties, it is felt that the presentation of a brief discussion on nucleation is essential here.

In the last three decades or so, numerous publications including many excellent books, review papers, and proceedings of international conferences on crystallization of glass, in general, and nucleation, in particular, have appeared in the literature. Some of the most important ones that can be recommended for further reading are: McMillan (1979), Strnad (1986), Höland and Beall (2002), James (1982, 1985, 1989), Beall (1993), Zanotto (1994), and Uhlmann and Uhlmann (1993).

In a homogeneous liquid or glass, crystallization starts with nucleation, i.e., the formation of crystal-like fluctuations/clusters that are able to grow further. Owing to the excess free energy of the crystal/liquid interfacial layer, the free energy of the smallest fluctuations is positive and has a maximum as a function of size. The clusters, smaller than the corresponding critical size, decay with a greater probability than the larger ones, which grow and reach a macroscopic size. The free-energy maximum represents the thermodynamic barrier the system has to pass before forming the crystalline phase.

The appearance of crystalline fluctuations may be catalyzed by foreign particles, container walls, and oxide layers on a surface, a process known as "heterogeneous nucleation."

In the preparation of glass articles, where the formation of nuclei and their subsequent growth to crystals must be avoided, knowing the nucleation rate of crystalline solid phases in glass-forming melts at different temperatures has great practical importance. On the other hand, in preparation of glass ceramics, and especially nano-glass ceramics, which is to be fabricated with tailor-made properties, only achievable by the strict control of microstructure, the knowledge on nucleation mechanism and kinetics is even more critical.

1.1.1 Homogeneous Nucleation

1.1.1.1 Theoretical Background

Homogeneous nucleation supposes the same probability of critical nucleus formation in any given volume or surface element of the system under study.

Tammann (1925) in his classical studies on the crystallization of supercooled liquids showed that below the equilibrium melting temperature, there exists a metastable temperature interval in which nuclei do not form at a detectable rate. In this zone, however, crystals can grow if nuclei are provided, i.e., if the melt is seeded.

At temperatures below this region, the crystallization process is controlled by the rate of formation of nuclei and the crystal growth rate. Melts exhibiting rapid increase in viscosity during cooling show maxima in nucleation and crystal growth rates. This is because of the fact that at the lower temperatures, the high viscosity hinders the atomic rearrangements and diffusion processes which are necessary for nucleation and crystal growth (McMillan, 1979). These melts normally show high glass-forming tendencies. Curves representing nucleation and crystal growth rates for viscous melts have the forms shown in Figure 1.1.

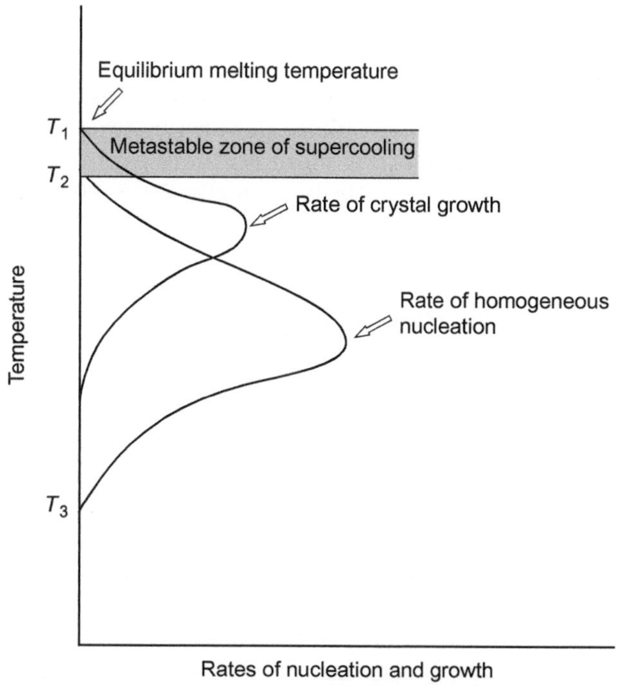

Figure 1.1 Rates of homogeneous nucleation and crystal growth in a viscous liquid.

According to this figure, in order to produce the largest possible number of small crystals, e.g., in nano-glass ceramics, nucleation should occur at or near to the temperature at which the maximum nucleation rate occurs.

It could be noted that there is also a temperature, T_3, below which the homogeneous nucleation rate is zero due to the high viscosity of the melt (Figure 1.1).

In the "classical nucleation theory" (CNT), which is the oldest and the best-studied theory used for calculation of the nucleation kinetics, it is assumed that the nuclei formation and growth take place via single molecule attachment/detachment.

According to the above theory, the rate of homogenous steady-state nucleation (I) at an absolute temperature (T) in a one-component supercooled liquid can be expressed by Eq. (1.1) (Christian, 1975).

$$I = A\exp\left[-\frac{W^* + \Delta G_D}{kT}\right] \tag{1.1}$$

where W^* is the thermodynamic free-energy barrier to nucleation and ΔG_D is the kinetic barrier, k is the Boltzmann constant, and A is a preexponential factor, given to a good approximation by:

$$A = n_v\left(\frac{kT}{h}\right) \tag{1.2}$$

where n_v is the number of atoms or, strictly, "formula units" of the crystallizing component phase per unit volume of the liquid and h is Planck's constant. A more accurate expression for A is given by Christian (1975). It is typically 10^{41}–$10^{42}/\text{m}^3/\text{s}$ and may be taken as effectively constant over the temperature range of nucleation measurements.

The thermodynamic barrier to nucleation, W^* (or ΔG^*), can be calculated in the following way.

The net change in energy, W, for the formation of a spherical nucleus of radius r is given by:

$$W = 4/3\pi r^3\Delta G_v + 4\pi r^2\sigma \tag{1.3}$$

In the first energy term ($4/3\pi r^3\Delta G_v$), ΔG_v, is the change in the free energy in transforming a unit volume of the liquid to a unit volume of the crystal, known as the free energy of crystallization (which is negative below T_m, providing the driving force for crystallization). The second term is given by $4\pi r^2\sigma$, where σ is the surface energy of the interface per unit area. This term is positive and acts as an obstacle for crystallization. Obviously, W is a function of r and the maximum value of it, W^* (thermodynamic barrier to nucleation) can be obtained from the W versus r plot.

The radius of the nucleus at the maximum point, r^*, defined as the critical-sized nucleus, can be determined by differentiating W with respect to r and setting equal to zero:

$$dW/dr = 4\pi r^2 \cdot \Delta G_v + 8\pi r\sigma = 0 \quad \text{when} \quad r = r^*$$

Hence,

$$r^* = -2\sigma/\Delta G_v \tag{1.4}$$

Substituting Eq. (1.4) into Eq. (1.3) yields the maximum value, W^* as

$$W^* = 16\pi\sigma^3/3(\Delta G_v)^2 \tag{1.5}$$

Crystal embryos below the critical radius (r^*) are unstable and will tend to dissolve. Embryos with radii exceeding r^* become stable nuclei and continue to grow.

The critical radius increases with rise in temperature, becoming infinite at the melting point.

The free-energy driving force ΔG (per mole) can be calculated if accurate thermodynamic data, including ΔC_p, the difference in specific heats at constant pressure between the crystal and liquid phases, are available as a function of temperature (James, 1985). An approximate expression, only applicable for small undercoolings below the melting point T_m and in which ΔC_p is taken as zero, is

$$\Delta G = -\frac{\Delta H_f(T_m - T)}{T_m} \tag{1.6}$$

where ΔH_f is the heat of fusion per mole and $\Delta G = \Delta G/v_m$ (v_m is the molar volume).

On the other hand, the kinetic barrier to nucleation, ΔG_D, can be expressed in terms of an effective diffusion coefficient D given by

$$D = \left(\frac{kT\lambda^2}{h}\right)\exp\left(-\frac{-\Delta G_D}{kT}\right) \tag{1.7}$$

where λ is a quantity of the order of atomic dimensions ("jump distance"). Various authors have related D to the viscosity of the liquid η by the Stokes–Einstein relation

$$D = \frac{kT}{3\pi\lambda\eta} \tag{1.8}$$

Alternatively, ΔG_D can be obtained from crystal growth measurements at similar temperatures, assuming the kinetic barriers for nucleation and for growth are the same (Höland and Beal, 2002).

The steady-state nucleation rate, as discussed above, is concerned with the linear part of the curve relating the number of nuclei at a fixed temperature with time (shown schematically in Figure 1.2).

Obviously, there is an initial period during which the nucleation rate is either zero or very small and after prolonged nucleation a saturation effect is reached after which no further nucleation occurs. Before the occurrence of saturation, the nucleation rate can be expressed by

$$I = I_o \exp\left(\frac{\tau}{t}\right) \tag{1.9}$$

where τ is the incubation time and I_o is the steady-state nucleation rate.

The calculation of nucleation rates in glasses using the above equations would require accurate values as a function of temperature for ΔG, ΔG_D, and σ.

Unfortunately, at present accurate data for the aforementioned values exist only for a limited number of glasses. However, even in these cases, the theoretical values should be evaluated by comparing them with experimentally obtained values. Fortunately during the last three to four decades, many experimental methods have been developed for determination of crystal nucleation rates in glasses. The most important methods will be discussed later.

1.1.1.2 Experimental Studies of Homogeneous Nucleation

James (1989), Zanotto (1987), and Zanotto and Weinberg (1989) based on numerous experimental nucleation rate data for several

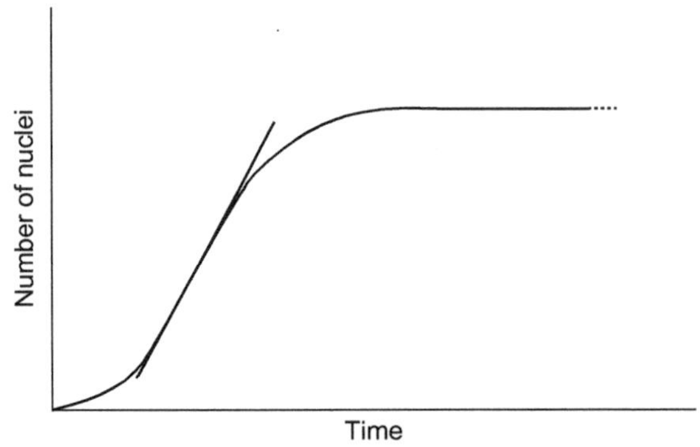

Figure 1.2 Number of nuclei per unit volume versus time at a fixed temperature.

silicate glasses, suggested that glasses having a reduced glass transition temperature, $T_{gr} = T_g/T_m$, higher than ~0.58–0.60, display only surface crystallization; while glasses possessing values T_{gr}<0.58–0.60, show volume (homogeneous) nucleation. However, since at temperatures $T<T_m$, the nucleation rate is always positive, the absence of volume nucleation for glasses having T_{gr}>0.60 merely indicates undetectable nucleation on laboratory time/size scales. When the glass transition range is reached at relatively high temperatures, T_{gr}>0.60, the work of critical cluster formation is still too large to allow the occurrence of measurable internal crystallization (since the critical cluster size and the work of critical cluster formation decreases with the difference $[T_m-T]$).

Since the work of critical cluster formation and the viscosity may be lower than bulk values on or close to interfaces, the surface crystallization is more commonly observed than internal crystallization (Fokin et al., 2003).

The transition from glasses demonstrating only surface crystallization (T_{gr}>0.60) to glasses exhibiting volume nucleation (T_{gr}<0.58) may be qualitatively explained by an increase in nucleation rate with decreasing T_{gr}.

The increase of the nucleation rate with decreasing T_{gr} was in fact demonstrated by Deubener (2000) for eight silicate glass systems: Li_2O-SiO_2, Na_2O-SiO_2, $BaO-SiO_2$, $CaO-SiO_2$, $Li_2O-Na_2O-SiO_2$, $Na_2O-CaO-SiO_2$, $Li_2O-BaO-SiO_2$, and $MgO-Al_2O_3-SiO_2$ and comprise both stoichiometric and nonstoichiometric compositions. For nonstoichiometric compositions, the liquidus temperatures, T_l, were used to calculate the reduced glass transition temperature.

In compositions belonging to the above silicate systems, in a relatively narrow range of T_{gr} (ranging from 0.47 to 0.58), the nucleation rate drops typically by about ~17 orders of magnitude!

Fokin et al. (2003) collected the available data for a number of silicate glasses that exhibit internal nucleation, extending the above list published by Deubener. Their work provided additional experimental evidence for the above-mentioned correlation between nucleation rate and T_{gr}. The existence of an analogous correlation between the temperature of maximum nucleation rate (T_{max}), the induction time at T_{max}, and the reduced glass transition temperature were also demonstrated.

The experimental results on internal homogeneous crystal nucleation in silicate glasses obtained in the last four decades were also reviewed (Fokin et al., 2006).

In this study, particular attention was paid to the analysis of the critical nuclei properties because of its decisive role in the determination of nucleation kinetics.

The authors especially discussed the very important problems that arise in the quantitative description of nucleation rates when using the CNT, e.g., the marked discrepancy observed between calculated and measured nucleation rates; the high value of the crystal nuclei/melt surface energies in comparison with the expected values; the increase of surface energy with increasing temperature.

The authors presented direct experimental evidence for the compositional changes occurring during the formation of critical nuclei and their growth from critical to macro-sizes and showed that during this process, both the thermodynamic driving force for nucleation and the critical nuclei/liquid interfacial energy decrease, in comparison with the respective properties of the macro-phase. On the basis of these findings, it was concluded that "in contrast to Gibbs' description of heterogeneous systems underlying CNT, the bulk thermodynamic properties of the critical nuclei generally differ from those of the corresponding macro-phase, resulting simultaneously in significant differences of the surface properties as compared with the respective parameters of the planar interface."

Finally, the authors concluded that, "despite the significant progress in understanding crystal nucleation in glasses in the past four decades, many problems still exist and this is likely to remain a highly interesting subject for both fundamental and applied research for a long time" (Fokin et al., 2006).

1.1.2 Heterogeneous Nucleation

1.1.2.1 Theoretical Background

1.1.2.1.1 General Considerations

Since the initiation of homogeneous nucleation in glass ceramics, as discussed above, is restricted to a quite limited number of glass systems, in most practical cases the fabrication of useful glass ceramics necessitates the development of volume (bulk) heterogeneous nucleation in glasses by the incorporation of some special nucleation agents into the base glass that act as catalysts for nucleation.

For nucleation on catalytic substrates, the classical theory considers the formation of a spherical-cap nucleus, making a contact angle θ with the substrate. The critical work of nuclei formation for heterogeneous nucleation, $\Delta G_{\mathrm{h}}^{*}$, is expressed as (James, 1982):

$$\Delta G_{\mathrm{h}}^{*} = \Delta G^{*} f(\theta), \text{ where } f(\theta) = (2 + \cos\theta)(1 - \cos\theta)^2/4 \qquad (1.10)$$

Based on Eq. (1.10), the nucleation barrier depends on θ and is very small if the nucleant substrate is completely wetted by the nucleus and the contact angle θ is close to 0. Heterogeneous nucleation is particularly effective if there is epitaxy between the nucleus and substrate. The criterion which determines whether epitaxy occurs is the value of disregistry between similar low-index planes for the foreign crystal and the precipitated crystals. It is generally agreed that a maximum disregistry of 15% for spacing in index planes of the catalyst and the nucleated phase can be tolerated in effective nucleation catalysis. With the increasing mismatch of crystal lattices, greater degrees of supercooling should exist before nucleation takes place.

The effect of preexisting surfaces in a supersaturated solution or a supercooled melt is to reduce the value of ΔG^*, the free energy for homogeneous nucleation, by decreasing the interface energy, σ which appears in Eqs. (1.3) and (1.4). The important feature of heterogeneous nucleation is that the interfacial tension between the heterogeneity and the nucleated phase must be low. Therefore, the influence of the catalyzing surface is determined by the contact angle θ at the substrate–melt–precipitate junction. The activation energy of heterogeneous nucleation can be expressed as

$$\sigma_{HL} = \sigma_{SH} + \sigma_{SL} \cos\theta \tag{1.11}$$

where σ_{HL} is the interfacial free energy between the heterogeneity and the melt, σ_{SH} is the interfacial free energy between the heterogeneity and primary crystal phase, and σ_{SL} is the interfacial free energy between the crystal phase and the melt.

For any contact angle (θ) less than 180, the free-energy barrier is less for nucleus formation on the surface of the heterogeneity than for homogeneous nucleation. As a result, heterogeneous nucleation will occur wherever possible in preference to homogeneous nucleation.

The presence of heterogeneities which have the effect of reducing ΔG^* will also reduce the incubation time τ. The incubation time, τ_H, for heterogeneous nucleation is given by

$$\tau_H = \tau f_1(\theta) \tag{1.12}$$

where τ is the incubation time for the "pure" glass, and $f_1(\theta)$ is given by (James, 1982):

$$f_1(\theta) = 1 - (\cos\theta)/2 - (\cos^2\theta)/2 \tag{1.13}$$

Hence, with increasing the efficiency of nucleating particles, the values of θ, $f_1(\theta)$ and the incubation time would be reduced.

1.1.2.1.2 Epitaxy in Heterogeneous Nucleation of Glasses

Epitaxy is a well-known phenomenon of crystallization. It is common in minerals which often exhibit special orientation relationships of their lattices. The epitaxy is naturally preferred because the occurrence of crystal nucleation at certain relative orientations and particular interface orientations minimize the lattice mismatch and facilitate the nucleation process in minerals (Habelitz et al., 1999). Heteroepitaxy is a special case of heterogeneous nucleation in which a distinct crystallographic relationship exists between the orientations of substrate and the material which is deposited on it.

It was originally supposed that epitaxy was always the result of a geometric match of the dimensions and shape of the substrate and deposit surface acts as a minimum-energy template to align the atoms of depositing material into a unique crystallographic orientation. However, this assumption is not always true and the term "epitaxy" includes a wide range of phenomena involving a variety of growth mechanisms (Sarma et al., 1997).

As Habelitz et al. (1999) stated, several authors suggested that not the lattice parameters, but the real structure of the substrate crystal faces is the origin of epitaxial crystallization (graphoepitaxy). For example, the occurrence of steps, kinks, and dislocations may strongly facilitate the nucleation of another crystalline phase as reported by Stoneham and Harding (1998).

Despite the critical role played by epitaxial considerations in the nucleation and growth of crystals in glass, relatively little information has been published on this subject. Headley and Loehman (1984) were the first investigators who reported the direct observation of epitaxial nucleation and growth in a $Li_2O{-}Al_2O_3{-}SiO_2$ glass in the presence of intentionally added nucleating agents.

They showed that Li_2SiO_3, $Li_2Si_2O_5$, and cristobalite crystallized epitaxially on lithium orthophosphate (Li_3PO_4) crystallites, which themselves were formed upon the addition of P_2O_5 nucleating agent. Although their results were specific to the particular composition chosen for investigation, they believed that the mechanism could generally be applicable for the crystallization of lithium aluminum silicate glasses that contain P_2O_5 nucleating agent. Commenting on the matter of selecting a nucleating agent that will form crystalline nuclei suitable for epitaxial overgrowth of a desired phase, the authors stressed that a simple consideration based on similarity of lattice spacing between low-index planes of prospective host and guest lattices is insufficient and can be misleading. According to these authors, e.g., the previous suggestion concerning the crystallization mechanism of $Li_2Si_2O_5$ in lithium silicate glasses, containing P_2O_5, possibly was not correct. Previously it was supposed that $Li_2Si_2O_5$ was possibly crystallized epitaxially on preformed Li_3PO_4 crystals because of the existing similarity in spacing between {111} planes of $Li_2Si_2O_5$ and either (111) or (210) planes of Li_3PO_4. Calculated misfits for either alignment showed only a small fraction of a percent. However, stereographic analysis carried out by the authors using the two orientation relations for $Li_2Si_2O_5$ growing epitaxially on Li_3PO_4 showed that the planar alignments suggested previously did not occur (Figure 1.3A and B) The orientation relations derived from the diffraction patterns are given in Figure 1.3B along with calculated misfits of aligned atomic planes and the Li_3PO_4 crystal face on which epitaxy occurs. As was shown in Figure 1.3B, $Li_2Si_2O_5$ is formed with either of the orientations listed.

Therefore, it was suggested that epitaxy is more likely controlled by proper matching of fundamental structural units between the two lattices, such as alignment of edges and/or faces of SiO_4 and PO_4 tetrahedra.

Since the approval of epitaxial crystallization in glass ceramics is difficult due to the random distribution of crystal orientations, Habelitz et al. (1999) suggested the utilization of oriented crystals obtained by the generation of textures, as an appropriate means to prove epitaxial growth. They found that upon extrusion of $SiO_2-B_2O_3-Al_2O_3-MgO-Na_2O-K_2O-F$ glasses, the solid solution of mullite and aluminum borate crystals became aligned with their elongated axes parallel to the extrusion direction. During subsequent heat treatments, fluorophlogopite platelets crystallized in part epitaxially on these star shape solid solution crystals.

Utilizing the texture goniometry and electron microscopy, it was also shown that the mica basal planes were orientated either parallel or perpendicular to the extrusion direction. Using selected-area electron diffraction in a transmission electron microscope (TEM), epitaxial orientation relationships between the lattices of the fluorophlogopite and the solid solution crystal were determined.

In samples cut perpendicular to the direction of extrusion star-like, solid solution crystals, owing to their excellent orientation, can be easily tilted into their [001] zone axes within the available tilting range of the microscope.

This was performed in several cases, and as shown in the TEM overview in Figure 1.4A and illustrated by the lattice-plane image depicted in Figure 1.4B, phlogopite (001) planes were

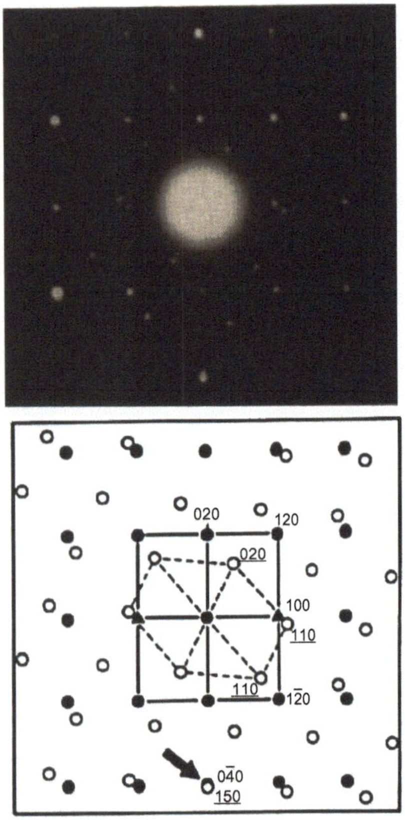

● hkl — lithium phosphate

○ <u>hkl</u> — lithium disilicate

▲ — double diffraction, lithium phosphate

(A)

Crystallographic Orientation Relations and Lattice Misfits for Epitaxial Growth of Cristobalite, Lithium Metasilicate, and Lithium Disilicate on Lithium Orthophosphate*

Affinitive lattices	Lithium orthophosphate face	Lattice periods (nm)†	Misfit (%)‡
	Cristobalite on	Li_3PO_4	
$(111)_{CR} \parallel (120)_{LP}$ $[\bar{1}10]_{CR} \parallel [001]_{LP}$	{120} only	$d(111)_{CR} = 0.4134$ $d(120)_{LP} = 0.3978$	3.8
		and	
		$d(\bar{1}10)_{CR} = 0.5063$ $d(001)_{LP} = 0.4923$	2.8
	Li metasilicate on	Li_3PO_4	
$(010)_{LS} \parallel (010)_{LP}$ $[001]_{LS} \parallel [001]_{LP}$	{120} only	$d(010)_{LS} = 0.5395$ $d(020)_{LP} = 0.5237$	2.9
		and	
		$d(001)_{LS} = 0.4675$ $d(001)_{LP} = 0.4923$	–5.3
	Li disilicate on Orientation I	Li_3PO_4	
$(1\bar{5}0)_{LS_2} \parallel (1\bar{1}0)_{LP}$ $[001]_{LS_2} \parallel [001]_{LP}$	{120}	$d(1\bar{5}0)_{LS_2} = 0.2619$ $d(2\bar{2}0)_{LP} = 0.2641$	–0.8
		and	
		$d(001)_{LS_2} = 0.4790$ $d(001)_{LP} = 0.4923$	–2.8
	Orientation II		
$(1\bar{5}0)_{LS_2} \parallel (0\bar{1}0)_{LP}$ $[001]_{LS_2} \parallel [001]_{LP}$	(010)	$d(1\bar{5}0)_{LS_2} = 0.2619$ $d(0\bar{4}0)_{LP} = 0.2619$	0
		and	
		$d(001)_{LS_2} = 0.4790$ $d(001)_{LP} = 0.4923$	–2.8

*CR, cristobalite; LS, lithium metasilicate; LS_2, lithium disilicate; LP, lithium orthophosphate, †Derived from unit-cell parameters given in Ref. 10, ‡Misfit (%) = 100(*b*–*a*)/*a*, where *a* = lattice spacing of substrate crystal and *b* = lattice spacing of overgrowth crystal.

(B)

Figure 1.3 (A) Electron diffraction pattern and indexed schematic of lithium disilicate crystal growing epitaxially in orientation II (shown in Figure 1.3B) on lithium orthophosphate crystal. Beam direction parallel to [001] zone axis in lithium orthophosphate and [001] zone axis in lithium disilicate. Arrows in the schematic point to overlapping reflections that identify parallel planes in the host and guest crystal lattices. (B) Crystallographic orientation relations and lattice misfits for epitaxial growth of cristobalite, lithium metasilicate, and lithium disilicate on lithium orthophosphate. (After Headley and Loehman, 1984. Reproduced with permission of the American Ceramic Society.)

frequently found to be perpendicular to the $[1\bar{1}0]$ direction of the aluminum borate-mullite solid solution. Hence, the introduction of texture tremendously facilitated the study of epitaxial relationships of the crystalline phases.

In the authors' opinion, the combination of texture formation and epitaxial growth would open new ways of microstructural design in extruded glass ceramics.

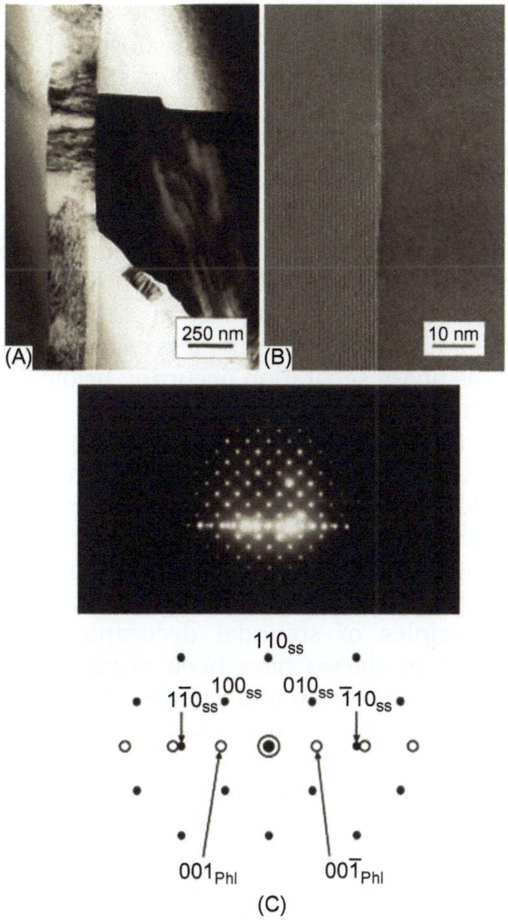

Figure 1.4 (A) TEM overview exhibiting a phlogopite plate epitaxially grown on a $(1\bar{1}0)$ facet of a star-like ss-crystal. (The latter was tilted to a [001] zone-axis orientation.) (B) Lattice-plane image of the interface of ss-crystal (ss, exact [001] orientation)-fluorophlogopite (*c*-axis perpendicular to the boundary plane). (C) Selected-area diffraction pattern of the interfacial region together with an indexing scheme. (After Habelitz et al., 1999. Reproduced with permission of the Acta Metallurgica Inc.)

Stoneham and Harding (1998) commented on the interesting issue of epitaxy involving noncrystalline solids. In the authors' opinion, although in this case, the idea of epitaxy could not be applied in a simple form, there are, however, constraints at the actual interface that can affect the process of nucleation. For example, in the case of two crystalline media separated by an amorphous layer, sometimes it seems that an epitaxial relationship exists between the two crystalline components. The authors tried to find an answer for the question "how can this happen, when the amorphous layer is about 10 nm thick?" Obviously, the conventional mechanisms are unlikely to be operative in this case.

According to the authors, this effect may possibly be explained by a phenomenon known as graphoepitaxy. The steps on the substrate may carry crystallographic information, i.e., the amorphous material could have some memory with ridges corresponding to those on the substrate. This probably could influence the epitaxy of the final crystalline layer.

1.1.2.1.3 The Effect of Glass-in-Glass Phase Separation on Heterogeneous Nucleation

Many glass-forming systems may be separated into two (or more) glassy phases during cooling to room temperature, or their subsequent heating to high temperatures.

If this phase separation takes place in the melt, at a temperature above the liquidus temperature, it is called the stable immiscibility, whereas phase separation occurring below the liquidus is termed the metastable immiscibility, or glass-in-glass phase separation. The metastable immiscibility is generally far more probable than the stable immiscibility in glass ceramics. A general review of liquid immiscibility in glass-forming systems has been given by James (1975).

There are two possible routes for phase separation: a nucleation and growth process or a spinodal decomposition process. The basic principles of spinodal decomposition process in phase separation of glasses have been discussed by Cahn and coworkers (Cahn, 1961; Cahn and Charles, 1965).

The glass-in-glass phase separation often plays a decisive role in crystal nucleation in glass. The occurrence of phase separation prior to crystal nucleation could have a significant effect upon the kinetics of the subsequent crystal nucleation processes in glass. Various ways in which phase separation may influence the kinetics of crystal nucleation have been dealt with by many investigators.

The most significant effects of the liquid–liquid phase separation on the crystal nucleation in glass-forming systems have been summarized by Uhlmann (1970) as follows:

i. In the absence of necessary driving force for the crystal nucleation in a homogeneous liquid, the occurrence of phase separation can provide it.

ii. The separation results in emergence of interfaces between the phases which can provide preferred sites for the nucleation of crystalline phases because of the associated surface energy surplus.

iii. The separation can result in one of the liquid phases having relatively higher atomic mobility at high undercoolings than the parent homogeneous phase.

iv. The interfacial regions between the separated phases may be enriched in some component, providing a locally higher driving force or mobility for nucleation.

According to the author, the third possibility seems to represent the most likely origin of the nucleation catalysis.

Since the principal difference between liquids in their ability either to form glasses or to follow the path of crystallization lies in their relative mobility, hence the provision of relatively high molecular mobility in a region of sizeable undercooling can be considered as the decisive role of phase separation in promoting the nucleation process.

Though the nucleation within the separated phase may sometimes be catalyzed by heterogeneities, but the occurrence of a region of relatively high molecular mobility would still be important (Uhlmann, 1970).

James (1982) has suggested that, in addition to the aforementioned effects proposed by Uhlmann, in some situations, e.g., for glass-ceramic systems in which the nucleation process was carried out at large undercoolings, the phase separation may enhance the nucleation rate by increasing the driving force of crystallization (thermodynamic effect).

The latter author also noted that the crystal–liquid interfacial energy (σ) could also be lower for one of the separated glass phases than for the parent glass, which also could result in an increase in the crystal nucleation rate.

It has further been suggested that a less soluble component might crystallize out at the interface and promote heterogeneous nucleation in the bulk of the glass (Scholes, 1970). Tomozawa (1972) suggested an alternative situation in which the preferential nucleation may occur in the diffusion zones, around liquid phase droplets.

1.1.2.2 Experimental Studies of Heterogeneous Nucleation

The process of heterogeneous nucleation of crystalline phases in glass is the most complicated process occurring in glassy materials in their conversion into glass ceramics. There are a number of diverse mechanisms involved in this type of nucleation. In the last three to four decades, numerous publications have appeared in the literature dealing with various aspects of the heterogeneous nucleation in glass. These include the mechanism of nucleation, the role of the nucleating agents, determination of the most effective agents, the optimum quantity of nucleants to be used, determination of the optimum temperature and time of nucleation process, and the kinetics of nucleation.

Here, some of the most important aspects of the topic from the practical point of view are discussed.

1.1.2.2.1 Heterogeneous Nucleation in the Presence of Phase Separation in Glass

A classical example for the effect of glass-in-glass phase separation on crystal nucleation in glasses is the $MgO-Al_2O_3-SiO_2$ (MAS) glass system, containing TiO_2 as a nucleation agent to produce cordierite glass ceramics. On heat treatment of these glasses, first a fine-scale glass-in-glass phase separation occurs. Next, the primary crystal nuclei, usually Mg_2TiO_5, are formed within the droplets of one of the separated glass phases, rich in TiO_2 and MgO. These nuclei then initiate heterogeneous nucleation of a metastable phase, β-quartz solid solution that at elevated temperatures may be transformed to the stable cordierite phase (Devekey and Majumdar, 1975).

More recently, Marghussian et al. (2009) suggested the occurrence of a similar, though less complicated nucleation path, in cordierite glass ceramics in the presence of V_2O_5. According to these authors, if V_2O_5 assumes a network-forming role in silicate glasses, the positive charge of V^{5+} as compared with neighboring Si^{4+} ions may induce an instability in the glass structure. In this condition, V^{5+} which is a cation of relatively high ionic field strength value ($z/a^2 = 1.40$) triggered a marked tendency for separation and greatly promotes the occurrence of microphase separation in glass.

Therefore, in V_2O_5-containing cordierite glasses, in the same way as TiO_2-containing glasses, after the occurrence of a metastable glass-in-glass phase separation, and formation of V_2O_5-rich droplets, a vanadate phase ($Mg_2V_2O_5$) is formed within the droplets (which were also enriched in MgO and Al_2O_3), catalyzing the crystallization of α-cordierite.

In some less complicated cases, the nuclei of the final crystalline phase may directly be formed within one of the separated glass phases.

For example, Kord et al. (2009a) reported that the calcium–titanium phosphate crystals were initiated from the phase separated regions of a $CaO-TiO_2-P_2O_5$ glass (Figure 1.5). The most critical step in the preparation of these glass ceramics is the phase separation process. The occurrence of phase separation by the spinodal decomposition mechanism in the glass specimen provides the required conditions for the formation of interconnected crystalline collections of the main phases, $Ca_3(PO_4)_2$ and $CaTi_4(PO_4)_6$ in the phase-separated regions of glasses, enriched in CaO and TiO_2, respectively. The $Ca_3(PO_4)_2$ phase is then leached out, leaving behind a porous skeleton of $CaTi_4(PO_4)_6$ (Kord et al., 2009a).

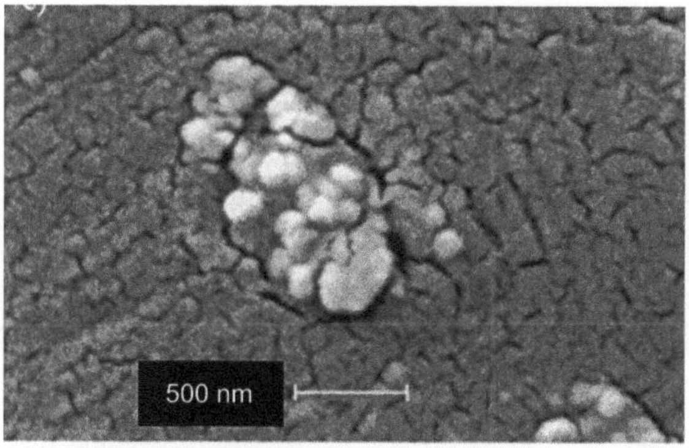

Figure 1.5 SEM micrographs for the specimen nucleated at 710°C for 24 h and subsequently heated at 765°C for 45 min, depicting the onset of crystal growth within the phase-separated glass specimen. (After Kord et al., 2009a. Reproduced with permission of Elsevier Science Publishing Company.)

The phase separation plays a very decisive role in controlling the crystallization process, and the microstructure of the aforementioned glass ceramics. It has been shown that without the formation of a fine and interconnected microstructure during the phase separation process, the whole process of leaching and fabrication of these highly porous bodies possessing very fine nano-size open porosity (Figure 1.6) would be impossible (Kord et al., 2009b).

Figure 1.7 is another example depicting the direct crystallization of a $BaFe_{12}O_{19}$ magnetic phase within the droplets of a phase-separated $BaO-Fe_2O_3-B_2O_3-SiO_2$ glass, in the presence of a ZrO_2 nucleating agent (Mirkazemi et al., 2007).

Radonjíc et al. (2004) prepared a transparent, ferroelectric glass ceramic, exhibiting high dielectric constant, based on sodium niobate crystals. The glass composition was located in the $NaNbO_3-SiO_2-Al_2O_3$ system. In order to achieve high nucleation rates for the crystalline phase, which is essential for making nanostructured, transparent samples, the phase separation process, preceding the nucleation of $NaNbO_3$ crystals in the glass, should be controlled. The authors lowered the phase separation and nucleation temperatures by addition of alumina to the parent silica-based glass. According the authors, the boundaries of phase-separated regions in glass served as appropriate heterogeneous sites for the fast nucleation of sodium niobate crystals. After a 15 min hold at 660°C, a phase separation, in a scale of 10–15 nm, appeared in the glass and after 15 min at 700°C a $NaNbO_3$ glass ceramic, exhibiting fine and homogenous microstructure in a rather fine scale (~40–60 nm) was obtained.

O'Flynn and Stanton (2010) studied the crystallization morphology and activation energy of fluorapatite formation, in apatite–mullite glass ceramics. They demonstrated that increasing levels of fluorine in the base glass led to a

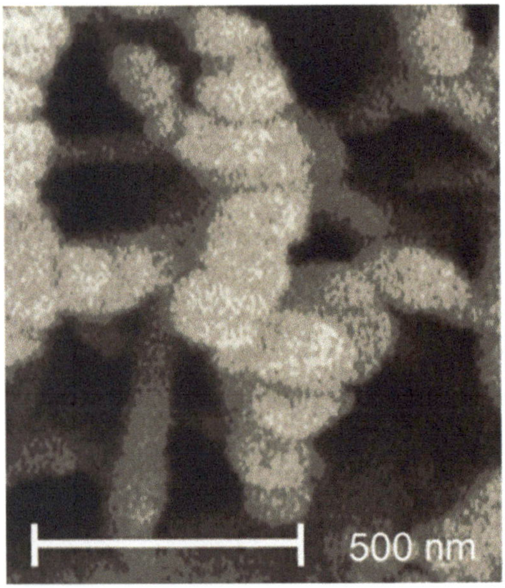

Figure 1.6 SEM micrograph exhibiting the mesh of nanometric pores located on the walls of large pores in the porous specimen. (After Kord et al., 2009b. Reproduced with permission of Elsevier Science Publishing Company.)

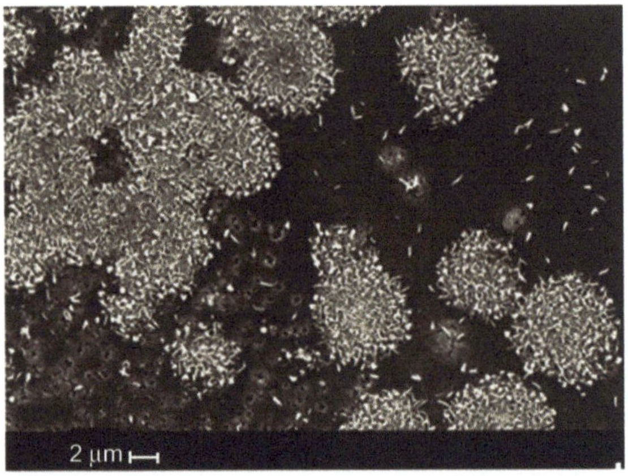

Figure 1.7 SEM micrograph of a magnetic glass-ceramic specimen, showing the nucleation and crystallization of $BaFe_{12}O_{19}$ particles within the droplet phase of the phase-separated glass. (After Mirkazemi et al., 2007. Reproduced with permission of Elsevier Science Publishing Company.)

corresponding increase in the activation energy for fluorapatite crystallization, and a decrease in fluorapatite crystal size. Apatite–mullite glass ceramics are known to phase separate and nucleate prior to crystallization. According to the authors, the phase-separated regions, which are enriched in certain elements, brings the composition closer to that of the resulting crystals. In addition, the surfaces of the phase-separated regions can act as heterogeneous nucleation sites, further reducing the barrier to nucleation. They postulated that, in general, nucleation will occur more easily in the separated droplet phase, rich in CaO, P_2O_5, and F, as their composition is similar to that of crystalline apatite. Also, in this case, crystal growth beyond the size of the droplet phase would be more difficult. Since the presence of higher concentration of nuclei within the droplet phase (which consumed fluorine) caused the glass surrounding the droplets to become fluorine deficient, the glass network connectivity in this region increased, and the latter region acts as a barrier to crystal growth (Figure 1.8). Hence, in the specimens of higher initial fluorine content, in which far more nuclei are formed during the crystallization process, the

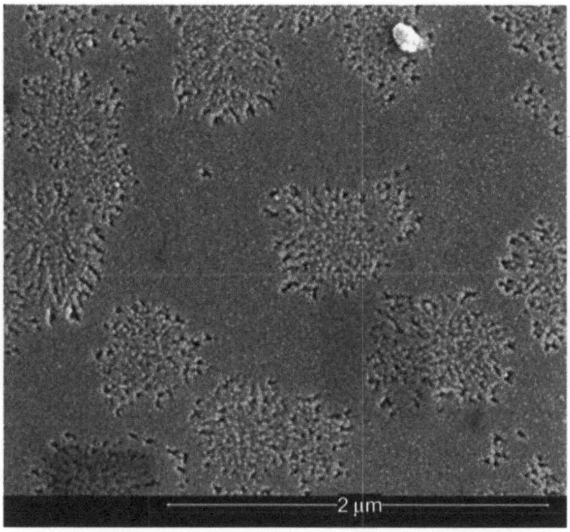

Figure 1.8 SEM micrograph of an apatite–mullite glass ceramic, exhibiting the crystallization of fluorapatite crystals within the separated droplet phase of the glass specimen after nucleating at 700°C for 1 h and subsequent heating to 1,050°C. (After O'Flynn and Stanton, 2010. Reproduced with permission of American Chemical Society.)

residual glass would become more fluorine deficient, as compared with the specimen of lower fluorine content. Therefore, it is reasonable to assume that in the specimens richer in fluorine, the barrier to crystallization of fluorapatite (the activation energy) be higher, and the formed crystal size smaller (Figure 1.8).

Höland et al. (1999) explained a rather complicated case of crystallization of apatite in the SiO_2–Al_2O_3–K_2O–CaO–F–P_2O_5 system, with the aim of producing a needle-like apatite–leucite glass ceramic, as a biomaterial for restorative dental applications.

It was demonstrated that leucite was formed on the surface of the monolithic glass by the surface nucleation and crystallization mechanism, whereas the crystallization of apatite proceeded by a complicated bulk nucleation mechanism.

The process of apatite formation in these glass ceramics was summarized as follows: The base glass underwent a phase separation process during the cooling of the melt, producing very small CaO–P_2O_5-rich droplets (50–250 nm in diameter, which led to the formation of $NaCaPO_4$ crystals, as the primary phase, after heat treatment at 580°C for 15 min. Consequently, another unknown "X-crystal phase" is precipitated at 640°C. Furthermore, $NaCaPO_4$ crystals were no longer detected; they were dissolved and the new unknown crystal phase was formed. The crystallization of apatite was finally observed after the precipitation of the X-phase, during heat treatment at 700°C for 8 h. It is interesting to note that in this condition, a high density

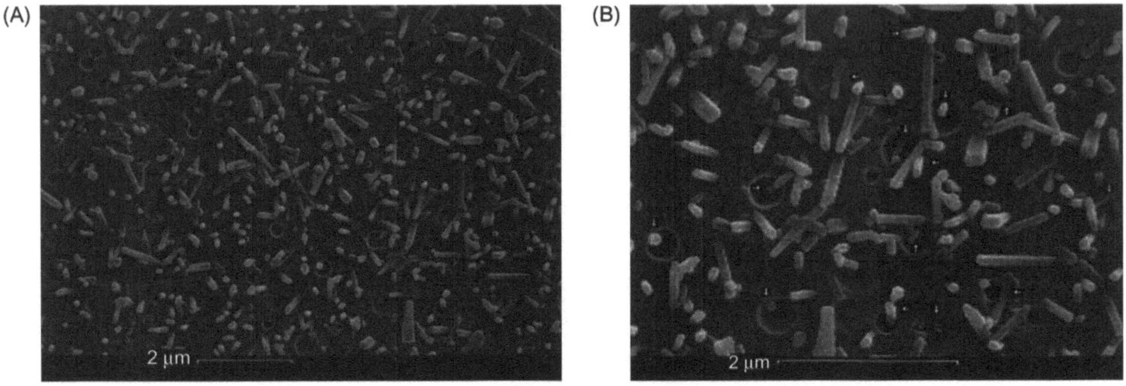

Figure 1.9 Crystallization of needle-like apatite in the volume of the glass ceramic after heat treatment at 700°C/8 h and additionally at 1,050°C/2 h. (A) SEM micrograph of a representative microstructure. (B) Excerpt from (A) with arrows indicating the needle-like apatite and its special circled environment. (After Höland et al., 1999. Reproduced with permission of Elsevier Science Publishing Company.)

of fluoroapatite and X-phase crystals was reached, with the maximum diameter of the crystallites ranging from 100 to 200 nm. At this stage, the primary fluoroapatite, $Ca_{10}(PO_4)_6F_2$, did not show a needle-like habit. However, if the sample was subsequently heat-treated at 1,050°C for 2 h, a needle-like apatite was appeared (Figure 1.9).

The authors proposed two mechanisms for the final nucleation of these apatite crystals, either a heterogeneous reaction of the primary crystals ($NaCaPO_4$ or X phase) with the glassy matrix or an interface reaction of the glass droplet–glass matrix (heterogeneous nucleation on the droplet surface).

Hill et al. (2007) performed real-time small-angle neutron scattering (SANS) studies on a calcium fluorapatite ($Ca_5[PO_4]_3F$) (FAP) glass ceramic. Figure 1.10A and B shows the SANS study of the glass heat-treated from 648°C to 780°C.

It can be seen that the FAP glass ceramic exhibited two scattering peaks; a sharp peak, corresponding to a spacing of 25–27 nm, plus a lower intensity peak at very low q.

The authors attributed this to the occurrence of APS in the glass by two different mechanisms, a nucleation process that gave rise to the scattering at low q and a spinodal decomposition process that gave rise to the well-pronounced maximum in scattering at higher q.

Upon elevation of temperature, the peak at higher q grew in intensity and moved to a lower q, consistent with the coarsening of the finer-scale phase-separated structure.

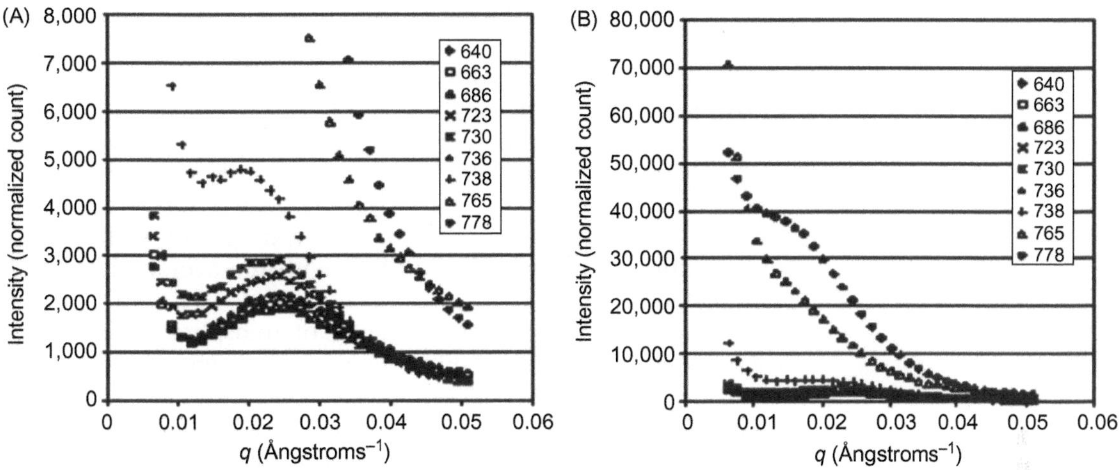

Figure 1.10 (A) SANS patterns (×10) for glass specimens heated at various temperatures in 648–780°C range. (B) (SANS) patterns (×1) for glasses heated at higher temperatures of the range. (After Hill et al., 2007. Reproduced with permission of the American Ceramic Society.)

The intensity of scattering at lower q was also increased with temperature.

The authors explained the main role of APS on crystal nucleation as provision of an internal surface for heterogeneous crystal nucleation or lowering the activation energies for homogenous crystal nucleation.

Nanoscale APS which often promotes crystal nucleation may also serve to hinder crystal growth and coarsening. Obviously, it will be difficult for a crystal to grow into the second amorphous phase, depleted in certain species that served the formation of the crystal phase in the first amorphous phase.

In fact, in this study, a strong correlation was found between the size of FAP crystals growing within the droplet phase and the size of the droplets, suggesting that the FAP crystals did not grow beyond the boundaries of the droplet phase.

A study, with the aim of exploring and explaining the phenomena of nano-crystal formation in oxyfluoride glasses was also carried out by Hill et al. (2010).

Nanostuctured glass ceramics have attracted considerable attention in recent years and are expected to play a major role in various optical applications. In particular, the so-called ultra-transparent oxyfluoride glass ceramics containing metal fluoride nano-crystals, sized between 5 and 100 nm, possess a huge application potential since the rare-earth-doped metal fluoride nano-crystals exhibit enhanced fluorescence, luminescence, and up-conversion characteristics. A more detailed account of the composition, properties, and application of oxyfluoride glass ceramics is presented in the following chapters.

Hill et al. (2010) investigated the crystallization of a starting glass of composition: SiO_2 29.7, Al_2O_3 16.1, P_2O_5, 11.0, CaO 29.7, CaF_2 2.0, and MgO 5.6 (wt%). The glass samples were heat-treated for 8 h at 850°C, followed by rapid cooling to room temperature.

The X-ray diffraction (XRD) patterns of the heat-treated glasses showed the presence of FAP, with the average crystallite size of ~11–21 nm.

Though, SANS showed no evidence of amorphous phase separation, ^{31}P Magic angle spinning nuclear magnetic resonance (MAS NMR) spectra suggested that the glass had undergone APS.

Also, ^{19}F MAS NMR spectra showed that F was below the detectable limits in the amorphous environment, after heat treatments at 850°C and 950°C, indicating that all the fluorine was incorporated into the FAP crystal phase.

According to the authors, the formation of nano-sized crystals in oxyfluoride glass ceramics requires the existence of a rapid nucleation rate with a low crystal growth rate in the system. The authors also summarized the design rules for such a system, among which the key rule is the occurrence of nanoscale amorphous phase separation in the glass system, where one of the separated glass phases is close to the stoichiometry of the fluorine-containing crystal phase.

More recently, Pablos-Martín et al. (2011) investigated the crystallization of a $55SiO_2{-}20Al_2O_3{-}15Na_2O{-}10LaF_3$ (mol%) glass. LaF_3 crystals with a maximum crystal size of 14 nm were obtained by heat treatment of the glass above glass transition temperature. It was shown that the glass was phase-separated prior to the nucleation of LaF_3 crystals. According to the authors, there is a difference in the details of the nucleation and growth processes in phase-separated and nonphase-separated glasses. In nonphase-separated glasses, such as the CaF_2- or BaF_2-containing glasses, fluoride crystals precipitate from a homogeneous glass; therefore, the increase in viscosity due to the silica enrichment around the formed crystals limits further crystal growth. On the other hand, the heat treatment of phase-separated glasses, such as the studied glasses, results in the growth of LaF_3 crystals inside the lanthanum and silicon-rich droplets. As a result of this process, silicon redistribution occurs around the crystals, forming a shell, which then inhibits further crystal growth. The TEM micrograph of the base glass showed the formation of droplets enriched in elements of high atomic number (strong scattering), due to the occurrence of a

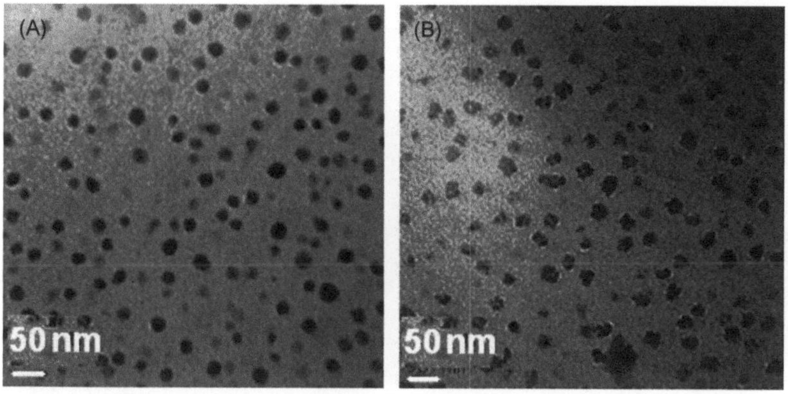

Figure 1.11 Zero-loss filtered bright-field transmission electron microscopic image of (A) the base glass sample, exhibiting phase-separated droplets enriched in elements of high atomic number. (B) The glass ceramic obtained after heat treatment at 620°C for 40 h, exhibiting less spherical droplets as a result of the formation of LaF_3 crystallites. (After Pablos-Martín et al., 2011. Reproduced with permission of the American Ceramic Society.)

liquid–liquid phase separation in glass. The droplets were fully amorphous and had an average diameter in the range of 30–40 nm (Figure 1.11A).

The average size of the phase-separated droplets did not change significantly after heating for 40 h at 620°C (Figure 1.11B), but the shape of the droplets became less spherical as a result of the formation of LaF_3 crystallites.

1.1.2.2.2 Crystal Nucleation in the Absence of Phase Separation in Glass

The occurrence of phase separation however is a common prerequisite for glass crystallization, but it is never an essential condition for crystallization of all glasses. In some conditions, the crystalline phases acting as heterogeneous nucleation centers for the crystallization of the final phases can be nucleated homogenously from a glass without prior glass-in-glass phase separation.

For example, it has been shown (Rezvani et al., 2005a) that $MgCr_2O_4$ spinel crystallites served as heterogeneous nucleation sites for crystallization of an aluminian diopside phase. The spinel crystallites were previously precipitated in a SiO_2–Al_2O_3–CaO–MgO–(R_2O) glass, in the presence of Cr_2O_3, Fe_2O_3, and TiO_2 nucleants without any prior phase separation.

Riello et al. (2001) investigated the nucleation and crystallization behavior of a transparent glass-ceramic glaze of low

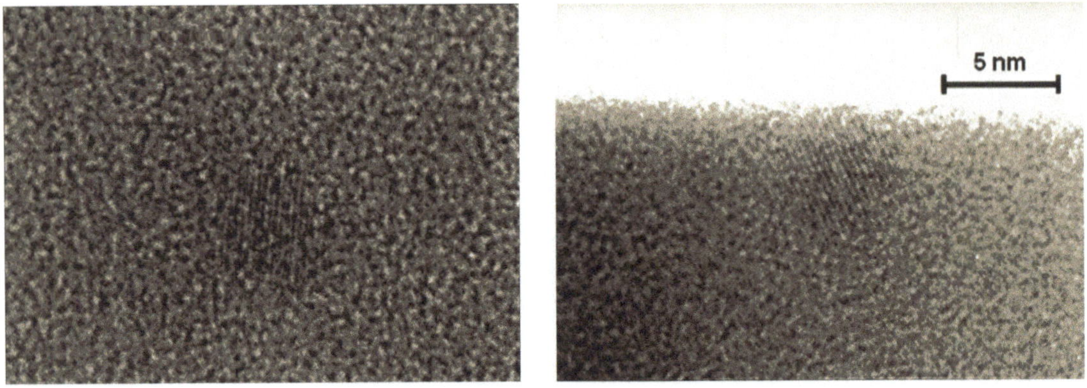

Figure 1.12 Digitalized HRTEM images showing two spherodal crystalline nuclei of $ZrTiO_4$ embedded in the glassy matrix (sample nucleated at 953 K for 20 h). (After Riello et al., 2001. Reproduced with permission of Elsevier Science Publishing Company.)

thermal expansion in the $Li_2O-Al_2O_3-SiO_2$ system in the presence of various oxide nucleants such as TiO_2, ZrO_2, and P_2O_5. They showed that during the nucleation treatment of 20 h at 953 K, some ~4 nm-sized crystallites were first precipitated from the glass, presumably by a homogeneous mechanism. Figure 1.12 shows the high-resolution TEM (HRTEM) micrographs of the nucleated sample. The XRD examination utilizing the Rietveld analysis (Rietveld, 1969) revealed that the crystallites shown in Figure 1.12 are the single crystals of $ZrTiO_4$.

It was also shown that ~0.28 nm apart parallel lines, observed inside the crystallites, are in fact the traces of (111) planes of $ZrTiO_4$ structure, with an interplanar distance of 0.2927 nm. These $ZrTiO_4$ crystallites then served as heterogeneous nucleation sites for crystallization of the desired β-eucryptite phase, at higher temperatures.

Kukkonen et al. (2001) studied the crystallization behavior of a transparent Er^{3+}-doped oxyfluoride glass ceramic. The studied glass composition was located in the $SiO_2-Al_2O_3-CdF_2-PbF_2-ZnF_2-ErF_3$ system. HRTEM images of an as-prepared glass prior to heat treatment revealed that the glass was not phase-separated. However, after a heat treatment of 6 h at 394°C, which was very close to T_g (glass transformation temperature), some ordered regions (0.5 nm) appeared in the glass that were associated with the beginning of nucleation of a fluorite phase (Figure 1.13). A heat treatment of 168 h at 394°C resulted in a transparent glass ceramic, containing PbF_2 crystallites with the mean size of ~18 nm as its main crystalline phase.

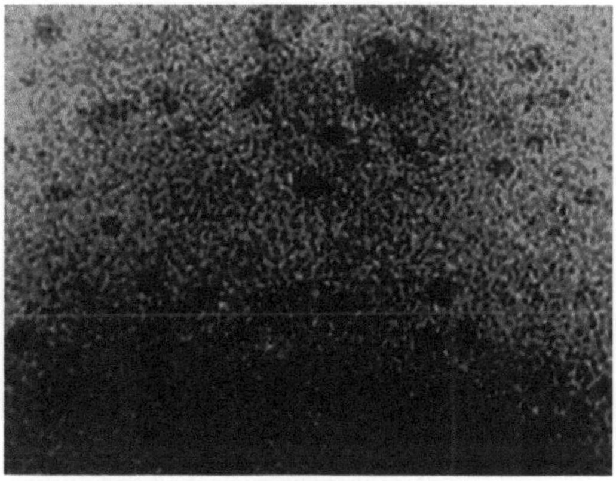

Figure 1.13 HRTEM showing the appearance of ordered regions (0.5 nm) in a glass of SiO_2—Al_2O_3—CdF_2—PbF_2—ZnF_2—ErF_3 system due to the onset of nucleation of PbF_2 crystals after heat treatment for 6 h at 394°C. (After Kukkonen et al., 2001. Reproduced with permission of Elsevier Science Publishing Company.)

1.1.2.2.3 Secondary Crystallization

In the cases in which more than one phase are crystallized from a glass, it is likely that the phases which crystallized first play a role in the crystallization process of others. For example, it is possible that a stable phase nucleates heterogeneously on the previously formed metastable or stable crystallites, or the metastable phase transforms to the stable phase with time.

Figure 1.14 depicts the crystallization of $BaFe_{12}O_{19}$ (barium hexaferrite) platelets, growing on the surface of previously crystallized BaB_2O_4 ellipsoids in a B_2O_3—BaO—Fe_2O_3 glass (Marghussian et al., 2003).

The micrograph of a cordierite–mullite glass ceramic (Figure 1.15) shows how the previously formed plate-like particles of cordierite served as the heterogeneous nucleation sites for tiny mullite crystals, after 15 min heat treatment at differential thermal analysis (DTA) exo-peak temperature of 1,056°C (Marghussian et al., 2009).

1.1.2.2.4 Determination of the Optimum Nucleation Temperature and Time

Marotta et al. (1981) in their pioneering work in the field of glass crystallization proposed a fast DTA method for evaluating the influence of temperature and time of heat treatments on bulk nucleation in glass. In this method, first the DTA exothermic peak temperature of the non-pretreated glass sample, T'_p, is determined, then the DTA peak temperatures of the glass samples heat-treated at different "nucleation" temperatures, T_p, are measured.

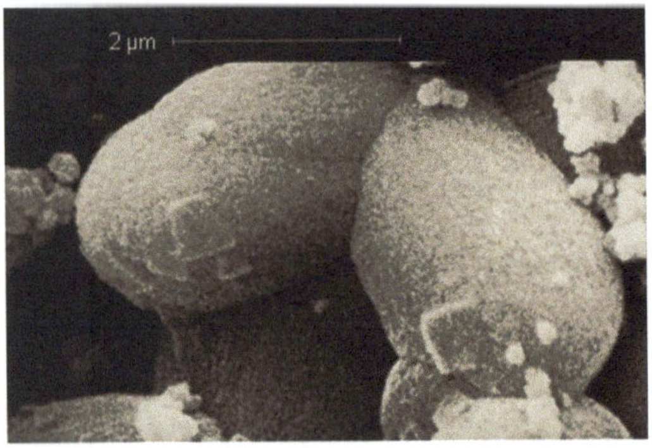

Figure 1.14 SEM micrograph exhibiting the crystallization of $BaFe_{12}O_{19}$ (barium hexaferrite) platelets, growing on the surface of previously crystallized BaB_2O_4 ellipsoids in a B_2O_3—BaO—Fe_2O_3 glass. (After Marghussian et al., 2003. Reproduced with permission of Elsevier Science Publishing Company.)

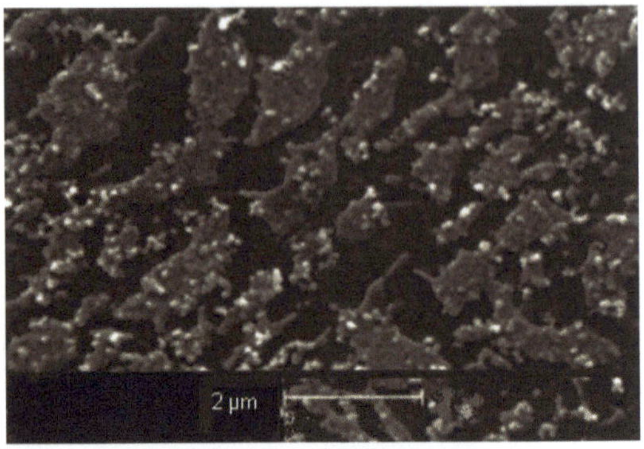

Figure 1.15 SEM micrograph depicting the heterogeneous nucleation and crystallization of mullite tiny crystals on the surface of the previously formed plate-like particles of cordierite, after the heat treatment of 15 min at 1,056°C. (After Marghussian et al., 2009. Reproduced with permission of Elsevier Science Publishing Company.)

The plot of $\Delta T_p = (T'_p - T_p)$ versus the pretreatment temperature of samples gives a curve which shows the same shape and maximum as a nucleation rate–temperature curve.

Such a curve exhibiting the crystal nucleation of a Li_2O–$2SiO_2$ glass is shown in Figure 1.16. The maximum of this curve (455°C) shows a good match with the previously determined temperature of maximum nucleation rate for the same glass as reported by James (1974).

If the samples are held for different times at this so-determined temperature of maximum nucleation rate, then a plot of $(T'_p - T_p)$ against the times of the heat treatments gives an isothermal nucleation curve, which shows how the nuclei number-order of magnitude is increased by the duration of the heat treatments.

Although, Marotta et al. used the already known suitable nucleation temperatures in their study, such "nucleation" temperatures can be selected in the temperature range between the glass transition temperature, T_g, and dilatometric softening point of glass, T_d.

Ray and Day (1990) also proposed a similar DTA method for determination of maximum nucleation temperature in a lithium disilicate glass.

In this method, **the DTA exothermic peak heights** of glass specimens were used as the criterion of nuclei concentration present in a glass.

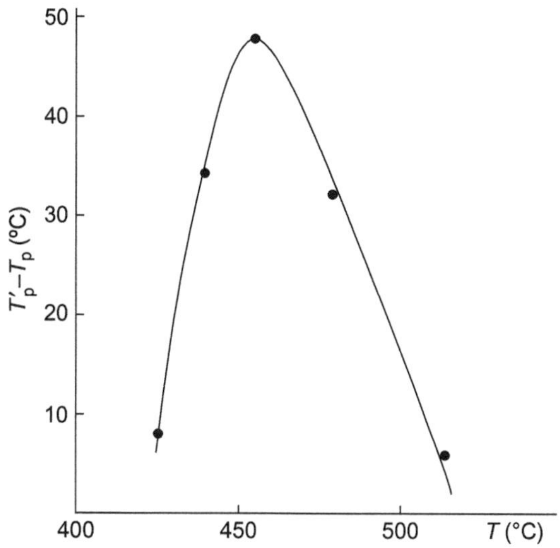

Figure 1.16 $T'_p - T_p$ plotted against temperature of nucleation pretreatment for $Li_2O—2SiO_2$ glass. (After Marotta et al., 1981. Reproduced with permission of Springer Science + Business Media, Inc.)

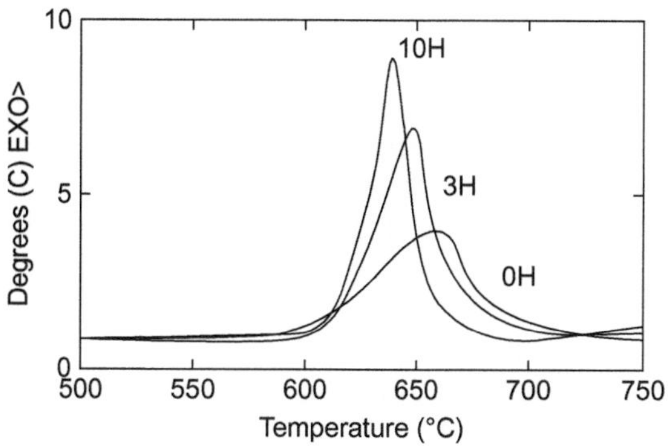

Figure 1.17 DTA crystallization traces for $Li_2O—2SiO_2$ glasses. (0H) as-quenched, (3H) nucleated at 453°C for 3 h, and (10H) nucleated at 453°C for 10 h. (After Ray and Day, 1990. Reproduced with permission of the American Ceramic Society.)

A lithium disilicate glass was investigated, because the thermodynamic and crystallization data were available for it in the literature.

As the DTA crystallization peaks for the glasses previously nucleated for 3 or 10 h (specimens 3H and 10H) at 453°C, the temperature corresponding to the maximum nucleation rate, along with that of the as-quenched sample (0H), are shown in Figure 1.17. For the same sample weight, the peak height, $(\delta T)_p$, clearly increases, while the DTA peak temperature (T_p) decreases with increasing nucleating time.

With increasing time at the nucleation temperature (453°C), the concentration of nuclei in the glass increases, which, in turn, increases the crystallization rate during the DTA run, as indicated by the higher $(\delta T)_p$ and lower T_p, as shown in Figure 1.17.

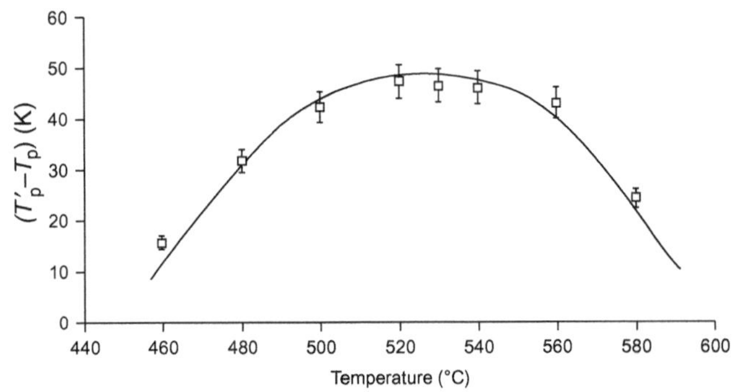

Figure 1.18 Function of $\Delta T_p = (T'_p - T_p)$ versus temperature of isothermal treatment. T'_p temperature of the first exothermic DTA (DSC) peak of the non-pretreated reference sample and T_p, temperature of the first exothermic peak of the pretreated sample. (After Höland et al., 2006. Reproduced with permission of Elsevier Science Publishing Company.)

The increase in $(\delta T)_p$, and decrease in T_p, which are considered to be the direct result of increasing concentration of nuclei in the glass, are expected to continue until the glass becomes saturated with nuclei.

Therefore, if the changes in $(\delta T)_p$ and T_p are plotted against nucleating time (at the suitable nucleation temperature), it is expected to observe a "leveling off" in the curves, indicating the saturation of glass with nuclei.

Since an increasing concentration of nuclei causes a decrease in T_p (before saturation), plots of T_p versus nucleation temperature can be used to determine the optimum nucleation temperature. Obviously, the optimum nucleation temperature (causing the maximum nucleation rate) would be located at the minimum of the above T_p versus T_n curves.

Figure 1.18 depicts the ΔT_p (DTA peak temperature difference between the as-prepared and prenucleated glass samples) versus the nucleation pretreatment temperature (T_n) plot for a glass derived from the SiO_2–Li_2O–Al_2O_3–K_2O–ZrO_2–P_2O_5 system, in the presence of P_2O_5 as a nucleating agent (Höland et al., 2006).

The broad maximum observed at ~525°C indicates the temperature of the maximum nucleation rate, as discussed above.

The main stages of the rather complicated crystallization path of the aforementioned glass were studied, utilizing high-temperature XRD (HT-XRD), and explained by the authors as follows: the aforementioned maximum temperature at ~525°C was attributed to the nucleation of Li_3PO_4 as the first crystalline phase formed, which according to the authors, most probably

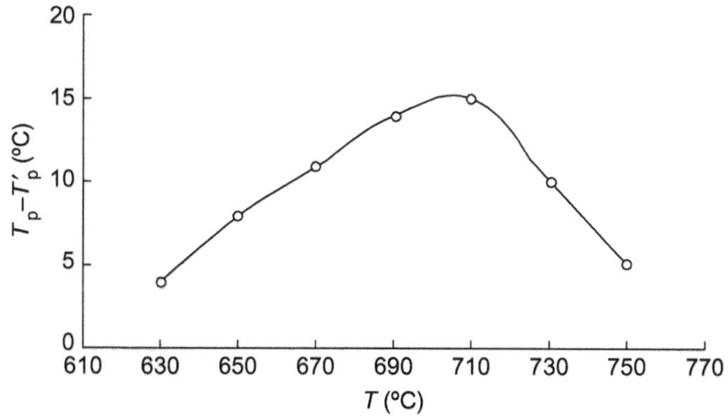

Figure 1.19 ($T_p - T'_p$) plotted against temperature of heat treatment for a glass of SiO_2–CaO–MgO–Na_2O system, containing a combination of $Fe_2O_3 + Cr_2O_3$ nucleating agents. T_p and T'_p are the crystallization peak temperatures of the as-quenched and previously nucleated samples, respectively. (After Alizadeh and Marghussian, 2000a. Reproduced with permission of Elsevier Science Publishing Company.)

acted as a heterogeneous catalyst, for the crystallization of both Li_2SiO_3 and $Li_2Si_2O_5$ phases.

The latter phase grew as agglomerated nano-crystals but remained in a relative small amount. By contrast, lithium metasilicate (Li_2SiO_3) grew rapidly and decomposed at 780–820°C with the result of a drastic increase of the lithium disilicate phase. The authors finally concluded that the crystal growth of lithium disilicate in multicomponent systems can be controlled by the nucleation process and by the preliminary phase formation process of lithium metasilicate. If the formation of the latter phase is maximized, the best possible conditions are created for the formation of high amounts of lithium disilicate, possessing fine crystallites.

Alizadeh and Marghussian (2000a) also obtained a similar curve (Figure 1.19) for the nucleation of a glass of composition 59.68 SiO_2, 12.23 CaO, 15.02 MgO, 5.08 Na_2O, 4.00 Fe_2O_3, and 4.00 Cr_2O_3 (wt%), located in the SiO_2–CaO–MgO–Na_2O system, containing a combination of $Fe_2O_3 + Cr_2O_3$ nucleating agents. In this way, the maximum nucleation temperature was found as 710°C. The main crystalline phase of the resulting glass ceramic was diopside.

1.1.2.2.5 Determination of the Type and Amount of Effective Nucleating Agents by DTA

A DTA method can also be utilized to evaluate the effectiveness of the nucleating agents in promoting bulk crystal nucleation in glasses (Leonelli et al., 1991). In this method, samples with different specific surface areas and containing various nucleants are studied by DTA. The increase of the specific surface area does not significantly influence the glass crystallization temperatures of the specimens with a marked tendency toward bulk (volume) nucleation and crystallization, whereas for surface nucleation mechanism, the crystallization peak maxima shift relatively toward lower temperatures. Therefore, the ΔT values, the differences between exothermic crystallization peak temperatures for glasses in two particle size ranges, can be used as a criterion to verify the effectiveness of the added nucleants. The glass samples containing the more effective nucleating agents usually show relatively small differences in DTA peak positions and are anticipated to undergo bulk nucleation.

It should be noted that the criterion of ΔT values, discussed above, shows only the effectiveness of a given nucleating agent to induce bulk nucleation in a glass and tells nothing about the actual nucleation rate in the presence of that nucleant. Therefore, it should be used only as a *first approximation* to differentiate between various nucleating agents and should never be used as the final, decisive criterion to select the most effective bulk nucleation agent in a glass. The other parameters, discussed in Section 1.1.2.2.4, i.e., the variation of T_p and $(\delta T)_p$ (the maximum of DTA crystallization peak temperature and the height of the peak) with the type and amount of nucleating agents should also be considered in conjunction with the peak shift.

Many authors have utilized the above method to identify the effective nucleants. Alizadeh and Marghussian (2000b) investigated the crystallization behavior of wollastonite–diopside glass ceramics in the ternary SiO_2–MgO–CaO system containing different nucleating agents. With attention to the ΔT parameter, as discussed above, and considering other parameters, including the microstructure of the resulting glass ceramics, they concluded that if $\Delta T < 40°C$, the occurrence of a certain degree of bulk nucleation could be expected in the glass samples, whereas for specimens with $\Delta T > 40°C$, surface crystallization could mainly be observed.

Table 1.1 gives the summary of ΔT values for some glasses containing various nucleating agents and Table 1.2 depicts the chemical composition of the glasses.

In the above work, in conformity with the aforementioned criterion, $V_2O_5 + MoO_3$ and $CaF_2 + MoO_3$ pairs had the ability to induce bulk nucleation, whereas $WO_3 + Fe_2O_3$ pair was relatively ineffective in this respect (Table 1.1). The authors also stated that in addition to the ΔT parameter, the sharpness of crystallization peak, and the maximum peak temperature are other factors showing the occurrence of bulk nucleation and the susceptibility of samples to volume crystallization.

Table 1.1 Difference in Crystallization Peak Temperatures (ΔT Parameter) due to the Variation of Particle Sizes for Some Glass Samples (the Glass Compositions Are Given in Table 1.2)

Sample No.	ΔT (C)
1	55
2	52
3	40
4	22
5	26
6	26
7	23
8	31
9	30

Table 1.2 Chemical Compositions of the Glass Samples Given in Table 1.1 (wt%)

Sample No.	SiO_2	CaO	MgO	Na_2O	Fe_2O_3	WO_3	V_2O_5	MoO_3	CaF_2	MoO_3
1	59.68	18.23	9.02	5.08	6	2	–	–	–	–
2	59.68	12.23	15.02	5.08	6	2	–	–	–	–
3	59.68	6.23	21.02	5.08	6	2	–	–	–	–
4	59.68	18.23	12.02	5.08	–	–	4	4	–	–
5	59.68	12.23	15.02	5.08	–	–	4	4	–	–
6	59.68	6.23	21.02	5.08	–	–	4	4	–	–
7	59.68	18.23	9.02	5.08	–	–	–	–	4	4
8	59.68	12.23	15.02	5.08	–	–	–	–	4	4
9	59.68	6.23	21.02	5.08	–	–	–	–	4	4

In this respect, it was shown that $CaF_2 + MoO_3$ pair was the best nucleating agent for the studied glasses, resulting in glass-ceramic materials exhibiting higher crystallization values.

Rezvani et al. (2005a) investigated the effects of Cr_2O_3, Fe_2O_3, and TiO_2 nucleants on the crystallization behavior of

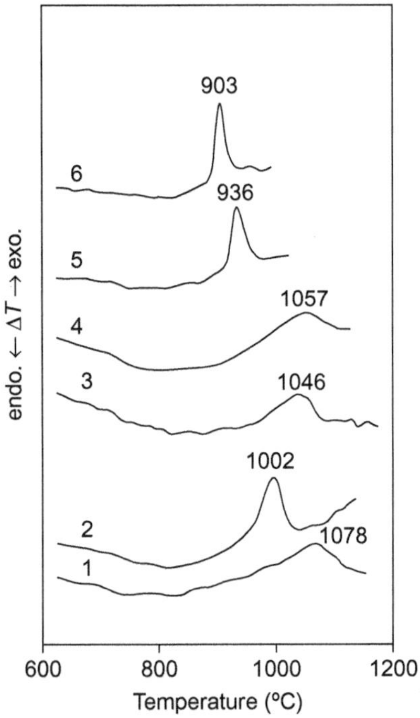

Figure 1.20 DTA traces of the glass samples containing various nucleants. The composition of glasses is given in Table 1.3 as AR (1), AR-Cr_3 (2), AR-Fe_5 (3), AR-Ti_5 (4), AR-Cr_3Fe_5 (5), and AR-$Cr_3Fe_5Ti_5$ (6). (After Rezvani et al., 2005a. Reproduced with permission of Elsevier Science Publishing Company.)

glass compositions in the SiO_2–Al_2O_3–MgO–CaO–(R_2O) system. In this investigation, Cr_2O_3, Fe_2O_3, and TiO_2 were added in various amounts and combinations to a base glass of composition SiO_2 55.05, Al_2O_3 13.61, CaO 24.42, MgO 6.92, Na_2O 2.82, and K_2O 3.02 (weight ratio).

Figure 1.20 shows the DTA results for the base glass as well as for the most promising specimens, exhibiting the sharpest and lowest temperature exothermic peaks among the whole specimens investigated in this work. Table 1.3 summarizes the DTA results for different particle sizes.

It was deduced that the base composition (specimen AR), showing a weak and broad exothermic peak in DTA (Figure 1.20), was not capable of inducing effective bulk nucleation and crystallization in the glass.

According to the results given in Table 1.3 and Figure 1.20, the specimen AR-$Cr_3Fe_5Ti_5$ containing $3Cr_2O_3$, $5Fe_2O_3$, and $5TiO_2$ (wt%) additives resulted in the sharpest peak, and the lowest DTA exothermic peak temperature exhibiting an almost 180°C reduction in peak temperature in comparison with the base glass.

Table 1.3 Summary of DTA Crystallization Peak Temperatures (T_c) for the Glass Samples of Two Different Particle Sizes*

Specimen	Particle Size	T_c (°C)	Crystallization Peak Shift
AR-Cr_3	A	1002	−18
	B	984	
AR-Fe_5	A	1046	−75
	B	971	
AR-Ti_5	A	1059	−71
	B	988	
AR-Cr_3Fe_5	A	936	+4
	B	940	
AR-$Cr_3Fe_5Ti_5$	A	903	+7
	B	910	

*A, particle size, 0.45–0.55 mm; B, particle size, <63 μm. *Suffixes Cr, Fe, and Ti denote Cr_2O_3, Fe_2O_3, and TiO_2, respectively, and the numbers following the suffixes indicate the wt% of each oxide.

This is a clear indication of the occurrence of more efficient bulk nucleation and crystallization in the presence of the triple nucleating agents in the glass composition. It was also observed that AR-Cr_3Fe_5 and AR-$Cr_3Fe_5Ti_5$ samples show very low exo-peak shifts in comparison to other specimens (Table 1.3), verifying the above-mentioned effective bulk crystallization mechanism in these specimens. Meanwhile, the latter samples exhibited unusual positive exo-peak shift values. The authors attributed this unusual behavior to the sintering and coarsening of the glass powders, i.e., with increasing the temperature during the DTA run, the above powder specimens, because of their high sinterability, rapidly became agglomerated and acted as coarse grain samples. Therefore, the prediction of surface/bulk mechanism of crystallization by the comparison of DTA peak shifts of the samples with different particle sizes in the case of powders with high sintering tendencies could be misleading.

1.1.2.2.6 Determination of Crystal Nucleation rates

Particle Counting Method The kinetics of volume nucleation in some "simple" stoichiometric glass compositions, in which the crystallizing phase has the same composition as the parent glass, have been determined quantitatively using optical microscopy

(Kalinina et al., 1977; James, 1985). Two methods have been used. At higher temperatures, above the nucleation maximum, a single-stage heat treatment can be used. After nucleation, the number per unit volume, N_v, can then be obtained using the micrographs of polished and lightly etched sections of the samples taken by optical microscopy (James, 1974).

For lower nucleation temperatures (T_N), a two-stage heat treatment is usually used. After the nucleation treatment, first the glass is heated up very quickly to a higher "growth" or "development" temperature (T_D), held at that temperature for a short time then rapidly quenched to ambient temperature. In this way, the previously nucleated particles are grown to observable dimensions suitable for the microscopic examination. The criterion of choosing the T_D temperature is the existence of a rapid growth rate, along with a negligible nucleation rate. The validity of this procedure depends on the assumption that on heating to the second stage, the large majority of the formed nuclei actually grow into small crystals and do not redissolve. In this way, the number of particles observed after the growth treatment should be very close to the number of nuclei formed at T_N (James, 1982).

James (1974) stated that, if this assumption was incorrect, then varying the growth temperature T_D should significantly affect the number of crystals observed after a given nucleation heat treatment. The measurements conducted on glasses with the compositions near the $Li_2O{-}SiO_2$ stoichiometric composition showed that for the fixed T_N temperatures, the variation of T_D did not significantly affect the N_v versus t curves. Hence, the assumption that the large majority of the nuclei do not redissolve during the second-stage treatment was justified for the nucleation temperatures used.

However, Kalinina et al. (1976, 1977) showed that when lower nucleation temperatures were chosen for the same glass composition, the aforementioned "dissolution effect" could increase the measured or apparent induction time if the growth temperature T_D was high enough. In this case, the whole nucleation curve could be shifted parallel to the time axis. However, the value of the steady-state nucleation rate, I, determined from the curve, might not be affected, provided the heat treatment times at T_N were sufficiently long.

According to James (1982), the double-stage method is capable of giving accurate results provided that when using lower nucleation temperatures, where the induction times are relatively longer, the growth temperature is not too high and the nucleation times are not too short.

In some cases, especially the existence of high nucleation rates, electron microscopy may be required to observe and count the particles, and in some conditions, it is possible to use single-stage growth treatment, provided that the growth rate is high enough at the nucleation temperature.

Thermal Analysis Methods A much faster and precise DTA method was also introduced for nucleation and growth rate measurements in glasses that require a smaller amount of sample and need no special sample preparation (Ray et al., 2000, Ranasingheh et al., 2002). According to this method, the nucleation rate can be determined using the following equation:

$$\frac{A_1}{A_2} = \frac{M_1\left[1-\frac{\pi}{3}(I_N t_N + I_G t_{G1} + N_q)U_G^3 t_{G1}^3\right]\left[(I_N t_N + I_G t_{G1}) + N_q\right]}{M_2\left[1-\frac{\pi}{3}(I_N t_N + I_G t_{G2} + N_q)U_G^3 t_{G2}^3\right]\left[(I_N t_N + I_G t_{G2}) + N_q\right]} \quad (1.14)$$

where I_N is the nucleation rate at a given nucleation temperature (T_N), t_N is the nucleation time at T_N, A_1 and A_2 are the areas of DTA peaks after crystal growth heat treatment at crystallization temperature (T_G) for times t_{G1} and t_{G2}, respectively, and M_1 and M_2 are the sample masses for two DTA runs. The quantity N_q is the number of nuclei quenched in the specimen before the onset of nucleation heat treatment at T_N, and I_G and U_G are the steady-state nucleation and growth rates at T_G, respectively.

For measuring the nucleation rate according to the above-mentioned method, the number of nuclei ($I_N t_N$) formed during the nucleation heat treatment at T_N should be determined as a function of t_N from two DTA measurements for each t_N. A plot of ($I_N t_N$) versus t_N should be a straight line, its slope yielding the value of I_N at T_N. Determining ($I_N t_N$) using Eq. (1.14) requires that three unknown parameters be known, namely, N_q, I_G, and U_G. The value of U_G can be determined by the conventional methods. To determine N_q and I_G, three DTA runs for an as-quenched glass ($I_N t_N = 0$), with different heat treatment times (t_{G1}, t_{G2}, t_{G3}) at T_G are required. If A_1, A_2, and A_3 are the respective DTA peak areas, then three equations can be developed for A_1/A_2, A_1/A_3, and A_2/A_3 from Eq. (1.14) and simultaneous solutions of any two of the three equations will yield the values for N_q and I_G, which can then be used in Eq. (1.14) as constants for determining ($I_N t_N$).

The aforementioned authors successfully applied the above DTA method to a $Na_2O \cdot 2CaO \cdot 3SiO_2$ glass, whose I and U curves were known to overlap significantly. Hence, they concluded that this DTA method provides a rapid alternative means for determining the nucleation and crystal growth rates in glasses, even when these rate curves overlap on the temperature scale.

More recently, Rezvani et al. (2011) also utilized this DTA method to calculate the nucleation rate of diopside crystals in a $SiO_2-Al_2O_3-CaO-MgO-(R_2O)$ glass in the presence of Cr_2O_3, Fe_2O_3, and TiO_2 nucleating agents. The optimum nucleation rate of the aluminian diopside crystals was determined for the most promising glass specimen, containing all three nucleant, as $3.10 \pm 0.18 \times 10^{18}$ m^3/s. This is one of the highest nucleation rates reported so far for glass ceramics. Therefore, the authors' previous claims (Rezvani et al., 2005b) concerning the marked effectiveness of the triple nucleants in inducing bulk nucleation in the glasses of this system were confirmed.

These claims were previously substantiated by reduced values of the activation energy for crystallization and lower ΔT values (DTA exo-peak shifts upon changing the glass particle sizes) (see Section 1.1.2.2.5).

Fokin et al. (2010) critically assessed the validity of DTA–DSC (Differential Scanning Calorimetry) methods for studying the nucleation kinetics in glasses. These authors, employing two silicate glasses with the stoichiometric compositions $Li_2O \cdot 2SiO_2$ (L_1S_2) and $Na_2O \cdot 2CaO \cdot 3SiO_2$ ($N_1C_2S_3$), as "model" glasses, studied the DSC/DTA methods against the traditional microscopy method, to estimate the temperature range of the nucleation rate maximum, and the number of supercritical crystal nuclei.

The authors concluded that the dependence of the DSC/DTA crystallization peak position on the number of preexisting nuclei is not linear. This fact, as well as the non-steady-state nucleation, is always expected to occur at some temperature around the glass transition range, distorts the real temperature dependence of the nucleation rate, when the shift of crystallization peak method was employed.

Furthermore, since the ratio between volume and surface crystallization depends not only on sample size but also on the number of density of internal crystals, the role of surface crystallization would be more pronounced when the number of crystals in the volume is small, e.g., in the case of athermic crystals. Therefore, particle size alone is not sufficient to give a satisfactory explanation of the relative importance of surface crystallization. On the other hand, crystal growth in the nonisothermal regime can have a significant effect on the calculated crystal number density and nucleation rates. To make the conditions far more complicated, the formation of nonequilibrium phases or the presence of elastic stresses can also have a significant influence on the area of DTA/DSC crystallization peaks.

All these factors can prevent obtaining an accurate analysis of nucleation kinetics by DSC/DTA methods.

Finally, the authors concluded that in order to accurately determine the nucleation kinetics by DTA/DSC methods, a preliminary knowledge of the nucleation and growth rates of the studied glass is essential to take into account all the above effects. If all these conditions were met properly, such nonisothermal methods can give useful kinetic information, which includes the crystal number density, but they could be as laborious as the traditional microscopy methods!

1.1.3 Nonclassical Theories of Crystal Nucleation in Glass

1.1.3.1 General Considerations

In the CNT, the "droplet" approximation is adopted to calculate W^*, and r^*, the nucleation barrier and critical radius, respectively. As stated previously (Eqs. [1.5] and [1.4]), they are formulated as $W^* = 16\pi\sigma^3/3(\Delta G_v)^2$ and $r^* = \frac{2\sigma}{\Delta G_v}$, where ΔG_v is the bulk free-energy change in crystallization, per unit volume, and σ is the interfacial free energy of the equilibrium (planar) crystal/liquid interface for unit area. Here, the nuclei are considered as particles which show bulk properties. This would necessitate the assumption of existence of extremely sharp interfaces between crystal and liquid phase interfaces. In contrast, computer simulations and more advanced theories predict interfaces of several molecular layers thick. Hence, the validity of the classical approach seems to be questionable. Indeed, many investigators demonstrated that the measured and calculated nucleation rates differ by several orders of magnitude (see Section 1.1.1.2 and Weinberg, 1999; Fokin et al., 2006).

The computer simulations also indicate that the surface tension of the crystal/liquid interface depends on the size of the crystalline clusters for small sizes (Gránásy, 1995, 1996). Since the critical size depends on the undercooling, the curvature effect leads to an apparent temperature dependence of the interfacial free energy. Owing to the general lack of independent data for the free energy, σ, of the undercooled crystal/liquid interface, a direct comparison between the experimental and theoretical nucleation rates is not possible. Nevertheless, a method, introduced by Turnbull to test the CNT, showed that a temperature-independent interfacial free energy is incompatible with the measured nucleation rates (Gránásy and James, 1999).

Considerable efforts have been made in recent years to overcome the difficulties of the classical theory and to represent a more accurate and satisfactory theory of crystal nucleation in glass. In these "nonclassical cluster theories," two main approaches have been adopted, namely, phenomenological and molecular. The phenomenological theories start from the thermodynamic data reflecting true molecular interactions. Therefore, provided that the curvature effect on the interfacial free energy is incorporated in them, these theories can generally be applicable.

The molecular theories, on the other hand, deduce the thermodynamic and interfacial properties from the molecular interaction, giving detailed information on simple model systems.

1.1.3.2 Phenomenological Models

Several phenomenological cluster models have been proposed for the nucleation process.

For example, Gránásy and James (1999) introduced the diffuse interface theory (DIT) as the simplest model that takes into account the diffuseness of the interface. In this theory essentially the same information is used, as in the classical theory. The above-mentioned theory had previously been applied to vapor condensation and liquid crystallization processes (Gránásy, 1995, 1996).

Assuming a size-independent interface thickness, and bulk properties in the interior of nuclei, a size-dependent interfacial free energy was predicted. The theory without using adjustable parameters represented a more reasonable explanation of the condensation experiments on liquids of disordered surface. The treatment was also extended to crystal nucleation in oxide glasses. It was shown that the DIT is consistent with a variety of experimental data on crystal nucleation in glasses. A method was also outlined and discussed to distinguish between homogeneous and bulk heterogeneous processes (Gránásy et al., 1996).

1.1.3.3 Density Functional Theory

The molecular theories of freezing that offer theoretically well-established means to describe diffuse interfaces are based on the modern density functional techniques (DFT). These theories describe the crystal as an inhomogeneous liquid with spatially varying density that displays the symmetries of the crystal. This approach relies on the density functional theorems of classical systems, according to which the thermodynamic quantities such as the Helmholtz free energy, *F*, are unique functional of the density (Gránásy and James, 1999). The equilibrium structure is the one that minimizes the grand potential functional.

It has been shown that the DFT is a reasonable and convenient tool for studying various model systems. Its application to real substances, however, needs the inclusion of realistic molecular interactions. Though this method had already achieved some success in describing the condensation of nonpolar substances, it failed to reproduce reasonable results when applied to oxide glasses. In the calculations regarding the nucleation rates of eight stoichiometric oxide glasses utilizing DFT models, with different approximations, large deviations from the physically reasonable values were observed. Moreover, most of these molecular theories failed to reproduce the positive temperature coefficient for the interfacial free energy of the critical nuclei, while the phenomenological and semiempirical, Cahn–Hilliard-type models produced acceptable results (Gránásy and James, 1999). According to the authors, it is expected that the interfacial information provided by computer simulations and semiempirical calculations will play a decisive role in the future advance of the field.

1.2 Crystal Growth in Glass

After occurrence of an efficient nucleation in glass, the subsequent growth of nuclei is the second decisive step in the crystallization process of glass ceramics. The two steps, nucleation and growth of crystals, together determine the microstructural details, hence the properties of glass-ceramic materials. Therefore, it is essential to have a brief discussion on theoretical, as well as experimental aspects of the topic, **"crystal growth in glass."**

1.2.1 Theoretical Background

For stoichiometric glass compositions that do not undergo compositional changes during crystallization, i.e., polymorphic crystallization, long-range diffusion is not essential for crystal growth; thus, interfacial rearrangements are likely to control the crystal growth process. In this case, the focus of theoretical treatments is directed at the nature of the interface. Three standard models are used to describe the crystal growth process in glasses, which are based on different views of the nature of the crystal/liquid interface (Uhlmann and Uhlmann, 1993). These models are:

i. The normal or continuous growth model
ii. The screw dislocation model
iii. The two-dimensional (2D) surface-nucleated growth

All the above models are based on the assumption that the release of latent heat does not substantially alter the crystal–melt interface temperature.

1.2.1.1 Normal Growth Model

In the framework of the normal (or continuous) growth model, the interface is pictured as rough on an atomic scale and all the sites on the interface are assumed to be equivalent growth sites. The growth rate is expressed by:

$$U = \upsilon\lambda\left[1 - \exp\left(-\frac{\Delta G}{RT}\right)\right] \tag{1.15}$$

where υ is the frequency of atomic jumps at the interface, λ is the distance advanced by the interface in a unit kinetic process (usually taken as a molecular diameter), ΔG is the thermodynamic driving force for crystallization, i.e., the difference between the free energies of the undercooled melt and crystalline phase per mole, T is the absolute temperature, and R is the gas constant.

Assuming that the molecular motion required for crystal growth is similar to that involved in transport in the bulk liquid, the jump frequency factor, υ, can be represented through the Stokes–Einstein equation:

$$\upsilon = \frac{K_B T}{3\pi\lambda^3\eta} \tag{1.16}$$

where η is the shear viscosity, k_B is the Boltzmann constant, and λ is the jump distance.

1.2.1.2 The Screw Dislocation Growth Model

The screw dislocation growth model views the interface as smooth, but imperfect on an atomic scale, with growth taking place at step sites provided by screw dislocations, intersecting the interface. The crystal growth rate U is given by

$$U = f\upsilon\lambda\left[1 - \exp\left(-\frac{\Delta G}{RT}\right)\right] \tag{1.17}$$

$$\upsilon = \upsilon_0 \exp\left(-\frac{\Delta G_D}{RT}\right) \tag{1.18}$$

where υ is the frequency of atomic jumps at the interface, υ_o is the vibrational frequency of the growth-controlling atoms, and ΔG_D is the activation free energy for diffusion across the interface. The fraction of sites on the interface where atoms can preferentially be added or removed, f, is given by

$$f = \frac{\lambda \Delta G}{4\pi\sigma T_m} \approx \frac{(1 - T_r)}{2\pi} \tag{1.19}$$

where

$$T_r \equiv \frac{T}{T_m} \tag{1.20}$$

σ is the specific surface energy of the liquid/crystal interface and T_m is the thermodynamic melting point.

1.2.1.3 2D Surface-Nucleated Growth

In this model, growth occurs by the formation and lateral growth of 2D nuclei on the interface and the crystal growth rate is given by

$$U = C\frac{D_u}{\lambda^2}\exp\left(\frac{-B}{T\,\Delta G}\right) \tag{1.21}$$

$$B = \frac{\pi\lambda V_m \sigma^2}{3k_B} \tag{1.22}$$

$$C = \frac{\sqrt[3]{\pi N_s \lambda^{\frac{5}{3}}}}{\Gamma\left(\frac{4}{3}\right)}\left[1 - \exp(-\Delta G/RT)\right] \tag{1.23}$$

where D_u is the effective diffusion coefficient (m^2/s) of the molecular species that control atomic or molecular attachment at the liquid/crystal interface, V_m is the molar volume (m^3/mol), k_B is the Boltzmann constant, σ is the surface energy (J/m^2), N_s is the number of growth sites per unit area (m^{-2}), and Γ is the gamma function.

1.2.1.3.1 Jackson's Model for the Interface

Jackson (1958a,b, 1967) proposed a successful approach based on the characteristics of materials interface morphology and on the entropy of fusion, ΔS_f. In his model, materials with small entropies of fusion ($\Delta S_f < 2R$), such as GeO_2 and SiO_2, crystallize and melt with nonfaceted interface morphologies. They show small growth rate anisotropies, i.e., little differences in growth rate for different orientations and exhibit little asymmetry in the curves of growth and melting rates versus driving force in the vicinity of the melting point. Their interface site factors, i.e., fraction of preferred growth sites on the interface, are also independent of undercooling and superheating, hence, exhibit the form predicted by the normal growth model.

The materials characterized by large entropies of fusion ($\Delta S_f > 4R$), such as most oxide melts, on the other hand, exhibit faceted interface morphologies in growth and nonfaceted morphologies in melting, show large growth rate anisotropies, and have interface site factors that increase with increasing undercooling (Klein and Uhlmann, 1974).

Klein and Uhlmann (1974) represented a quite interesting method to determine the mechanism of crystal growth in glasses. In this method, the temperature dependence of the interface site factor can be determined by constructing a reduced growth rate, U_R, versus undercooling relation. The reduced growth rate is defined as

$$U_R = U\eta\left[1 - \exp\left(-\frac{\Delta H_f\,\Delta T}{RTT_E}\right)\right]^{-1} \tag{1.24}$$

where U is the growth rate, η is the viscosity, ΔH_f is the heat of fusion, ΔT is the undercooling ($\Delta T = T_E - T$), (T_E is the melting point and T is the temperature), and R is the gas constant.

According to the authors, a horizontal U_R versus ΔT relation indicates that the fraction of preferred growth sites on the interface do not change with temperature. In this case, the normal growth model provides a useful description of the crystallization process, in which atoms can be added to, or leave from, any site on the crystal–liquid interface. This model would be consistent with the experimental findings if the interface is rough on an atomic scale, i.e., possess a large fraction of step sites, where atoms can preferentially be added or removed.

If the U_R versus ΔT relationship exhibits a straight line of positive slope passing through the origin, growth by the screw dislocation mechanism is probably operative. By this mechanism, growth takes place preferentially, at step sites provided by screw dislocations, intersecting the interface. For the model to provide a useful representation of growth process, the interface must be rather smooth on an atomic scale and must be imperfect.

When the U_R versus ΔT relationship exhibits a positive curvature and a plot of log ($u\eta$) versus $1/T\,\Delta T$ is a straight line of negative slope, the kinetic model termed as "surface-nucleated growth" is probably operative. In this model, crystallization takes place at step sites provided by 2D nuclei formed on the interface. For the model to correspond to reality, the interface must be smooth on an atomic scale and must be free of intersecting screw dislocations.

1.2.2 Experimental Studies of Crystal Growth in Glass

The kinetics and mechanisms of crystal growth have been extensively studied for several inorganic glass-forming systems, within the last 30–40 years.

For example, the crystallization behavior of anorthite from a melt of the same composition was determined over the ranges of temperature between 1,173 and 1,273 K and between 1,523 and 1,773 K (Klein and Uhlmann, 1974). It was found that the triclinic form invariably appeared as the crystallization product, the interface morphology was faceted under all conditions of growth, and the growth took place preferentially in the *c* direction.

As was indicated previously, the nature of the crystal growth process in a material can be elucidated from the variation of reduced growth rate with undercooling. The reduced growth rate versus undercooling relationship constructed for the above material exhibited positive slope and positive curvature (Figure 1.21) indicating an interface site factor which increases continuously with increasing undercooling. This was consistent with the observed faceted interface morphologies and also indicated the existence of sizable anisotropy in the growth rate. These morphologies were similar to those observed in the growth from the melt of other materials, characterized by large entropies of fusion, predicted by Jackson (1958a,b).

As was stated earlier, the positive curvature of the U_R versus ΔT relationship (Figure 1.21) is probably suggestive of the occurrence of growth by a surface nucleation mechanism. To ascertain this possibility, the $\log(u\eta)$ versus $1/T\Delta T$ relationship was also constructed (Figure 1.21). The form of this relationship was not the straight line predicted by the standard models for such growth, but appreciable curvature was noted. Such curvature was not in accordance with that of the standard models for surface nucleation growth, which predict a straight line of negative slope, but was consistent with computer simulations of crystal growth in materials with large entropies of fusion.

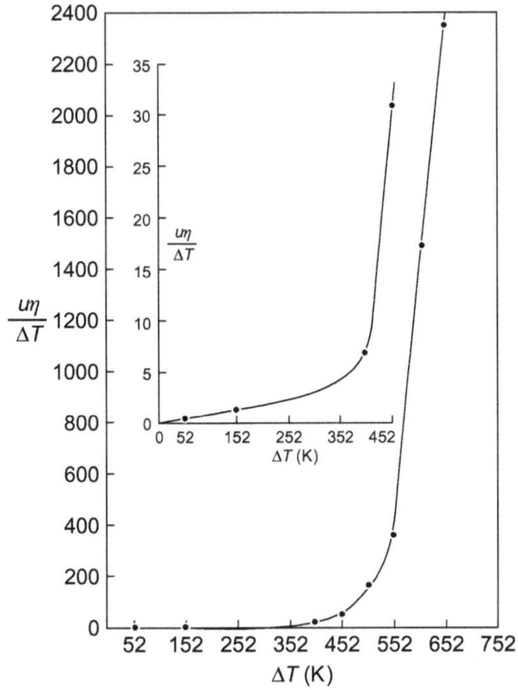

Figure 1.21 Reduced growth rate versus undercooling relation for anorthite. Insert shows small-undercooling data on expanded scale. (After Klein and Uhlmann, 1974. Reproduced with permission of John Wiley & Sons Publishing Company.)

Uhlmann (1982) reviewed the advances in the area of crystal growth in glass-forming systems. Summarizing the results of experimental data during the 1970–1980 period, it was commented that the computer modeling had supported the earlier suggestion of Jackson concerning the importance of fusion entropy (ΔS_f) in the characterization of the growth process in various materials. Moreover, the author noted that the standard models for crystal growth of materials, possessing high values of fusion entropy, represented limiting cases for the expected behavior and defects such as screw dislocations were unimportant in the growth of low ΔS_f materials and most important for the growth of high ΔS_f materials at modest undercoolings. The experimental data concerning the materials of high ΔS_f (to which the great majority of the studied glasses belonged) also revealed the existence of faceted morphologies or at least gave evidence of highly anisotropic growth. It was also observed that the fraction of preferred growth sites on the interface increased with increasing undercooling. At small undercoolings, the imperfect crystals seemed to grow with screw dislocation mechanism, while the perfect crystals of the same material exhibited surface-nucleated growth mechanism (Uhlmann, 1982).

The experimental data concerning the crystallization of materials in the condition, in which large composition changes existed, were in general accord with the expectations of diffusion-controlled growth. Growth rates independent of time were frequently observed; such growth rates correlated closely with interface morphologies which were fibrillar or dendritic in form. It was also concluded that phase separation generally does not have a significant effect on the crystal growth rate (Uhlmann, 1982).

Burgner and Weinberg (2001a), by means of analyzing crystal growth data obtained for several different inorganic glasses, concluded that the determination of growth mechanism from the temperature dependence of reduced growth rate is a complex process and the correct mechanism cannot be inferred from a mere inspection of the reduced growth rate versus undercooling plot. Moreover, they showed that the crystal growth mechanism of a given system can change several times and depend upon the region of undercooling.

The authors also stated that in order to establish the correct growth mechanism, a good agreement must be found between both the temperature dependence and magnitude of experimental and calculated growth rates, and it should also be ensured that the parameters entering into the equations are physically meaningful.

Burgner and Weinberg (2001b) also assessed the applicability of standard models of crystal growth to the rates of lithium disilicate crystal growth in a binary lithium silicate glass. It was found that the reduced growth rate versus undercooling graph consisted of three regimes. For undercoolings $<140°C$, the reduced growth rate curve suggested either 2D surface nucleation or screw dislocation growth. Although, for the 2D surface nucleation model, it was found that calculated growth rates were in excellent agreement with experimental results, with respect to both magnitude and temperature, the screw dislocation model provided an equally adequate description of experimental data, in the region of small undercooling. However, the values of the interface site factor obtained for this model were not physically meaningful. Moreover, the surface energy value obtained by interpreting the experimental data with respect to the screw dislocation model was unduly small.

For undercoolings $>400°C$, the reduced growth rate plot showed that the operative crystal growth mechanism is 2D surface nucleation, but detailed calculations cast doubt upon this conclusion. In the intermediate undercooling range, possibly a transitional behavior was operative to which none of the standard models appeared to be applicable.

According to the authors, although the interface site factor seemed to be relatively independent of temperature in this intermediate regime, it was highly unlikely that this region corresponded to a transition to the normal growth model. This deduction was based on the observation of spherulites and faceted surface crystalline layers in the microstructure of the glass specimens heat-treated at a temperature within the apparent transitional regime, namely, 635°C.

The authors concluded that crystal growth over a broad ΔT range is complex with many possible mechanism changes due to phase transitions, compound crystal growth, and other morphological changes. Therefore, the usual phenomenological laws cannot be applicable over the entire temperature range of measurable growth, but only over limited temperature regimes.

Švadlák et al. (2006) studied the crystal growth kinetics of Sb_2S_3 in $(GeS_2)_{0.2}(Sb_2S_3)_{0.8}$ glass by DSC and optical microscopy. Antimony trisulfide (Sb_2S_3) with an orthorhombic crystalline structure is a semiconductor that has many potential applications, such as solar energy conversion, optoelectronics in the IR region, and photocatalysis.

2D crystal growth was observed for the Sb_2S_3, the sole crystallizing phase, by optical microscopy at the temperature range of 525 K $\leq T \leq$ 556 K.

Also, the reduced growth rate plot as a function of supercooling gave a curve exhibiting positive curvature (Figure 1.22) and the plot of $\log(u\eta)$ versus $1/T\Delta T$ exhibited a straight line of negative slope. These observations indicated the occurrence of an interface-controlled 2D surface-nucleated growth mechanism for the crystallization of Sb_2S_3 crystals in the above glass.

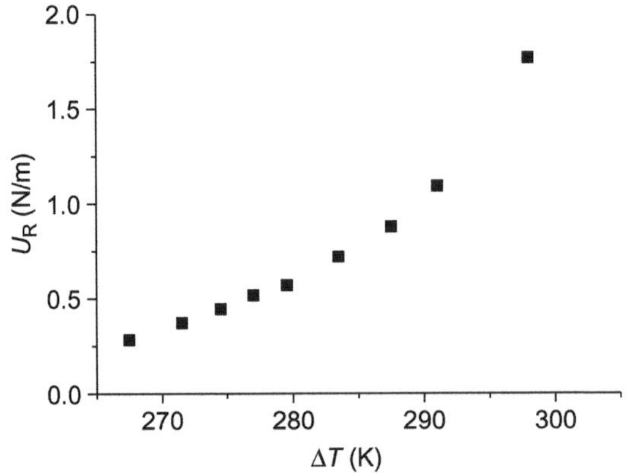

Figure 1.22 Reduced growth rate versus undercooling for crystallization of Sb_2S_3 in undercooled $(GeS_2)_{0.2}(Sb_2S_3)_{0.8}$ melt. (After Švadlák et al., 2006. Reproduced with permission of Elsevier Science Publishing Company.)

Reinsch et al. (2008) measured and collected literature data for the crystal growth rate, $u(T)$, of μ-cordierite and diopside in their isochemical glass-forming melts. The data covered the temperature ranges of 800–1,350°C for cordierite and 750–1,378°C for diopside. The maximum of $u(T)$ was observed at ~1,250°C for both systems. For μ-cordierite, a smooth shoulder was observed at ~970°C.

$U(T)$ was fitted using standard crystal growth models, utilizing the measured and collected viscosity data. For diopside, the experimental $u(T)$ fitted well both to the 2D surface nucleation and the screw dislocation growth models. However, the screw dislocation model yielded parameters of more significant physical meaning. For cordierite, also the experimental growth rates could be fitted well to these two models, but the best fittings of $u(T)$, including the observed shoulder, were attained for a combined mechanism. The good fittings indicated that viscosity can be used to assess the transport mechanism that determines the crystal growth in these systems, from the melting point T_m, to the vicinity of the glass transition temperature, T_g.

Barták and Málek (2012) investigated Sb_2S_3 crystal growth kinetics in $(GeS_2)_x(Sb_2S_3)_{1-x}$ thin films ($x = 0.4$ and 0.5) in the temperature range of 575–623 K. It was observed that the thin films crystallized as complex crystalline structures that were composed of bundles of submicrometer thin Sb_2S_3 crystals. The reduced crystal growth rate versus ΔT and $\ln(u\eta)$ versus $1/(T\Delta T)$ plots, for both compositions (with $x = 0.4$ and 0.5 values), showed linear dependence with a positive and negative curvature, for the respective plots. Therefore, an interface-controlled 2D surface-nucleated growth mechanism was suggested for the thin films. The calculated crystal growth rate for this model was in good agreement with experimental data. It was also observed that while the crystal growth rates of bulk glass and thin film are similar for $x = 0.4$ composition, the crystal growth rate is considerably higher in thin film than in bulk glass for $x = 0.5$ composition.

1.3 Alternative Mechanisms of Glass Crystallization at Low Temperatures

Though the aforementioned mechanisms of nucleation and growth for glass crystallization are now well established for numerous glass-forming systems, they have usually been substantiated by experiments carried out at relatively high temperatures, well above the glass transition temperature (T_g). The mechanisms of glass crystallization in the range of low temperatures, on the other hand, are not sufficiently studied. More recently, it has been proposed that near the glass transition temperature (T_g), the crystallization of glass may take place through the rearrangement of the glass structural elements. Metastable, intermediate phases of defected structure are then formed, and their composition and structure are rebuilt gradually with the temperature increase (Stoch, 2003, 2005). Glass crystallization in such conditions can be considered as **disorder–order transition**, analogous to the mechanism of the decomposition of solid solutions in crystal phases or the polymorphic transition of solids.

Stoch (2005) studied the early stages of crystallization of a glass located in the SiO_2–Al_2O_3–MgO system ($47SiO_2$–$22Al_2O_3$–17MgO–2ZnO–$12TiO_2$, wt%) near the temperature T_g (TiO_2 was used as a nucleant).

According to the author, the obtained results showed that the TiO_2 activation of crystallization of the above glasses, beginning near T_g, proceeded directly by successive displacement and ordering of atoms of the glass structure, similar to the transitions of disorder–order type, occurring in the crystalline solid bodies. Since the bonding of TiO_2 in silicate network is relatively weak, Ti^{4+} is mobile enough to form new crystal phases such as Mg,Al-titanates, as the first crystalline phase, at low temperatures around T_g. The silicate phases, including the so-called high quartz solid solution, then crystallized successively at higher temperatures around titanate phases. The higher the heating temperature of the glass, the stronger was the characteristic XRD peak of high quartz$_{ss}$ and above 825°C, it was the main crystal phase; above 1,000°C, cordierite ($2MgO \cdot 2Al_2O_3 \cdot 5SiO_2$) crystallized.

The investigation of the initial stage of crystallization of the above glass by HRTEM and electron diffraction also confirmed the formation of dendrite aggregates about 10–100 nm at 725°C (Dutkiewicz et al., 2003; Stoch, 2003).

Figure 1.23 shows a high-resolution micrograph, presenting a nucleus (upper right corner) of ~5 nm thickness and less than 20 nm length surrounded by an amorphous matrix in a glass crystallized at 725°C. In certain crystallite orientations, the phase contrast allows to differentiate lattice planes of ≈0.5 nm spacing being close to (020) planes of $MgTi_2O_5$ phase (Dutkiewicz et al., 2003). This observation is in accord with the XRD results regarding the formation of a titanate phase, as the first crystalline phase formed in the initial stage of crystallization of this glass.

It is interesting to note that the phenomenon of glass-in-glass phase separation, as a prerequisite of crystallization of

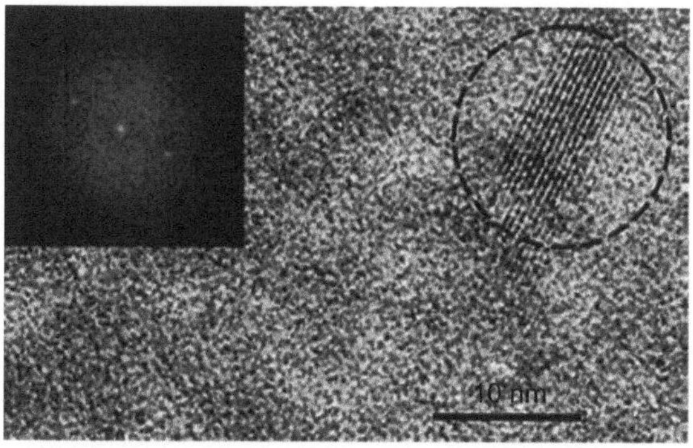

Figure 1.23 HRTEM and inserted Fourier transform from circled area taken from a glass heat-treated at 725°C for 4 h. (After Dutkiewicz et al., 2003. Reproduced with permission of Elsevier Science Publishing Company.)

these glasses, as suggested by several authors and discussed previously, was not observed in this study.

According to the authors (Dutkiewicz et al., 2003; Stoch, 2003), at temperatures near T_g, the glass still retained the properties of a rigid body, hence the rate of crystals growth was limited and the titanate grains did not exceed the nanometric dimensions. At higher temperatures, Ti is incorporated into the crystallizing Mg silicate structure and does not act directly as a nucleator in the heterogeneous glass crystallization process. Upon the elevation of temperature, the decomposition of the glass structure may selectively increase the mobility of certain components, which determines the composition and order of the subsequent formation of compounds crystallizing. However, if ZrO_2 is replaced by TiO_2 as a nucleating agent, the crystallization mechanism is radically altered (Stoch, 2005). In this case, the T_g and crystallization temperatures are both increased. In a two-stage crystallization, first ZrO_2 is precipitated, then acted as a heterogeneous nucleating agent for the crystallization of the final product, cordierite. Obviously, in this case, the intermediate metastable phases of Mg–Al titanate and high-temperature quartz solid solution do not appear.

This is an interesting example for the phenomenon that the type of the chosen nucleation agent may alter the crystallization mechanism and the products of glass crystallization.

1.4 Overall Glass Crystallization Kinetics

1.4.1 Theoretical Background

The kinetics of overall crystallization process in glass is usually described by a theory derived independently, in the late 1930s by Kolmogorov (1937), Johnson and Mehl (1939), and Avrami (1939,

1940, 1941) best known as Johnson–Mehl–Avrami–Kolmogorov theory (JMAK). If the conditions on which the JMAK theory was derived are observed, the volume fraction transformed as a function of time, $\alpha(t)$, in isothermal conditions, is given by

$$\alpha(t) = 1 - \exp\left[-g\int_0^t I(t')\left(\int_{t'}^t U(t)\mathrm{d}t\right)^3 \mathrm{d}t'\right] \tag{1.25}$$

where g is a shape factor, which is equal to $4\pi/3$ for spherical crystals, and I and U are the crystal nucleation and growth rates, respectively.

This equation takes up two limiting forms. For constant crystal nucleation and growth rates throughout the transformation (homogeneous nucleation), Eq. (1.25) can be expressed as

$$\alpha = 1 - \exp\left(-\frac{gIU^3t^4}{4}\right) \tag{1.26}$$

On the other hand, if the number of growing crystals does not change, e.g., all nucleation centers are exhausted at an early stage of transformation, crystal growth occurs from a fixed number of sites and the above equation can be written as

$$\alpha = 1 - \exp(-gNU^3t^3) \tag{1.27}$$

where N is the number of sites per unit volume.

Avrami has proposed the following general relation for isothermal crystallization processes:

$$\alpha = 1 - \exp(-kt^n) \tag{1.28}$$

where α is the fraction of material crystallized after time t and n is a constant known as "Avrami exponent" that, depending upon the crystallization mechanism, typically has values between 1 and 4; k is reaction rate constant, which is primarily a function of nucleation and crystallization rates and related to temperature by the Arrhenius equation:

$$k = \upsilon \exp(-E/RT) \tag{1.29}$$

where E is the activation energy, υ is a frequency factor, and R is the universal gas constant.

Equation (1.28) is usually linearized and takes the following form:

$$\ln\{-[\ln(1-\alpha)]\} = \ln k + n \ln t \tag{1.30}$$

Therefore, the Avrami coefficient, n, can be calculated from the slope of a $\ln[-\ln(1-\alpha)]$ versus $\ln(t)$ plot. The linearity of such plot is taken as an indication of the validity of the JMAK equation for the case under study.

Equation (1.28) is only valid for isothermal crystallization, but it has been used frequently for the interpretation of nonisothermal DTA/DSC experiments. Although in some cases reasonable activation energies could be obtained, **but this procedure is not appropriate and could not be recommended**.

Furthermore, it is known that Eq. (1.28) must be modified (or is totally erroneous) for certain types of isothermal crystallization processes (Weinberg and Birnie, 1996).

Some examples of the above-mentioned situations can be encountered in systems where finite size effects are important or multiple crystallization mechanisms occur (Weinberg, 1991, 1992) and where the second phase particles are anisotropic (Shepilov and Baik, 1994).

Thermal analysis is a very useful tool for studying the kinetics of crystallization in glass for nonisothermal conditions, as a rapid and convenient means. Hence, many investigators used DTA and DSC techniques to determine kinetic parameters such as activation energy for crystallization of glass, and n and m constants which are related to the mechanism of crystallization and directionality of crystal growth, respectively.

For nonisothermal conditions, the crystallization kinetics can be studied by reheating the glass at a constant rate, $\varphi = \mathrm{d}T/\mathrm{d}t$, so that at any time t, the temperature T is given by

$$T = T_o + \varphi t \tag{1.31}$$

where T_o is the starting temperature.

Under this condition, k is no longer a constant but varies with time, and the Avrami equation is modified as

$$y = 1 - \exp\left[-\int_0^t k(t)\mathrm{d}t\right]^n \tag{1.32}$$

using Eq. (1.32) and assuming n as a constant, the following equation has been derived for the maximum rate of crystallization (Bansal et al., 1983).

$$\ln(T_p^2/\varphi) = \ln(E/R) - \ln \upsilon + E/RT_p \tag{1.33}$$

where T_p is the temperature for the crystallization peak at a heating rate φ.

Equation (1.33) is basically the same as the equation derived by Kissinger (1956), using different approaches as

$$\ln(T_p^2/\varphi) = E/RT_p + \text{constant} \tag{1.34}$$

Some authors used the above equation to examine the kinetics of crystallization in glass. Since this equation has been proposed for the determination of chemical reactions order, it is not suitable for the study of glass crystallization that proceeds by the nucleation and growth processes and the activation energies thus obtained would be physically meaningless.

According to Eq. (1.33), a plot of $\ln(T_p^2/\varphi)$ versus $1/T_p$ should yield a straight line. The activation energy E and the frequency factor υ can be calculated from the slope and intercept, respectively.

Matusita et al. (1984) also derived an equation for the determination of kinetic factors of nonisothermal crystallization process in glass, as follows:

$$-\ln(1-x) = (k\alpha^{-n})\exp(-1.052mE/RT) \qquad (1.35)$$

Here x is the volume fraction of crystals precipitated in a glass when it was heated up with a constant rate of α to the temperature of T. The value of x can be calculated from the ratio of the partial area at the temperature T to the total area of the exothermic crystallization peak, and m is the growth dimensionality of the precipitating crystals.

Similarly, for the variation of crystal volume fraction the equation was derived as

$$dx/dt = k(1-x)\alpha^{-(n-1)}\exp(-1.052mE/RT) \qquad (1.36)$$

Here, $n = m + 1$ for a quenched glass containing no nuclei and $n = m$ for a glass containing a sufficiently large number of nuclei. In the cases of 3D, 2D and 1D crystal growth, $m = 3$, 2, and 1, respectively. For example, in a thin film glass (2D growth), $m = 2$ and in the surface crystallization or fiber glass (1D growth) $m = 1$.

The n-value can be obtained from the plot of $\ln[-\ln(1-x)]$ versus $\ln \alpha$ at a specific temperature. Theoretically, the maximum value of n is 4 and minimum value is 1. Hence, the corresponding m-values must be 3 and 1, respectively.

In the cases in which $n = 2$, the corresponding m-value would be 2 or 1, and when $n = 3$, the corresponding m-value is 3 or 2.

It should be noted that the aforementioned values of n and m parameters have all been determined assuming an interface-controlled mechanism of crystal growth. As stated in Section 1.2.1, for the stoichiometric glass compositions that do not undergo compositional changes during crystallization, long-range diffusion is not essential for crystal growth, thus, interfacial rearrangements are likely to control the crystal growth process. Obviously, when considerable compositional differences exist between the parent glass and the growing crystals, long-range **diffusional processes** may control the growth rate of crystals.

The Avrami parameter, n, which is related to the mechanism of crystallization in glass, can also be determined from the shape of the DTA crystallization exotherm, using the following equation derived by Augis and Bennett (1978):

$$n = (2.5/\Delta T)(RT_p^2/E) \quad (1.37)$$

where ΔT is the width of the crystallization exotherm at half maximum.

According to this equation, a sharp peak (small ΔT, large n) implies bulk crystallization, while a broad peak (large ΔT, small n) signifies surface crystallization. The values of n and m for various crystallization mechanisms, as summarized by Donald (2004), are depicted in Table 1.4.

In order to obtain the activation energy, Eq. (1.35) was rewritten as

$$\ln \alpha = -1.052m/nE/RT - 1/n \ln[-\ln(1-x)] + \text{constant} \quad (1.38)$$

where E is the activation energy for crystal growth and n is the Avrami exponent.

Thus the plot of $\ln \alpha$ against $1/T$, where T is the temperature at which the crystal volume fraction reaches a specific value, gives a straight line and the slope gives the value of $1.052(m/n)E$. The activation energy can be obtained when the ratio m/n is known. Since this plot was very similar to the so-called Ozawa plot, Ozawa (1971) derived for chemical reactions as

$$\{d \ln[-\ln(1-x)]/d \ln\alpha\}_T = -n \quad (1.39)$$

the authors called it a "**modified Ozawa-type plot**."

Since it is known that the volume fraction of crystals at the peak temperature, T_p, in DSC or DTA curves is almost the same irrespective of α, this equation should apply for the peak temperature (Matusita et al., 1984).

The rate of increase of x reaches its maximum at a temperature T_p. Solving Eq. (1.36) for $d(dx/dt)/dt = 0$, the following equation was derived:

$$\ln(\alpha^n/T_p^2) = -1.052rnE/RT_p + \text{constant} \quad (1.40)$$

This is very similar to the so-called Kissinger equation (Eq. [1.34]) when $n = m = 1$, so they called it a **modified Kissinger-type equation**.

It should be noted that the existence of the above condition of $n = m = 1$ is the only condition in which the use of the Kissinger equation (Eq. [1.34]) is permissible for the interpretation of kinetics of glass crystallization in nonisothermal conditions.

Table 1.4 The Values of *n* and *m* Parameters Used in the Determination of Activation Energies and Mechanisms of Crystallization

Crystallization Mechanism	*n*	*m*
Bulk crystallization with a constant number of nuclei (i.e., a well-nucleated sample with the number of nuclei being independent of the heating rate)		
3D growth of crystals	3	3
2D growth of crystals	2	2
1D growth of crystals	1	1
Bulk crystallization with a constant number of nuclei (i.e., a well-nucleated sample with the number of nuclei being independent of the heating rate) with crystal growth rate proportional to $t^{-0.5}$ (diffusion controlled)		
3D growth of crystals	1.5	1.5
2D growth of crystals	1	1
1D growth of crystals	0.5	0.5
Bulk crystallization with an increasing number of nuclei (i.e., an as-quenched sample with the number of nuclei inversely proportional to the heating rate)		
3D growth of crystals	4	3
2D growth of crystals	3	2
1D growth of crystals	2	1
Bulk crystallization with an increasing number of nuclei (i.e., an as-quenched sample with the number of nuclei inversely proportional to the heating rate) with crystal growth rate proportional to $t^{-0.5}$ (diffusion controlled)		
3D growth of crystals	2.5	1.5
2D growth of crystals	2	1
1D growth of crystals	1.5	0.5
Surface crystallization	1	1

Source: After Donald (2004).

Usually, the change of $\ln T_p^2$ with α is negligibly small compared with the change of $\ln \alpha^n$, and, therefore, the modified Kissinger-type equation (Eq. [1.40]) is essentially the same as the modified Ozawa-type.

Marotta et al. (1982) also proposed a method of analysis to obtain the activation energy of crystal growth in glass from a single derivative differential thermal analysis (DDTA) curve. They derived the following equations for the calculation of activation energies for the bulk and surface crystallization of glass:

$$E/R(1/T_{f1} - 1/T_{f2}) = 1.59 \quad \text{(for surface crystallization)} \tag{1.41}$$

$$E/R(1/T_{f1} - 1/T_{f2}) = 0.64 \quad \text{(for bulk crystallization)} \tag{1.42}$$

where T_{f1} and T_{f2} are the temperatures of two inflection points of the DTA crystallization peak. If the DTA curve and its derivative are simultaneously recorded, the two inflection points, i.e., the maximum and the minimum slopes, of the DTA peak would correspond to the maximum and the minimum of the DDTA double peak, as shown in Figure 1.24. Therefore, the temperatures T_{f1} and T_{f2} can be easily and exactly detected on the DDT curve.

For studying the surface and bulk crystallization respectively, as-quenched powdered samples and well-nucleated bulk samples have to be used.

1.4.2 Experimental Studies of the Crystallization Kinetics in Glass

Numerous authors have utilized the aforementioned DTA techniques to determine the kinetic parameters of glass crystallization since the advent of these methods.

For example, Ray and Day (1990) determined the activation energy for crystallization, *E*, and the Avrami exponent, *n*, of a lithium disilicate glass, nucleated at different temperatures for 3 h, utilizing the Kissinger and Augis and Bennett equations (Eqs. [1.34] and [1.37], respectively).

As expected, *E* was practically independent of the nucleation temperature (and, hence, concentration of nuclei), and its value was consistent with the results of other investigators.

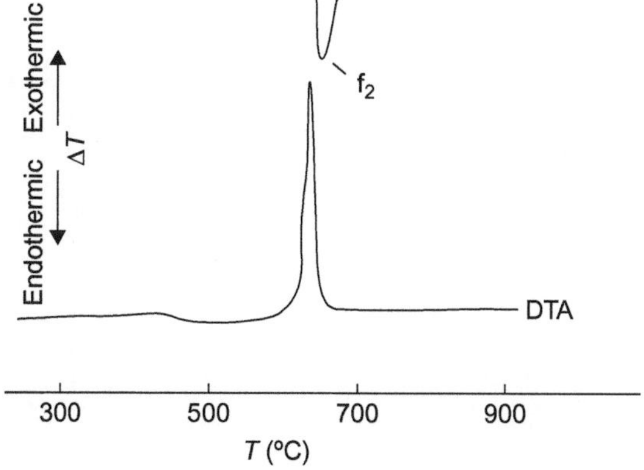

Figure 1.24 DTA and DDTA curves of a bulk sample of $Li_2O.2SiO_2$ glass. (After Marotta et al., 1982. Reproduced with permission of Springer Science + Business Media, Inc.)

But it is very interesting to note that the Avrami exponent, n, varied markedly with the alteration of nucleation temperature. Its value increased from 1.5 to 3.8 (Figure 1.25) as the concentration of nuclei reached a maximum at 453°C.

This indicates that the mechanism of crystallization (hence, of nucleation) changed from surface to bulk as the concentration of nuclei increased. Since, upon the increase of the nucleation temperature above 453°C, the concentration of nuclei in the glass decreased again, the bulk crystallization process became less dominant, and the n value dropped. Yekta and Marghussian (2001) studying the effect of P_2O_5, B_2O_3, and PbO additives on the sintering and crystallization behavior of β-quartz$_{ss}$ and gahnite glass ceramics determined the crystallization mechanism of the glasses (n parameter) by the Ozawa method. The aforementioned authors also obtained the crystallization activation energy of glasses utilizing the modified Kissinger, Matusita, and Marotta methods.

The glass compositions which were located in the $ZnO-Al_2O_3-SiO_2$ system are given in Table 1.5.

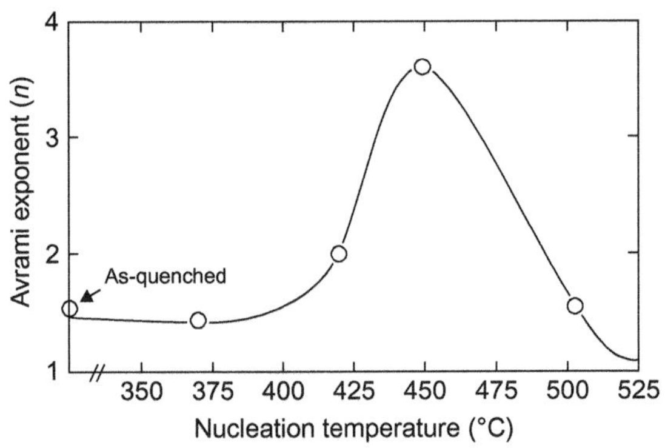

Figure 1.25 Dependence of Avrami exponent, n, on the nucleating temperature for the $Li_2O-2SiO_2$ glass (nucleation time = 3 h). (After Ray and Day, 1990. Reproduced with permission of the American Ceramic Society.)

Table 1.5 Chemical Composition of β-quartz$_{SS}$ and Gahnite Glass Ceramics (wt%)

Glass	SiO_2	Al_2O_3	ZnO	ZrO_2/TiO_2	Li_2O	As_2O_3	PbO
A (β-quartz$_{ss}$)	45.57	22.78	22.78	6.37(ZrO_2)	2	0.49	—
A PbO17.5	45.57	22.78	5.28	6.37(ZrO_2)	2	0.49	17.50
B (Gahnite)	46.50	23.25	23.25	6.50(TiO_2)	—	0.50	—
B PbO17.5	38.36	19.18	19.18	5.36(TiO_2)	—	0.41	17.50

The m and n constants and activation energy for crystallization of the glasses determined from the above-mentioned methods are summarized in Table 1.6.

It can be seen that while there is no difference between the activation energies for crystallization of the two gahnite glass ceramics (with PbO additive and without additive), the activation energy for crystallization of glass APbO17.5 is higher than the base glass A. In this way, the improved sinterability (accompanied with the suppressed crystallization tendency) of the lead-bearing β-quartz$_{ss}$ glass ceramics as compared with the lead-free ones could be explained. This suppressed crystallization tendency which was manifested by the increase in the DTA exothermic peak temperatures and the shortening and broadening of these peaks (Figure 1.26) was further verified by the observed decrease in n and m values (from 3 and 2 to 1 and 1) indicating a change in crystallization mechanism from 2D volume crystallization to surface crystallization, and an increase in the activation energy for the crystallization in PbO-bearing A glass.

It is also interesting to note, while in some cases there is a good match between the values of activation energies obtained using different methods, in some cases a marked inconsistency is observed in these values.

Rezvani et al. (2005b) investigated the crystallization mechanism and activation energy for crystallization of glasses in the SiO_2–Al_2O_3–CaO–MgO system, containing a mixture of Cr_2O_3, Fe_2O_3, and TiO_2 (in the single, double, and triple nucleant series) using various methods, based on DTA. The base glass composition was SiO_2 55.05, Al_2O_3 13.61, CaO 24.42, MgO 6.92, Na_2O 2.82, and K_2O 3.02 (weight part). The nucleating agents, Cr_2O_3, Fe_2O_3, and

Table 1.6 *m* and *n* Parameters and the Crystallization Activation Energies of Glasses with the Chemical Compositions Given in Table 1.5

Glass	*n*	*m*	Activation Energy (kJ/mol)		
			Marrota	*Matusita*	*Modified Kissinger*
A	3	2	301.6	303.5	370.5
A PbO17.5	1	1	373.7	309.5	392.0
B	3	2	259.6	323.0	381.0
B PbO17.5	3	2	260.0	359.0	381.0

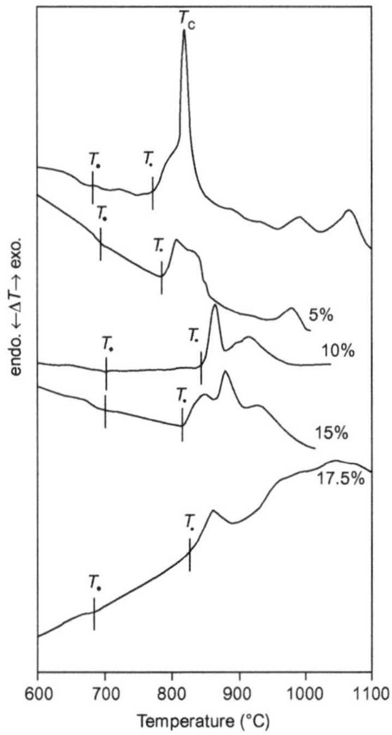

Figure 1.26 DTA traces of glass A containing various amount of PbO. The composition of glass is given in Table 1.5. (After Yekta and Marghussian, 2001. Reproduced with permission of Springer Science + Business Media, Inc.)

TiO_2 were added to the base glass composition in various combinations. The Avrami constants and activation energy of crystallization of glasses were determined by Matusita, modified Kissinger, and Marotta methods. Table 1.7 summarizes the results. According to this table, in each method, specimens containing three weight part of Cr_2O_3 exhibit the minimum crystallization activation energy within the single nucleant-containing series. But in the specimens containing a mixture of Cr_2O_3, Fe_2O_3, and TiO_2, generally a reduction of E and an increase of n value are observed. Considering the Avrami constants in each method and the differences of crystallization peak shifts for fine and coarse particles of different compositions (as discussed in Section 1.1.2.2.5 and given in Table 1.3), the authors deduced that the n values determined by the Marotta method are more reasonable than the others.

According to the data exhibited in Table 1.7, the minimum crystallization activation energy and maximum Avrami constants were obtained for the specimen AR-Cr3Fe5Ti5.

Regarding the values of $n = \sim 3$ and $m = \sim 2$, a 2D bulk crystallization mechanism was anticipated for this specimen, which

Table 1.7 Avrami Exponents and Crystallization Activation Energies Determined by Various Methods for the Prepared Glasses

Glass*	Avrami Constant, *n*		Activation Energy (kJ/mol)		
	Matusita	*Marotta*	*Matusita*	*Marotta*	*Modified Kissinger*
AR-Cr3	2.8	1.9	366.4	314.0	334.4
AR-Fe5	2.7	2.0	399.0	343.7	378.1
AR-Ti5	1.7	1.5	494.3	440.7	489.4
AR-Cr3Fe5	3.0	2.6	356.3	321.5	341.4
AR-Cr3Fe5Ti5	3.2	3.0	254.0	207.5	183.2

*Suffixes Cr, Fe, and Ti denote Cr_2O_3, Fe_2O_3, and TiO_2 respectively and the numbers following the suffixes indicate the wt% of each oxide.

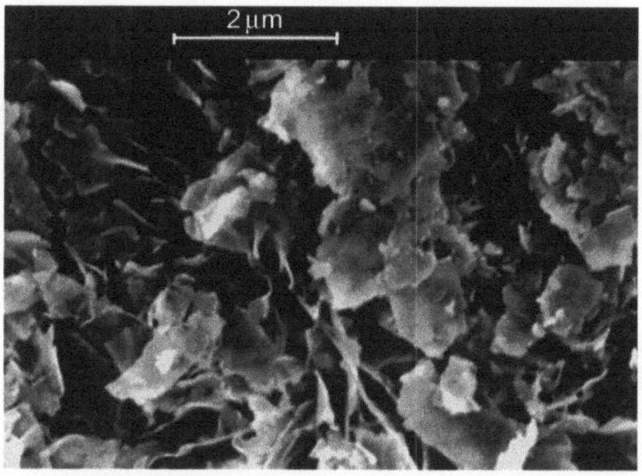

Figure 1.27 SEM micrograph of AR-$Cr_3Fe_5Ti_5$ glass ceramic nucleated at 740°C for 3 h and then crystallized at 885°C for 3 h. (After Rezvani et al., 2005b. Reproduced with permission of Elsevier Science Publishing Company.)

was in accord with the scanning electron microscopy (SEM) micrograph of the specimen (Figure 1.27).

The values of the activation energies obtained by different methods (Table 1.7), with some exceptions, show a reasonable consistency.

Clupper and Hench (2003) studied the crystallization kinetics of the tape-cast bioactive glass 45S5 of the composition $45SiO_2$, $24.5Na_2O$, 24.5CaO, $6P_2O_5$ (wt%) using DTA. $Na_2Ca_2Si_3O_9$ phase was crystallized during heating of the glass up to 1,000°C.

The activation energy for crystallization was determined by the modified Kissinger method and the Avrami exponent, n, was calculated utilizing both the Ozawa and Augis–Bennett methods. The obtained activation energy value was 350 kJ/mol and the n exponents were 0.96 and 0.94 for the Ozawa and Augis–Bennett methods, respectively.

Obviously, there is a close agreement between the values of the Avrami exponent, obtained using the Ozawa and Augis–Bennett methods. Such results ($n = \sim 1$) indicated that the glass has undergone diffusion-controlled surface crystallization during the heat treatment (see Table 1.4).

Considering the small average particle size of 3 μm for the initial glass specimen prepared by jet milling, and its large surface area relative to the total material volume, the dominant surface crystallization mechanism seems to be quite reasonable.

The calculated activation energy of 350 kJ/mol for this glass is relatively lower in comparison with the reported values of the activation energy of $CaO–P_2O_5–SiO_2$ glasses, which does not contain Na_2O. Given the relatively high content of Na in the composition of the studied glass, it is reasonable that it exhibited a relatively disrupted structure and lower viscosity. This resulted in a lower activation energy and higher crystallization rate in the glass.

Pablos-Martín et al. (2011) studied the crystallization behavior of a $55SiO_2–20Al_2O_3–15Na_2O$-10 LaF_3 (mol%) glass that resulted in the formation of LaF_3 nano-crystals. They also determined the Avrami exponent, n, using the Ozawa method and the m parameter using the Matusita method. The activation energy of the crystallization for the above glass was also determined utilizing several methods. These values of activation energy were in good agreement with each other.

The values of n and m parameters, obtained by the authors in the crystallization temperature range 660–680°C, were $m = n \cong 1$. This indicates the occurrence of a bulk crystallization mechanism with a constant number of nuclei (i.e., a well-nucleated sample in which the number of nuclei is independent of the heating rate).

It was shown that the phase-separated droplets present in the as-prepared glass contained not only lanthanum and fluorine but also silicon and oxygen. During the crystallization process, several LaF_3 crystals were nucleated within the droplets, but their growth was limited to some extent by an excess of silicon. In this condition, the growth rate was mainly controlled by the diffusion of fluorine. Hence, considering the $m = n \cong 1$ relation, according to the parameters summarized in Table 1.4, the authors rightly assigned a bulk crystallization mechanism to this glass, in which the crystals grow in 2D with a diffusion-controlled process.

Kržmanc et al. (2011) studied the nucleation and crystallization of $MgO-B_2O_3-SiO_2$ (MBS) glass by means of a nonisothermal, thermal analysis technique, XRD and SEM. The temperature range of the nucleation and the temperature of the maximum nucleation rate for MBS glass were determined from the dependences of the inverse temperature at the DSC peak ($1/T_p$) and the maximum intensity of the exothermic DSC crystallization peak (δT_p) on the nucleation temperature (T_n). For MBS glass, the nucleation occurred at 600–750°C, with the maximum nucleation rate at 700°C, whereas the nucleation and crystal growth took place simultaneously in the narrow temperature range 700–750°C. The analyses of nonisothermal data for the bulk MBS glass, using the most common models (Ozawa, Kissinger, modified Kissinger, Ozawa–Chen, etc.) showed that $n = m = 1.5$. Hence, it was deduced that the crystallization of $Mg_2B_2O_5$ from the MBS glass was 3D bulk with a diffusion-controlled crystal growth rate.

SEM investigation also revealed the presence of 3D, spherical, $Mg_2B_2O_5$ crystallites in the microstructure of the resulting glass ceramics, consistent with the calculated crystallization parameters, m and n. Moreover, different methods (Kissinger, Ozawa–Chen) yielded very similar values of activation energy (E), for the same type of sample (bulk or powder). However, higher values of E were obtained for the nucleated bulk glass than for the glass powder, which was attributed to the simultaneous occurrence of both surface and bulk crystallization. According to the authors, consistency of the E values obtained using different methods confirmed the reliability of the nonisothermal DSC methods for the study of the crystallization kinetics in the MBS glasses.

1.5 Concluding Remarks

As stated above, in preparation of glass ceramics, and especially nano-glass ceramics with tailor-made properties, the knowledge on nucleation mechanism and kinetics is very critical.

On the other hand, as stated above, the shortcomings of the CNT in explaining the precise process of the crystal nucleation in glass has been recognized in recent years. Therefore, it has been felt that further experimental and theoretical work should be conducted to obtain more extensive and precise information on the nucleation and crystallization in glass.

The following areas can be named as the most revealing and important areas of the current and future research in this field:

1. The research activities to overcome the difficulties of the classical theory and to represent a more accurate and satisfactory theory of crystal nucleation in glass through nonclassical theories. So far, these activities have mainly been based on the interfacial information provided by computer simulations and semiempirical calculations.
2. The experimental and theoretical work conducted to obtain more extensive and precise information on the role of the epitaxial relationship in the heterogeneous nucleation of glass systems. So far, relatively little information exists in this field and without such information, the issue of heterogeneous nucleation in glass would remain more or less unpredictable, mainly based on trial and error.
3. Exploration of the alternative ways of explaining the process of nucleation in nano-glass ceramics. The crystallization of these materials is preferably carried out in relatively low temperature ranges, near T_g, in which, as stated above, at times the mechanism of nucleation may not properly be described by the classic CNT theory. In this condition, the crystallization of glass may take place through the rearrangement of the glass structural elements, similar to the disorder–order transition. This point, which is very important in crystallization of the nano-glass ceramics, has been generally neglected and much further work is required to clarify the subject.

2

OPTICAL PROPERTIES OF NANO-GLASS CERAMICS

CHAPTER OUTLINE

The most important optical property of nano-glass ceramics is their transparency, i.e., the ability to transmit light (electromagnetic waves), in certain range of wavelengths according to their specific application. The transparency, however, is not the sole requirement to be fulfilled by these glass ceramics as promising candidates for various current applications, or potential applications in the near future. The great attention that has been attracted in recent years by nano-glass ceramic is mainly because of their ability to combine transparency with other desired properties, such as mechanical, thermal, chemical, and electromagnetic.

2.1 Theoretical Background of Transparency

There are two main mechanisms that may hinder the travel of light through a glass ceramic, which are as follows:

1. The light scattering due to the presence of two (or more) phases, the glass matrix and the dispersed crystalline particles, possessing different refractive indexes.

V.K. Marghussian, Nano-Glass Ceramics, ISBN: 978-0-323-35386-1.

DOI: http://dx.doi.org/10.1016/B978-0-323-35386-1.00002-5

2. The absorption of light by ionic/atomic species which are present in both the glass matrix and dispersed particles.

Effect of the first mechanism, the light scattering, which is the far more effective obstacle for the transmission of light through glass ceramics, can be minimized by (a) trying to achieve closely matched indexes of refraction between the two (or more) phases existing in the glass ceramic and low birefringence in the crystals or (b) by reducing the size of the dispersed crystalline particles to much smaller sizes than the wavelength of the incident light (Beall and Pinckney, 1999).

Mg–Zn stuffed β-quartz solid solution is an example for the criterion (a) in which, despite crystal sizes of up to 10 μm, good transparency can be achieved, whereas the nano-glass ceramics should satisfy the second criterion (small crystallite size).

Among several scattering theories, there are two theories which could better describe the mechanism of scattering in nano-glass ceramics. The first theory, known as Rayleigh–Gans model (Kerker, 1969) assumes the existence of widely separated independent scatterers in a glass matrix. In this case, σ_p, the total turbidity or attenuation due to scattering, is given as $\sigma_p \approx (2/3)NVk^4a^3(n\,\Delta n)^2$, where N is the particle number density, V the particle volume, a the particle radius, $k=(2\pi/\lambda)$ (where λ is the wavelength), n the refractive index of the crystal, and Δn the index difference between the crystal and the matrix. For practical purposes, transparency is achieved here with particle radii of <15 nm and a refractive index difference of <0.1 between the glass and the dispersed crystals (Beall and Pinckney, 1999).

The other scattering model assumes the existence of small particles that are more closely spaced; the distance between particles should be no smaller than the particle radius but can be up to 6 times the particle radius. In this condition, the turbidity is given by the equation as developed by Hopper (1985):

$$\sigma_c \approx \left[\left(\frac{2}{3}\times 10^{-3}\right)k^4\theta^3\right](n\,\Delta n)^2$$

where $\theta=[a+(W/2)]$ is the mean phase width, in which a and W are the particle radius and the inter-particle spacing, respectively. In this case, improved transparency is allowed with particle sizes <30 nm at larger refractive index differences, up to $\Delta n = 0.3$ (Beall and Pinckney, 1999).

2.2 Application of Optical Nano-Glass Ceramics

As stated above, nano-glass ceramics as good candidates for various diversified applications are expected to have the capability of combining good transparency with other desired properties regarding the given application. One of the classification methods of these materials is according to the main characteristics (other than transparency) that determine their application. In this way, the transparent nano-glass ceramics can be classified into several groups.

In this chapter, the processing, properties, and application of some of the most important transparent nano-glass ceramics are discussed.

2.2.1 Low Thermal Expansion Glass Ceramics

These glass ceramics usually combine the transparency with low thermal expansion and high mechanical strength. Originally developed for use in the high-precision optical applications such as telescope mirror blanks, these glass ceramics have become known and entered the domestic market in applications such as cooker tops, cookware, and as reflectors for digital projectors. The most important materials of this group are stuffed **β-quartz$_{ss}$ glass ceramics**, the detailed description of the structure and processing of which is discussed in the following sections.

2.2.1.1 Structure, Properties, and Application of Stuffed β-Quartz$_{ss}$ Glass Ceramics

Buerger (1954) was the first investigator who recognized that certain aluminosilicate crystals, composed of three-dimensional networks of SiO_4 and AlO_4 tetrahedra, are similar in structure to crystalline forms of silica. These so-called stuffed derivatives of silica polymorphs may be imagined of as derived from silica networks by replacement of Si^{4+} by Al^{3+}, accompanied by the filling of structural vacancies by monovalent cations.

β-Eucryptite, $LiAlSiO_4$, and β-spodumene $LiAlSi_2O_6$, which are respectively similar to β-quartz and keatite structures, are such stuffed derivatives.

Monovalent Li^{+}, divalent Mg^{2+}, and, to a smaller extent, divalent Zn^{2+} ions can also randomly fill the interstitial vacancies in the β-quartz structure when Al^{3+} replaces Si^{4+} producing stuffed β-quartz solid solutions (β-$quartz_{ss}$).

The aforementioned materials are well known for their very low thermal expansion coefficients over considerable temperature intervals. The two unique properties of these glass ceramics, the ultralow thermal expansion and the ability to be polished similar to a glass, allow these transparent glass ceramics to be very suitable for some optical applications such as mirror blank materials.

For such uses, the thermal expansion of the material is very important, since any change in ambient temperature in the neighborhood of the mirror during use may result in a change of focus and lost time on the telescope.

On the other hand, since most of the ceramic materials have low thermal conductivities, there is the possibility of large thermal gradient buildup during the dissipation of frictional heat. In order to prevent the distortion of glass bodies, often a long and tedious procedure should be used during the finishing process. The use of ultralow expansion materials would help in minimizing the distortion problems and accelerating the finishing process (Duke and Chase, 1968).

As stated previously, the very fine microstructure of these transparent glass ceramics is also an important property. Since glass ceramics are polyphase crystalline assemblages, with significant residual glass, polishing problems could arise due to differential hardness between the phases. It was found that when the grain size was kept smaller than the wavelength of visible light, an optical finish, similar to that possible with glasses, could be obtained with no evidence of relief polishing (Duke and Chase, 1968).

Their transparent nature also allows inspection of the mirror blanks for residual stress and quality.

2.2.1.2 Processing of Stuffed β-$Quartz_{ss}$ Glass Ceramics

The composition of transparent glass ceramics, comprising β-$quartz_{ss}$ crystallites as their main crystalline phase, is usually located in the system SiO_2–Al_2O_3–Li_2O with added TiO_2 and ZrO_2 as nucleation agents. The stuffed β-$quartz_{ss}$ crystals are usually precipitated in the parent glass at relatively low temperatures, in the range of 700–850°C. The resulting glass ceramics owe their transparency to the very small size of the dispersed crystallites in the matrix, and/or the existence of complete optical isotropy in their structure, i.e., a minimized refractive index difference between all crystalline and amorphous phases and small birefringence within any single crystalline phase.

Although nucleation is the key step in obtaining fine-grained glass ceramics, only a small number of possible nucleating agents are known today that induce precipitation of β-quartz phases with crystallite sizes well below 100 nm in $Li_2O-Al_2O_3-SiO_2$ (LAS) glasses. Among these, TiO_2, ZrO_2, and combinations of them are still considered to be the most prominent representatives (Wondraczekw and Pradeau, 2008).

Many investigators have shown that the heterogeneous nucleation in LAS glass ceramics is often associated with an initial amorphous phase separation, followed by the formation of crystalline phases containing nucleating oxides.

Doherty et al. (1967) used transmission electron microscopy (TEM) of thin sections, produced from bulk samples, to investigate the nucleation and crystallization of LAS glasses containing TiO_2. They showed that during cooling from the melt, phase separation occurred in the glass specimens, and on subsequent heat treatment a large number of titanium–aluminum crystals, ~50 Å in diameter, were formed in the specimens. These crystals served as heterogeneous nuclei for the crystallization of the remaining glass. Photomicrographs of various stages of crystallization showed the development of fine-grained glass ceramics.

Beall et al. (1967) showed that metastable solid solutions with the β-quartz structure could be crystallized from most glasses in the system $SiO_2-Mg(AlO_2)_2-LiAlO_2$, as well as the glasses containing the additional components $Zn(AlO_2)_2$, $Al(AlO_2)_3$, Li_2ZnO_2, and Li_2BeO_2. Internal nucleation was induced by the addition of ZrO_2 or TiO_2. Either transparent or opaque crystalline materials could be formed from glasses containing ~70% SiO_2. Transparency was realized due to the characteristics of the major stuffed β-quartz phase, exhibiting low birefringence and minute crystal size. The thermal expansion values varied from -20 to $+50 \times 10^{-7}/°C$.

Duke and Chase (1968) described the sequence of nucleation and crystallization of β-quartz glass ceramics in more detail. According to them, various glass compositions in the basic $Li_2Al_2O_4-SiO_2$ system, nucleated with titania and zirconia, can be crystallized uniformly to yield transparent, very low expansion β-quartz solid solution glass ceramics. The suitable compositions studied were in the range:

64–70SiO_2, 18–25Al_2O_3, 2–5Li_2O, 0–4MgO, and 0–8ZnO (wt%). Other oxides, such as Na_2O, K_2O, CaO, SrO, BaO, and B_2O_3 were also used in some specimens, in varying amounts.

The crystallization sequence of a typical bulk glass sample with the composition 65.0SiO_2, 23.0Al_2O_3, 3.8Li_2O, 1.8MgO,

1.5ZnO, $2.0ZrO_2$, $2.0TiO_2$, and $0.9As_2O_5$ (wt%), which was located in the above composition range, was described as follows (Beall and Duke, 1969; Duke and Chase, 1968):

Initially a glass-in-glass phase separation process occurred in the 700–750°C range, followed by the crystallization of either cubic zirconia or titanate crystals. These randomly dispersed crystals, then acted as nuclei for the crystallization of β-$quartz_{ss}$, with an average crystallite size of ~0.1, after heat treating the glass in the 800–850°C range.

The β-$quartz_{ss}$ phase formed in the above temperature range was metastable and at higher temperatures was transformed to a spodumene solid solution.

A typical time–temperature stability diagram for the above glass, plotted by Duke and Chase (1968), is shown in Figure 2.1. Such a plot could be useful in both processing and application of β-$quartz_{ss}$ glass ceramics.

The glasses were stable against uncontrolled internal devitrification and thus were suitable for casting into thick sections.

The transparent glass ceramic mirror blanks, with average expansion coefficients of $0 \pm 0.3 \times 10^{-7}$/°C (in 5–50°C temperature range) were fabricated.

Wares up to 15 cm thick were fabricated with strain levels of <10 mμ/cm and no change in the expansion or transparency through the blank (Duke and Chase, 1968).

In order to clarify the crystallization mechanism of LAS glasses, the commercial lithium aluminosilicate glass ceramics were studied utilizing high-resolution TEM (HRTEM) (Maier and Müller, 1987). After various heat treatments, it was found

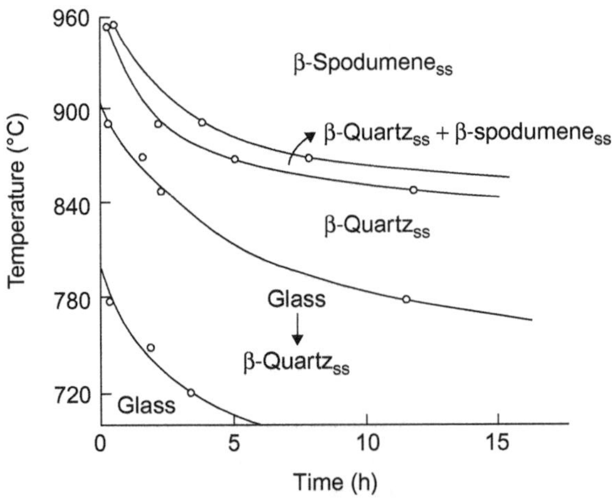

Figure 2.1 Time–temperature metastable phase diagram for a LAS glass ceramic.
(After Duke and Chase, 1968. Reproduced with permission of the Optical Society of America.)

that epitaxy of β-quartz crystals on $ZrTiO_4$-type crystallites was the critical step in crystallization of LAS glasses, containing ZrO_2 and TiO_2.

It was also shown that each β-quartz crystal, in addition to a $ZrTiO_4$ crystallite at its center, was also surrounded by a large number of peripheral $ZrTiO_4$ crystallites. It was suggested that these crystallites were initiated in the peripheral residual glass phase, which underwent a secondary phase separation during the crystallization of the parent LAS glass. Due to the occurrence of this phase separation, the residual glass was enriched in TiO_2 and ZrO_2.

The ZrO_2 and/or TiO_2-induced nucleation in LAS glass ceramics, and the effect of these nucleating agents' content and ratio upon the nucleation rate were also investigated (Schiffner and Pannhorst, 1987).

Besides ZrO_2 and TiO_2, other less common oxides such as Nb_2O_5, HfO_2, and Ta_2O_5 have also been utilized as effective nucleating agents for β-quartz$_{ss}$ glass ceramics, sometimes in combination with TiO_2 and ZrO_2 oxides.

For example, niobium pentoxide, Nb_2O_5, was used to promote crystallization in $Li_2O \cdot Al_2O_3 \cdot 6SiO_2$ glasses (Hsu and Speyer, 1991). It was shown that ≥2 mol% of the above nucleating agent-enhanced crystallization in these glasses. Glasses containing 4.0 and 8.0 mol% T–Nb_2O_5 (T form, orthorhombic system) exhibited a high tendency to form dispersed TT–Nb_2O_5 (monoclinic system), precipitates during the glass-quenching process. The crystallization process in glasses containing 2.0 or 4.0 mol% T–Nb_2O_5 occurred via microphase separation, followed by the formation of dispersed TT–Nb_2O_5 crystalline precipitates (760°C). These precipitates then acted as heterogeneous nucleation sites for the crystallization of β-quartz solid solution at 850–900°C. The latter phase transformed into β-spodumene$_{ss}$ at higher temperatures.

Wondraczekw and Pradeau (2008) studied the heterogeneous nucleation and crystallization of lithium aluminosilicate glasses, containing HfO_2 as a nucleating agent. It was suggested that hafnia can be used as a batch component in lithium aluminosilicate glasses, alone or in combination with other agents such as TiO_2 or ZrO_2, to nucleate nano-crystalline transparent glass ceramics with a low Coefficient of Thermal Expansion (CTE). It was also shown that the temperature of the onset of crystallization was increased due to the replacement of ZrO_2 by HfO_2 which, however, was not significant if <20–30% of ZrO_2 was replaced. It was shown that lattice parameters and, particularly, anisotropy of the nuclei phase depend on the extent of ZrO_2–HfO_2 substitution in a specific base glass. The lowest degree of anisotropy of nuclei crystallites was found when about 20–30% of ZrO_2 was replaced by HfO_2.

Considering the size of obtained β-quartz crystallites, the authors concluded that with decreasing the anisotropy, the nucleation efficiency increases.

The crystallization mechanism and microstructure of LAS glass ceramics with the addition of Ta_2O_5, as a nucleating agent, was also investigated (Li et al., 2010).

The following sequence of crystallization was found to occur with increasing temperature:

First, the occurrence of phase separation followed by the formation of precursor nuclei of Ta_2O_5 crystals, and finally, the crystallization of β-$quartz_{ss}$. By increasing the amount of Ta_2O_5 from 2 to 6 mol%, the activation energy for crystallization decreased from 297.73 to 218.66 kJ/mol, while the Avrami exponent increased from 1.76 to 3.39, indicating the change of crystallization mechanism from surface crystallization to bulk crystallization.

Figure 2.2 demonstrates the scanning electron microscopy (SEM) micrographs of the samples containing 4% Ta_2O_5; after heating at 850°C for 1 h (Figure 2.2A), relatively little amount of tiny sphere-shaped crystals, with the average size of 50–70 nm, were crystallized. With increasing temperature, the crystal size and crystallinity slightly increased (Figure 2.2B). After heat treating at 1,000°C, considerable crystallization occurred, and minor rod-shaped $AlTaO_4$ crystals also appeared (Figure 2.2C) coexisting with β-$quartz_{ss}$ and β-$spodumene_{ss}$. Hence, it was concluded that the amount of Ta_2O_5 had a significant effect on the crystal size, crystal shape, and crystallinity of the LAS glass ceramics. Moreover, by increasing the Ta_2O_5 addition, the crystallization mechanism changed from surface crystallization to bulk crystallization, confirming the above-mentioned results of the crystallization kinetics analysis.

Alekseeva et al. (2008) utilizing Raman spectroscopy studied the possibility of quantifying the SiO_2 content in the stuffed β-quartz phases, precipitated in lithium aluminosilicate glass ceramics. The parent glasses that were prepared by the conventional melt-quenching technique annealed and heat-treated in the temperature range of 660–850°C. In the Raman spectra, the one-mode behavior of complex mixed crystals of a series of stuffed β-quartz derivative phases, $x(LiAl)(1-x)SiO_2$, was determined. It was demonstrated that the linear dependence of the frequency of the high-frequency Raman band to SiO_2 content of

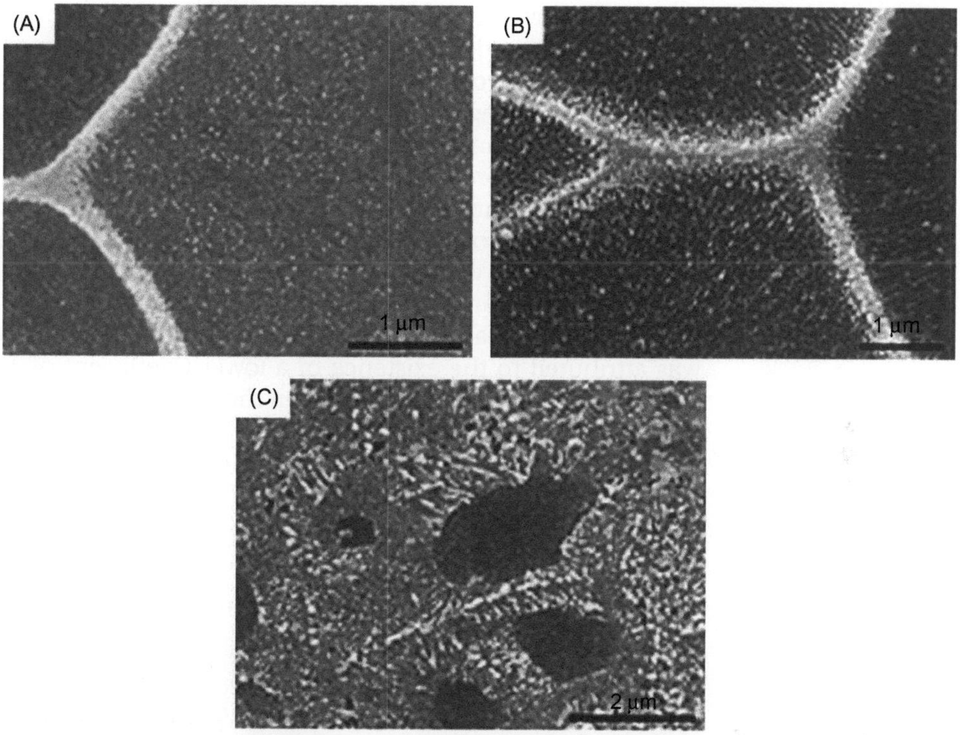

Figure 2.2 SEM images of the etched surfaces of the glass ceramics containing 4 mol% Ta_2O_5, heat-treated at different temperatures for 1 h. (A) 850°C, (B) 900°C, and (C) 1,000°C. (After Li et al., 2010. Reproduced with permission of Springer Science + Business Media, Inc.)

the glass samples can be used for quantifying the SiO_2 concentration in the crystalline phases of β-quartz structure.

The quantitative determination of SiO_2 content in glass ceramics, utilizing the Raman spectra, enabled the researchers to ascertain the previously proposed idea that the composition of β-quartz phase was independent of the type and amount of the nucleating agent. The previous suggestions concerning the enrichment of the β-quartz phase in SiO_2 with time and temperature of heat treatment was also confirmed.

More recently, Wu et al. (2012) studied the effects of B_2O_3 and P_2O_5 additives, and the heat-treatment conditions, on the microstructure and mechanical strength of bulk lithium aluminosilicate glass ceramics nucleated with TiO_2 and ZrO_2. The phase-development sequence observed by the authors was the usual sequence of phase separation, precipitation of the precursor phase, the nucleation and crystal growth of the main phase, β-quartz$_{ss}$, and its transformation to β-spodumene$_{ss}$ at higher temperatures. The changes of local environment of TiO_2 and ZrO_2

during the devitrification process were also studied by the X-ray absorption near-edge structure (XANES). The experimental results showed that the coordination environment of Ti and Zr changed significantly, owing to the development of secondary phases, such as $TiZrO_4$ and $ZrSiO_4$, during the crystallization process, which was in accord with the suggestions of other researchers.

The phase-transformation temperature of β-quartz solid solutions to β-spodumene solid solutions was also shown to decrease ~100°C with the addition of B_2O_3 and P_2O_5.

Samples containing B_2O_3 also exhibited an increased flexural strength, in comparison with the samples doped with P_2O_5. This was attributed to the existence of a lower Coefficient of Thermal Expansion (CTE) mismatch between the constituent phases of the B_2O_3-doped glass ceramics, as well as the more uniform grain-size distribution, and lower viscosity of the glass matrix of the latter samples.

2.2.2 Luminescent Glass Ceramics

Theoretical Background

The term luminescence, introduced in 1888 by Eilhard Wiedemann, is defined as emission of light that usually occurs at low temperatures and is thus a form of cold body radiation. It can be caused by chemical reactions, electrical energy, electromagnetic radiation, stress, etc. on a crystal. This distinguishes luminescence from incandescence, which is light generated by high temperatures.

Photoluminescence (PL), one of many forms of luminescence (light emission), is distinguished by photoexcitation (excitation by photons) and is a process in which a substance absorbs photons (electromagnetic radiation) and then reradiates photons. Quantum mechanically, this can be described as an excitation to a higher energy state and then a return to a lower energy state, accompanied by the emission of a photon. The period between absorption and emission is typically extremely short, in the order of 10 ns. Under special circumstances, however, this period can be extended into minutes or hours.

Ultimately, available energy states and allowed transitions between states (and therefore, wavelengths of light preferentially absorbed and emitted) are determined by the rules of quantum mechanics. A basic understanding of the principles involved can be gained by studying the electron configurations, and molecular orbitals of simple atoms and molecules. **Some of the most promising applications of luminescent glass ceramics are lasers, optical amplifiers, solar cells, and photonic devices.**

Laser Applications A laser is a source of monochromatic radiation of high intensity, coherence, and directionality, in the ultraviolet (UV), visible, or infrared (IR) optical regions. The laser mechanism is based on the radiative process of stimulated emission. A requirement of laser action is the existence of a metastable excited state with a long lifetime, as the starting level of stimulated emission.

Another requirement is the existence of a larger population in this initial metastable state than in the final state, corresponding to the phenomenon of population inversion (Yamane and Asahara, 2002). Three or four energy-level schemes are used to reach these emission processes, responsible for laser action. The energy levels involved belong to active ions (rare-earth [RE] or transition metal ions). RE ions of the lanthanide series have been the most extensively used active ions for laser operation, because there are a large number of fluorescing states and wavelengths to choose among the 4f electron configurations, as illustrated in Figure 2.3, for some of the more commonly used RE ions. Among these ions, Er^{3+} is the most popular active ion recommended by the researchers of this field. Nevertheless, one

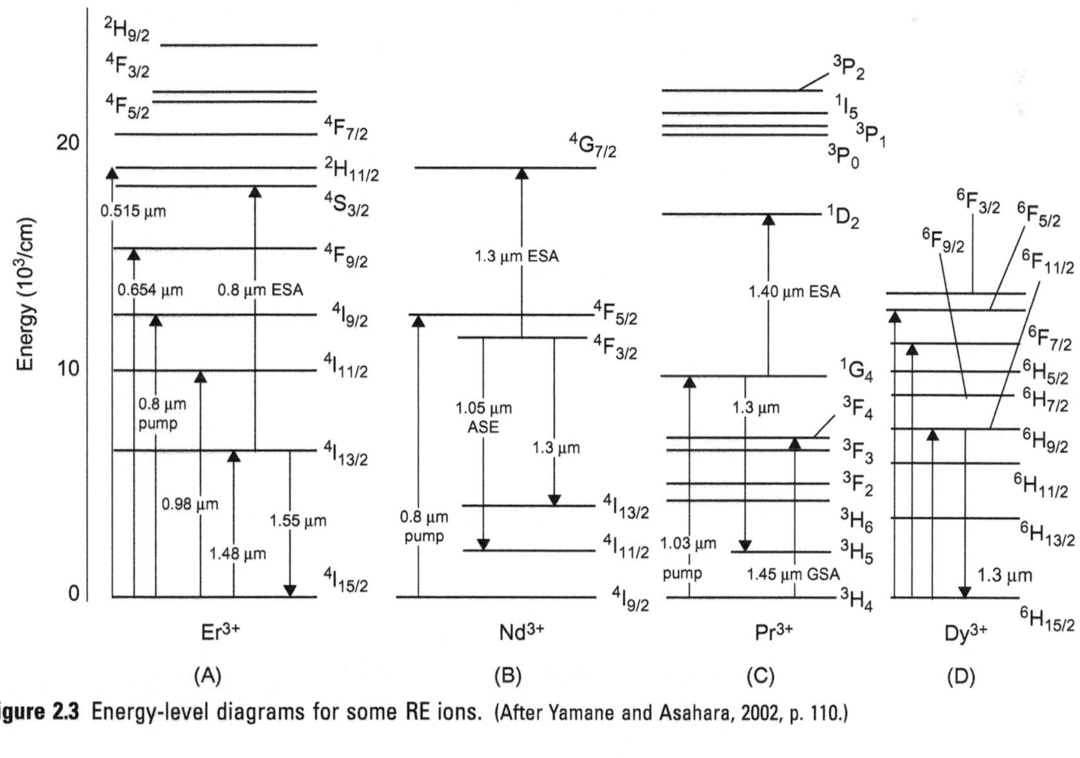

Figure 2.3 Energy-level diagrams for some RE ions. (After Yamane and Asahara, 2002, p. 110.)

of the obstacles to the use of RE ions can be a short lifetime of the excited states and nonradiative decay paths, which depend on the local environment of the ion and on the matrix vibrational energies. In glass, the nonradiative transitions of active ions are accomplished by a series of different mechanisms, namely, multiphonon relaxation, cross-relaxation, and cooperative up-conversion. Multiphonon relaxation occurs by a simultaneous creation of several phonons which suffice to equal the energy of the transition between the excited level and the next lower level. When this is larger than the highest vibrational energy of the solid matrix, several phonons may have to be created in order to bridge the energy gap. Since the probability for multiphonon decay decreases exponentially with the number of required phonons (Miyakawa and Dexter, 1970), it is desirable to surround the active ion by a matrix that possesses low vibrational energies. In glasses, the highest energy vibrations are stretchings of the anions against the glass-forming cations, whose frequency varies with the glass composition.

The other two dissipative processes, cross-relaxation and cooperative up-conversion, both involve active ion–ion interactions, wherein the intensity of fluorescence decreases as the concentration of the active ions increases. Cross-relaxation may take place between any two closed spaced active ions that happen to have two pairs of energy levels separated by the same amount. One of the ions, in an excited state, gives half of its energy to a ground state ion, so that both ions end up in the intermediate level. From this level, they both relax quickly to the ground state, via multiphonon relaxation.

Since the Er^{3+} ion has no intermediate states between the $^4I_{13/2}$ metastable level and the ground state, this cross-relaxation between an excited ion and one in the ground state cannot occur in materials doped with this ion (Yamane and Asahara, 2002).

The spontaneous cooperative up-conversion process may occur when two neighboring ions are in an excited state: one of them, A, thus promoting B to a higher level, while A relaxes to the ground state. From this higher energy level, the B ion relaxes rapidly, radiatively, or nonradiatively. This cooperative up-conversion process is believed to be the major cause of concentration quenching in Er^{3+}-doped glasses.

Frequency Up-Conversion At present, there is great interest in luminescent materials for efficient frequency conversion from IR to visible radiation, mainly because a visible source pumped by a near-IR (NIR) laser is useful for high-capacity data storage optical devices. This stimulated excitation process is designated by up-conversion which is a particular type of laser action. Hence, for increasing the efficiency of the process, both the fluorescence lifetime and the stimulated emission cross-section of the RE excited level should be maximized, whereas the nonradiative decay mechanisms should be minimized. The glassy host is required to possess a minimal absorption coefficient within the wavelength region of interest, plus the capability of incorporating large RE concentrations, low vibrational energies, and a high refractive index. Er^{3+} can provide up-converted visible fluorescence, in both fluoride crystals and glasses. In conventional oxide glasses, there is no report of the up-conversion phenomenon, because of large nonradiative losses, due to high-energy vibrations, e.g., Si—O ($\sim$1,100 cm^{-1}) and B—O ($\sim$1,400 cm^{-1}), which couple to the Er^{3+} ions.

Up-conversion fluorescence applications have therefore been based on nontraditional oxide glass systems, with lower vibrational energies. In the stimulated up-conversion processes, excited state absorption (ESA) and energy transfer involving RE ions in the solid matrix are important mechanisms. Much work has been devoted to up-conversion lasers involving Er^{3+} ions. The sensitization of Er^{3+} with Yb^{3+} may also be favorable for IR up-conversion (Mita, 1972; Auzel, 1990; Mita et al., 1993).

Amplification at 1.3 and 1.5 μm A driving force for research in RE-doped fibers and integrated optics waveguides has been their use for amplifying weak signals in optical communications systems at 1.3 and 1.5 μm, respectively. This may be achieved by simply splicing a section of RE-doped fiber into the transmission one and injecting pump light through a fiber coupler. The signal generated within the RE emission band stimulates emission of radiation at the same frequency, amplifying the optical communication signal with high gain, high efficiency, and low noise, which is highly advantageous for optical communications (Yamane and Asahara, 2002).

There are five main RE candidates for use as dopants in fiber or waveguide amplifiers for optical communications systems: Er^{3+}, Tm^{3+}, Nd^{3+}, Pr^{3+}, and Dy^{3+}.

As with the laser mechanism, the major problems concerning amplification efficiency are related to dissipative processes. In addition to all the nonradiative relaxation processes already mentioned, e.g., multiphonon relaxation, cross-relaxation and cooperative up-conversion, other dissipative processes may decrease the amplification efficiency. The phenomena of ESA and amplified spontaneous emission are two of the most compromising processes, which can be partially eliminated by changing the glass composition.

Solar Energy Applications In semiconductor solar cells, only photons with energy larger than the band gap are absorbed and therefore produce electron–hole pairs. Subsequently, the energy of these electron–hole pairs in excess over the band gap is lost by fast thermalization processes. These considerations, together with the shape of the solar spectrum, extended roughly from 350 to 2,500 nm, impose a limit for the efficiency of a solar cell. This limit is ~30% for silicon solar cells whose band gap presents an optimum value for this application of 1.12 eV (corresponding to photon wavelength of 1,100 nm).

The efficiency of these solar cells would be appreciably increased by a **down-conversion** layer placed at their front, which produces two photons with energy larger than the band gap from each high-energy incidents solar photon. For silicon solar cells, the specific aim would be to obtain two photons with wavelengths shorter than 1,100 nm from each incident solar photon in the range 350–550 nm. About 15% of the terrestrial solar energy is available to be used by silicon solar cells covered with a down-conversion layer (Richards, 2006; Strümpel et al., 2007).

The down-conversion processes, also known as quantum cutting, have attracted considerable attention and there are many recent publications devoted to study these processes in glass ceramics doped with lanthanide ions. In most of these papers, the studied systems were based on cooperative energy transfer from one Tb^{3+}, Pr^{3+}, or Tm^{3+} ion, with a narrow absorption peak at the blue-green, to a couple of nearby Yb^{3+} ions. The Yb^{3+} ions are interesting emitters because they only have an optical transition at ~1,000 nm, just over the silicon gap. These second-order down-conversion processes can be dominant over spontaneous emission and over first-order resonant energy transfer only for very short distances between the three involved ions, one donor and two acceptors. Therefore, high concentrations of Yb^{3+} ions are needed to get efficient cooperative energy transfer. On the other hand, there is an important constraint related to "concentration quenching," i.e., inhibition of the photon emission due to fast energy migration between Yb^{3+} ions taking to luminescence traps.

Most Common Luminescent Glass Ceramics

2.2.2.1 Transparent Mullite Glass Ceramics

Mullite is a nonstoichiometric compound whose composition generally lies between $3Al_2O_3 \cdot 2SiO_2$ and $2Al_2O_3 \cdot SiO_2$, and its crystal structure is derived from that of sillimanite, with chains of AlO_6 octahedra cross-linked by $(Si,Al)O_4$ tetrahedra. While sillimanite has silicon and aluminum cations ordered on the tetrahedral sites, mullite has excess Al^{3+} for Si^{4+}, with charge balance provided by oxygen vacancies. The chemical formula for mullite is often given by $Al_2(Al_{2+2x}\ Si_{2-2x})O_{10-x}$, where $x=0$ corresponds to sillimanite, $x=0.25$ corresponds to 3:2-mullite, and $x=0.4$ corresponds to 2:1 mullite (Duval et al., 2008). The thermodynamically stable modification of mullite with 60 mol% Al_2O_3 ($3Al_2O_3.2SiO_2$, i.e., 3:2-mullite) is one of the most important ceramic phases with low thermal expansion, low thermal conductivity, excellent creep resistance, good chemical stability, and high mechanical strength (Schneider et al., 1994). Although this solid solution phase is commonly found in human-made ceramics, only rarely does it occur as a natural mineral. Yet mullite is a major component of aluminosilicate ceramics and has been found in refractories and pottery dating back millennia (Duval et al., 2008). Recently new uses are being found for this ancient material in the areas of electronics and optics, as well as in high-temperature structural products.

Structure, Optical Properties, and Application of Transparent Mullite Glass Ceramics Transparent glass ceramics, including mullite and spinel glass ceramics, have attracted considerable attention in recent years as host media for luminescent transition metal ions. Among the transition metal ions, $\mathbf{Cr^{3+}}$ has particularly been studied as a suitable luminescent ion for potential use in tunable lasers and solar concentrators, since it has the ability to absorb broadly in the visible wavelength range, and fluoresces in the NIR (down-conversion). Although glasses have an even larger wavelength gap between Cr^{3+} absorption and fluorescence (Stokes shift), their quantum efficiency is generally poor, <15%. (Quantum efficiency is defined as the number of photons emitted by Cr^{3+} ions divided by the number of photons absorbed by these ions.) Although certain crystalline phases, such as alexandrite ($BeAl_2O_4$), have similar shifts with much higher efficiencies, they are not generally suitable for large-scale manufacturing in sheets which is necessary for some applications, e.g., solar collectors (Beall and Pinckney, 1999). Glass ceramics can combine the best of the crystal and glass properties.

Numerous studies using absorption and steady-state luminescence have shown that transition metal ions, e.g., Cr^{3+} ions, exhibit much better quantum efficiency of luminescence in glass ceramics, as compared with the glasses of the same composition. Mullite glass ceramics have been considered as suitable host media for the accommodation of Cr^{3+} ions, because the mullite crystal structure provides a suitable crystal-field environment for Cr^{3+}, where Cr^{3+} can be substituted for Al^{3+} in the octahedral sites. Furthermore, because mullite glass ceramics are self-nucleating, they do not require nucleating agents, the addition of which could have detrimental effects on optical properties. Mullite glass ceramics are also compatible with sheet manufacturing (Beall and Pinckney, 1999).

Therefore, chromium-doped mullite glass ceramics, combining good transparency with broad visible absorption, wide emission in the IR and high quantum efficiencies, are among the most suitable glass ceramics within this field.

Moreover, some studies indicated that the aforementioned characteristics of Cr^{3+} ions depend on the ligand field environment of them; the details of Cr distribution in the mullite structure may strongly affect their spectroscopic properties; and comprehensive work has been carried out, in the last 2–3 decades, on the site assignment of Cr in mullite structure and its effect upon the optical properties of Cr-doped mullites.

Although Cr^{3+} ions, because of their large octahedral stabilization energy, are expected to preferably replace octahedral site Al^{3+} ions, spectroscopic work suggested that the Cr distribution in the mullite structure is more complicated.

Knutson et al. (1989) utilizing optical and EPR (electron paramagnetic resonance) techniques studied the distribution and behavior of Cr^{3+} ions within a mullite ceramic host. The results indicated that the existence of low-field ions, which are likely to be located at sites in regions of disorder, accounts for most of the spectroscopic anomalies previously noted in these materials.

Previous work has shown that mullite possesses many properties characteristic of both crystalline hosts such as a large average crystal-field strength and strong R-line like emission and characteristic of low- or intermediate-field materials, generally resulting in $^4T_2$ emission. These materials also show large inhomogeneous broadening in both fields.

According to the authors, this anomalous behavior can be attributed to the presence of both ordered high-field-site (HFS) and disordered low-field-site (LFS) Cr^{3+} ions.

Rager et al. (1990) interpreted their EPR-spectroscopic data by incorporation of Cr in the octahedral M1 site as well as in an interstitial octahedral site. This distribution of Cr over two octahedral sites has also been confirmed by Ikeda et al. (1992) based on crystal-field spectroscopy. Also Nass et al. (1995) came to the same conclusion studying Cr-doped mullites by combined thermal, spectrometric, X-ray, and electron microscopic techniques.

Piriou et al. (1996) and Bauchspieβ et al. (1996) reported that Cr can be incorporated both in the M1 and interstitial sites in Cr_2O_3-rich mullites.

Piriou et al. (1996) studied the site assignment of Cr in mullite structure doped with Cr_2O_3 in concentrations ranging from 2 to 10 wt%, utilizing time resolved emission spectroscopy and site selective excitation. It was shown that in mullites with high Cr_2O_3 contents, Cr^{3+} ions were mostly located in LFS, whereas in Cr_2O_3-poor mullites, these ions were predominately incorporated in HFS, i.e., at octahedral M(1) positions replacing Al^{3+}. Upon increasing the Cr_2O_3 concentration, Cr^{3+} favorably entered the highly distorted LFS, which were attributed to interstitial lattice positions in the mullite structure.

Fischer and Schneider (2000) investigating a Cr-doped mullite specimen ($Cr_{0.5}Al_{3.92}Si_{1.58}O_{9.79}$) reported the predominant incorporation of Cr in the octahedrally coordinated M1 site. The mean M1–O distance of 1.935 Å was calculated from Al^{3+} and Cr^{3+} molar fractions obtained from the chemical analysis and structure refinement. The small displacement factors of the tetrahedral and octahedral sites indicated deficiencies in the electron densities which could be compensated by assuming additional Cr atoms. According to the authors, the predominant Cr incorporation into the M1 site caused an expansion of the octahedral bonds which was directly related to the observed lengthening of the *c* edge. The strong expansion of the long and elastic octahedral M1–Od bond in Cr-doped mullite, which would affect the *a* and *b* edges, was partly absorbed by a shortening of tetrahedral bonds.

Tkalčec et al. (2006) prepared and characterized the structure and properties of Cr-doped mullites from single-phase precursors containing up to 9.60 wt% Cr_2O_3 using a sol-gel technique followed by heat treatment. Five precursors with different chromium contents (0, 2.30, 4.74, 6.97, and 9.60 wt% Cr_2O_3) and with a constant atomic ratio $(Al + Cr)/Si = 3/1$ were prepared. Mullites in the prepared samples were orthorhombic with characteristic mullite structure. Unit-cell parameters linearly increased with Cr-doping level. The highest expansion was observed for *c* followed by *a* and *b*.

The structure of mullites was refined by the Rietveld method (1969) and location of Cr_2O_3 was performed by the EPR spectroscopy. At low chromium doping level (Cr_2O_3 content less than ~5 wt%) Cr^{3+} ions were substituted for Al^{3+} in the AlO_6 octahedra of the mullite structure (M1 sites). For higher doping level, Cr^{3+} ions were additionally substituted for Al^{3+} in the distorted octahedral AlO_6 sites of the second phase [θ-$(Al,Cr)_2O_3$ at 1,400°C, or α-$(Al,Cr)_2O_3$ at 1,600°C], which segregated in the system.

Figure 2.4A shows the typical spectroscopic spectra of a transparent Cr^{3+}-doped mullite glass ceramic (Beall and Pinckney, 1999). From the difference in absorption between M4 parent glass, with the chemical composition of $48SiO_2$, $25Al_2O_3$, $15B_2O_3$, $10ZnO$, $2K_2O$, $0.1Cr_2O_3$ (wt%), and its glass-ceramic counterpart, it can be deduced that the glass ceramic had better transparency in the key NIR region from 700 to 900 nm. The luminescence curves of the glass and glass ceramic are also shown in Figure 2.4B (the glass luminescence curve has been multiplied by 6 in order to make it more visible). The glass ceramic emitted widely from 680 to >1,000 nm. The glass ceramic was far more efficient, although the glass emitted further into the IR, and there is very little limited overlap between the glass-ceramic absorption and its emission (Beall and Pinckney, 1999).

Processing of Mullite Glass Ceramics Transparent mullite glass ceramics can be produced from simple binary glasses in the Al_2O_3–SiO_2 system without addition of any nucleating agents, because mullite glass ceramics are self-nucleating. The sequence of nucleation and crystallization of these glass ceramics in the SiO_2–Al_2O_3 and SiO_2–Al_2O_3–BaO systems has been described by Beall and Duke (1969). These authors reported the occurrence of a spontaneous phase separation in binary alumina-silica glasses, between ~7 and 55 mol% of alumina. The composition of the alumina-rich phase, usually the dispersed phase in the area of stable glass formation, is very close to mullite. For example, it was found that the phase separation was the essential step of nucleation of finely divided mullite crystals in quenched binary

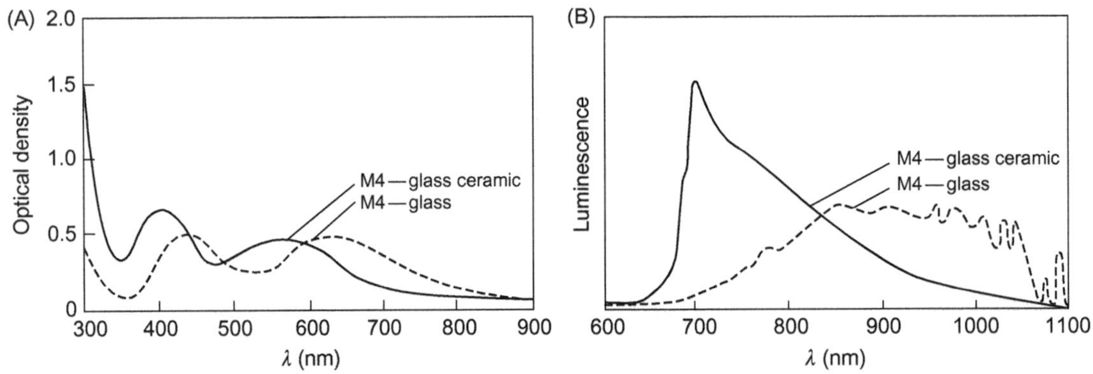

Figure 2.4 A. Absorption (optical density) and B. Luminescence as a function of wavelength for mullite parent glass (dashed line) and glass ceramic (solid line). Glass luminescence curve has been multiplied by 6. (Beall and Pinckney, 1999. American Ceramic Society.)

glass of composition 77% SiO_2–23% Al_2O_3 (wt%) upon heat treating above 1,000°C.

Figure 2.5 depicts the electron micrograph of a transparent mullite glass ceramic, formed by heat treatment of the glass of composition $50SiO_2$, $40Al_2O_3$, and 10BaO. Since the oxides such as BaO, K_2O, and Cs_2O form very stable aluminosilicate glasses that are resistant to cristobalite formation, BaO was added to the above glass composition to prevent the devitrification of the residual siliceous glass of the mullite glass ceramic to cristobalite.

However, the aforementioned oxides must be added in small amounts (<10 mol%), since they also suppress the basic phase separation of the glass which was essential for the crystallization of mullite.

Andrews et al. (1986) showed that glasses located in the SiO_2–Al_2O_3–B_2O_3–ZnO–K_2O system, containing >10 mol% Al_2O_3, phase-separated upon quenching, and on subsequent heat treatment, were converted to transparent glass ceramics, containing mullite nano-crystals. It was also shown that the Cr^{3+} ions, used as dopant, were mainly incorporated in the mullite crystalline phase. After the crystallization process, the luminescence quantum efficiency was increased from less than 1% to ~30%. The resulting glass ceramics also exhibited a low scattering loss of <0.01 cm^{-1}.

According to Beall and Pinckney (1999), the additions of components such as B_2O_3, MgO, ZnO, BaO, and alkaline oxides

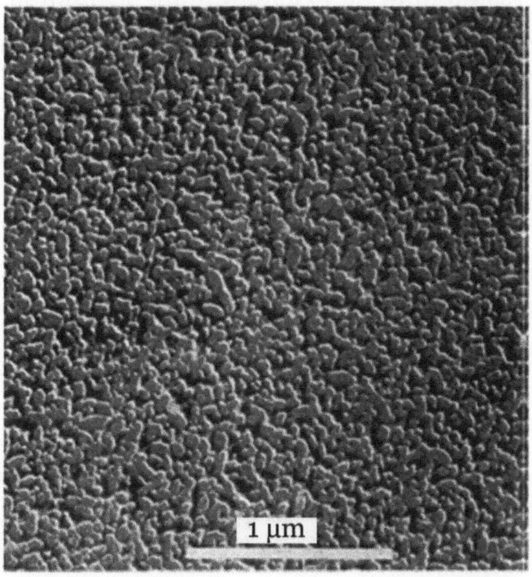

Figure 2.5 Electron micrograph of a transparent mullite glass ceramic, fabricated through the heat treatment of a $50SiO_2$–$40Al_2O_3$–10BaO (wt%) glass at 1,150°C for 8 h (white bar = 1 µm). (After Beall and Duke, 1969. Reproduced with permission of Springer Science + Business Media, Inc.)

can improve the melting characteristics of these otherwise refractory glasses, as well as suppressing their spontaneous phase separation. This would produce more stable glasses that phase separate controllably to Al_2O_3-rich droplets, within a continuous siliceous glass on reheating.

It was also shown that a glass of the composition $48SiO_2$, $29Al_2O_3$, $11B_2O_3$, 10ZnO, and $2K_2O$ (wt%), prior to its final heat treatment at 750°C for 4 h + 2 h at 800°C, which converted it into a transparent glass ceramic, has undergone a fine-scale phase separation. The resulting glass-ceramic contained rod-like mullite crystals of ~30 × 10 nm in size.

More recently, Hamedani et al. (2012) fabricated transparent mullite-based glass ceramics in the system SiO_2–Al_2O_3–B_2O_3–ZnO–K_2O bearing Cr additive and determined their crystallization behavior and optical properties.

The results showed that the optimum heat-treatment conditions for obtaining a transparent mullite glass ceramic were 750°C −4 h and 850°C −3 h, respectively, for the nucleation and crystallization steps. The activation energy for crystallization and the Avrami exponent, determined by the Matusita method, were 585 KJ/mol and 2 respectively.

The small differential thermal analysis (DTA) crystallization peak temperature difference of the coarse and fine glass particles and the Avrami constant both indicated that bulk crystallization was the main mechanism for crystallization of mullite.

Heat treatment of the glass under the optimum conditions led to the precipitation of 22 wt% mullite, exhibiting an average crystallite size of <50 nm.

The obtained glass ceramics showed an intense photoluminescence emission at 700 nm when excited by a 590 nm beam.

Table 2.1 summarizes the chemical compositions and the heat-treatment procedure of some representative transparent mullite glass ceramics.

2.2.2.2 Spinel Glass Ceramics

Spinels have cubic crystal structures with the general chemical formula AB_2O_4, where A is a tetrahedrally coordinated, typically divalent cation, such as magnesium, iron, or zinc, and B is an octahedrally coordinated, usually trivalent cation, such as aluminum, iron, or chromium. A large number of aluminates, ferrites, titanates, and solid solutions of them can be classified under the heading of spinel mineral group.

Table 2.1 Compositions of Some Representative Transparent Glass Ceramics Based on Mullite (wt%)

Oxide	Beall and Duke (1969)	Beall and Duke (1969)	Andrews et al. (1986)	Reisfeld and Kisilev (1987)	Beall and Pinckney (1999)	Beall and Pinckney (1999)
SiO_2	77	50	48	63	60	53
Al_2O_3	23	40	29	13.5	20	25
BaO	—	10	—	—	—	
B_2O_3	—	—	11	19.8	17.5	15
Na_2O	—	—	—	—	2.5	—
ZnO	—	—	10	—		—
K_2O	—	—	2	3.6		3
MgO	—	—	—	—		4
Heat-treatment temperature (°C)/time (h)	1000/10	1000/10	750/4 + 800 – 875/2	790/10 + 900/2		

The spinel crystals have attracted much attention in the last decade, as suitable host media for accommodation of transient metal ions.

Properties and Application of Transparent Spinel Glass Ceramics Due to the rapid development of the telecommunications industry, the demand to increase the transmission capacity of the wavelength division multiplexing (WDM) system has attracted considerable attention in recent years, and much effort has been devoted to materials and design for optical fibers, waveguide devices and optical fiber amplifiers to revolutionize the telecommunication systems.

Transition metal-doped transparent glass ceramics containing **spinel nano-crystals** have been extensively investigated in recent years for their potential applications in the two main fields in the lasing and telecommunication systems:

- **Broadband optical amplifiers and tunable lasers.**
- **Passive Q-switchers for 1.3 and 1.54 μm lasers.**

Broadband Optical Amplifiers and Tunable Lasers If ultra-broadband amplification with high-gain efficiency could be realized by the single wavelength pumping, a great revolution would be expected to occur in the WDM technology

(Suzuki et al., 2005). One of the ways of achieving this goal is by the incorporation of some transition metal ions into glasses or glass ceramics. In this connection, Cr^{4+}-doped glasses have extensively been investigated. However, as chromium ions can easily take multivalent states, such as Cr^{3+}, Cr^{4+}, and Cr^{6+}, strict valence control is required to obtain Cr^{4+}. In contrast, Ni^{2+} is extremely stable in almost all the hosts, and there is no need for the strict control of its valence states in the glass (Suzuki et al., 2005).

Hence, Ni^{2+}-doped transparent glass ceramics have received considerable attention for their potential applications in broadband optical amplifiers and tunable lasers.

Many researchers have studied the crystallization behavior, structure, and optical properties of the Ni^{2+}-doped zinc (magnesium) aluminosilicate glasses, containing spinel nano-crystals (Samson et al., 2002; Suzuki and Ohishi, 2004; Suzuki et al., 2005; Wu et al., 2008a,b, 2009a; Deng et al., 2011).

Suzuki and Ohishi (2004) obtained an efficient broadband NIR emission from Ni^{2+}-doped zinc aluminosilicate (ZAS) glasses after heat treatment and Suzuki et al. (2005) studied their structural and optical properties. The crystalline phase precipitated during heat treatment was identified as solid solutions of spinel compounds.

Figure 2.6A depicts an SAD (selected area diffraction) pattern and (B) a TEM image of the above glass ceramic, heat-treated at 680°C for 10 h. In the SAD image, ring patterns were clearly seen, indicating the presence of a polycrystalline phase.

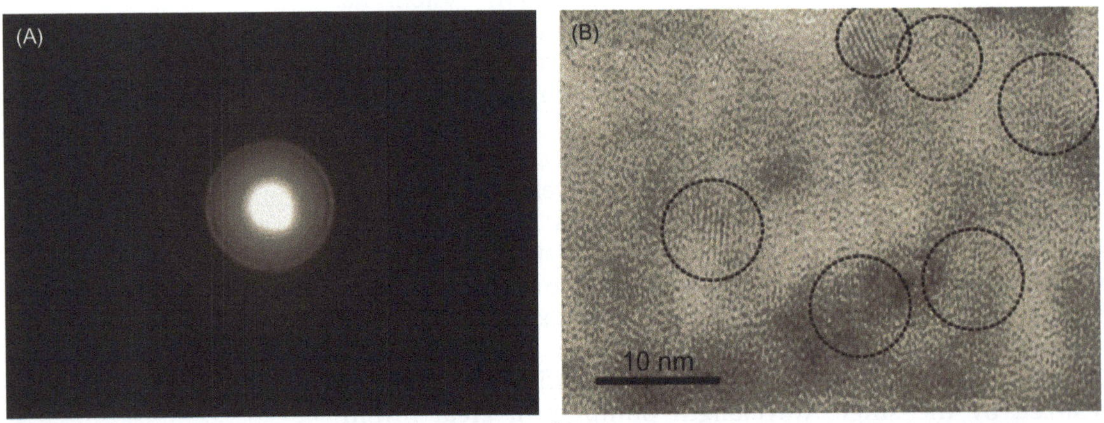

Figure 2.6 An SAD pattern (A) and TEM image (B) of a Ni^{2+}-doped ZAS glass, post-annealed for 10 h. Open circles in (B) show lattice images due to the presence of spinel crystalline phase. (After Suzuki et al., 2005. Reproduced with permission of Elsevier Science Publishing Company.)

Lattice images could also be seen in the open circles of TEM image in Figure 2.6B. The directions of the lattice images were randomly distributed, and the crystallite sizes could be estimated as 10–15 nm in diameter.

It was confirmed from optical absorption measurements that Ni^{2+} ions occupied fivefold sites in the as-cast glass and octahedral sites in the resulting glass ceramics.

The glass ceramics exhibited NIR emissions in which crystals containing Ni^{2+} ions acted as emission centers. The emission in this wavelength region was assigned to the downward $^3T_{2g}$ $(^3F) \rightarrow {}^3A_{2g}$ (^3F) transition of octahedral Ni^{2+} ions. It was shown that the emission intensity of the glasses increased with annealing time. The emission peak was located at ~1,440 nm for glass annealed for 15 min and moved to the shorter wavelength, located at ~1,350 nm, for glass annealed for 10 h. These observations indicated the existence of different types of octahedral Ni^{2+} sites under the different crystal-field strengths. Moreover, it was suggested that octahedral Ni^{2+} ions were formed in the low crystal field, at the early stage of annealing, and upon continuation of annealing shifted toward higher crystal fields.

It was also demonstrated that the emission intensity of Ni^{2+} in transparent glass ceramics of the aforementioned glass system, containing $ZnAl_2O_4$ nano-crystals, was improved ~8 times by Cr^{3+}/Ni^{2+}co-doping (Wu et al., 2008a). This enhanced emission was attributed to an efficient energy transfer from Cr^{3+} to Ni^{2+}. In fact, the NIR emission of Cr^{3+} with 532 nm excitation has a good overlap with the NIR absorption of Ni^{2+}. Hence, if Cr^{3+} and Ni^{2+} ions were introduced into the glass ceramics in combination, it could be expected that Ni^{2+} ions would be activated by Cr^{3+}-sensitizing ions. The energy transfer efficiency was estimated to be 57% and the detailed energy transfer mechanism was also discussed.

Wu et al. (2008b) also synthesized transparent Ni^{2+}-doped $MgO–Al_2O_3–SiO_2$ glass ceramics, without and with Ga_2O_3 addition. It was proved that the precipitation of spinel nano-crystals, identified as solid solutions in the glass ceramics, could be favored by Ga_2O_3 addition. The mean diameter of nano-crystals was reported as ~7.6 nm. The luminescent intensity of the Ni^{2+}-doped glass ceramics was largely enhanced by Ga_2O_3 addition. This was attributed to the increasing of Ni^{2+} in the octahedral sites and the reduction of the mean frequency of phonon density of states in the spinel nano-crystals of solid solutions. The full-width at half-maximum (FWHM) of emissions for the glass ceramics with different Ga_2O_3 content was all >200 nm. The emission lifetime was also increased with the Ga_2O_3 content and the longest lifetime was ~250 ms. The aforementioned Ni^{2+}-doped transparent glass ceramics with Ga_2O_3 addition could have potential application as broadband optical amplifier and laser materials.

The spectral properties of Ni-doped transparent MgO–Al_2O_3–Ga_2O_3–SiO_2–TiO_2 glass ceramics with the addition of Bi_2O_3 dopant was also investigated (Wu et al., 2009a). It was shown that the emission intensity of Bi/Ni co-doped glass ceramic was ~4 times stronger than Ni-doped glass ceramics, due to energy transfer from Bi to Ni (Figure 2.7). The Bi/Ni-doped glass ceramics containing 0.75 mol% Bi_2O_3 exhibited a NIR emission with FWHM of ~270 nm and a fluorescent lifetime of ~350 μs. Hence, they are very promising materials for applications in broadband optical amplifiers and tunable lasers.

Very recently, Alekseeva et al. (2014) synthesized transparent ZnO–Al_2O_3–SiO_2 glass ceramics, nucleated with TiO_2 and ZrO_2 and studied the influence of NiO addition (from 0.1 up to 3 mol %) on phase transformations and optical properties of them.

All parent glasses were X-ray amorphous and inhomogeneous (phase-separated), demonstrating a bimodal structure, according to small angle X-ray scattering (SAXS) data. Though SAXS intensity increased with increasing the NiO content in parent glasses, the distance between the inhomogeneous regions was independent of the NiO content implying that NiO did not play a role of additional nucleating agent.

The details of phase-transformation processes differed greatly for glasses at low and high NiO doping levels. In glasses

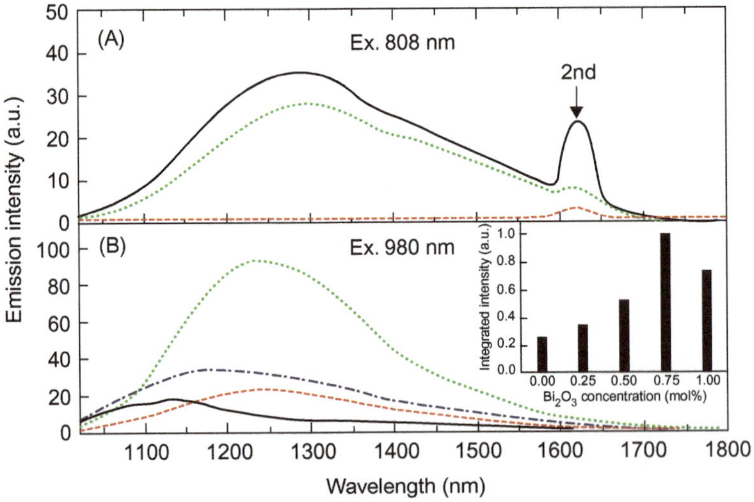

Figure 2.7 Emission spectra of Bi- (solid line), Ni- (dashed line), and Bi/Ni (dotted line)-doped (0.3 mol% NiO and 0.75 mol% Bi_2O_3) glass ceramics, excited by 808 nm (A) and 980 nm (B) Laser Diode (LD). The inset in (B) shows the dependence of the emission intensity of Ni on the Bi_2O_3 concentration and the dashed–dotted line is the combination of Ni and Bi luminescence. (After Wu et al., 2009a. Reproduced with permission of IOP Science Publishing Ltd.)

containing 0–0.1 mol% NiO, ZrTiO4, and Ni, Ti-doped gahnite crystals (with a normal spinel structure) precipitated simultaneously starting from heat treatment at above 730°C and traces of ZrO_2 tetragonal crystals were found in glass ceramics heat-treated in 1,000–1,200°C range.

In glasses doped with 1–3 mol% NiO, metastable nickel titanate-zirconate nano-crystals of 4–5 nm in size, with fluorite-type structure appeared during heat treatment in the temperature range of 730–800°C and its amount increased with the increase of the NiO content in the parent glass. The aforementioned phase decomposed at elevated temperatures above 800°C with the formation of Ni, Ti-doped gahnite, and ZrO_2 nanosized crystals which appeared at higher temperatures of 900–1,200°C.

Loiko et al. (2013) also investigated the optical absorption and luminescence properties of the aforementioned ZnO–Al_2O_3–SiO_2 glasses and glass ceramics for variable NiO content and heat-treatment conditions.

From the optical absorption data, it was evident that the as-cast glasses contained predominately five-coordinated Ni^{2+} ions embedded in amorphous glass matrix, along with a minor fraction of four-coordinated ions.

It was also proved that in the glass ceramics doped with 1–3 mol% NiO, Ni^{2+} ions were located in the octahedral sites of the fluorite-type nano-crystals appeared during heat treatment of the glasses at lower temperatures (730–800°C), whereas Ni^{2+} ions were located in both octahedral and tetrahedral sites in the Ni, Ti-doped gahnite nano-crystals formed after heat treatment at the higher temperature of 800–1,100°C.

No NIR luminescence was observed for as-cast glasses, whereas in gahnite-containing glass-ceramics intense luminescence was observed, associated with $^3T_{2g}$ (^3F) → $^3A_{2g}$ (^3F) transition for Ni^{2+} ions located in octahedral sites. The intensity of the aforementioned luminescence effect increased substantially with the increase of heat-treatment temperature or time. This was attributed to the increase of volume fraction of gahnite nano-phase and the decrease of the fraction of ions located at nano-crystal–residual glass phase interface.

Passive Q-switchers As indicated above, in the past decade, transparent glass ceramics, containing spinel nano-crystals doped with tetrahedral Co^{2+} ions, have also attracted much interest as good candidates for applications in **passive Q-switchers for 1.3 and 1.54 μm lasers** (Malyarevich et al., 2002; Duan et al., 2003, 2005; Chen et al., 2012, 2013).

Q-switching, sometimes known as **giant pulse formation**, is a technique by which a laser can be made to produce a pulsed output beam. The technique allows the production of light pulses with extremely high (gigawatt) peak power, much higher than would be produced by the same laser if it were operating in a continuous wave-constant output mode. Passively Q-switched solid-state lasers with high peak power and short pulse width are widely used in optical communications, pollution monitoring, material processing, medical surgery, and so on. The self-Q-switched laser materials, which combine the functions of the gain medium and the saturable absorber (SA), have attracted much attention, due to their compactness, low loss, and simplicity in the laser design and application (Chen et al., 2012).

As aforementioned applications in amplifiers and tunable lasers, the transition metal (especially Co^{2+})-doped zinc(magnesium) aluminosilicate glasses containing Mg/Zn spinel nano-crystals were also utilized in the Q-switching applications.

Malyarevich et al. (2002) studied the linear and nonlinear optical properties of cobalt-doped magnesium aluminosilicate transparent glass ceramics prepared under different conditions. It was shown that absorption and luminescence spectra and absorption bleaching of these glass ceramics were defined mainly by tetrahedrally coordinated Co^{2+} ions located in magnesium aluminum spinel nano-crystals. The lifetimes of the $^4T_1$ (^4F) and $^4T_2$ (^4F) excited states of the tetrahedral Co^{2+} ions were found to be in the ranges 25–40 and 120–450 ns, respectively, depending on the Co concentration.

Duan et al. (2003) prepared transparent glass ceramics in the SiO_2–Al_2O_3–ZnO–CoO system, containing zinc–aluminum spinel ($ZnAl_2O_4$) nano-crystals doped with tetrahedrally coordinated Co^{2+} ions, by the sol-gel method. Co^{2+} ions were incorporated in tetrahedral sites of $ZnAl_2O_4$ nano-crystals, which were in the size of 10–15 nm.

The optical absorption and luminescence of Co^{2+} ions located in the above glass ceramics were also studied. The absorption spectra of the sample heated at 900–950°C were very similar to that of Co^{2+} ions in bulk crystals, while the shape of the luminescence bands was completely different from Co^{2+} in bulk crystals.

Figure 2.8 depicts the emission spectra of the glass ceramic samples heat-treated at different temperatures/times.

The luminescence of Co^{2+} ions incorporated in $ZnAl_2O_4$ nano-crystals were assigned, in the visible and NIR spectral regions, to the $^2E(^2G) \rightarrow {}^4A_2(^4F)$ and the $^2E(^2G) \rightarrow {}^4T_2(^4F)$ transitions, respectively.

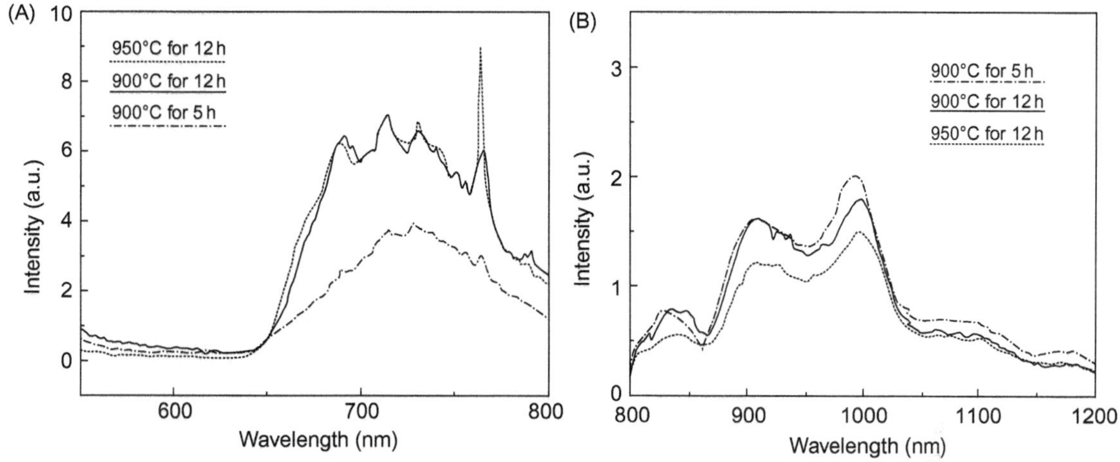

Figure 2.8 (A) Emission spectra of Co^{2+}-doped SiO_2—Al_2O_3—ZnO samples heat-treated at different temperatures and times; (B) The emission spectra for NIR region of Co^{2+}-doped SiO_2-Al_2O_3-Zno samples heat treated at different temperatures and times. (After Duan et al., 2003. Reproduced with permission of Elsevier Science Publishing Company.)

The difference in luminescence of Co^{2+} ions in glass ceramics and monocrystals arises from the different environments around Co^{2+} ions in these phases, since the local environment around the optically active ions has great influence on their luminescence. The luminescence of Co^{2+} ions in $ZnAl_2O_4$ monocrystal is mainly due to Co^{2+} ions located in the tetrahedral sites by substituting for Zn^{2+}. However, the presence of Co^{2+} ions, located in different crystalline and glassy sites in ZAS glass ceramics, led to the fine structure of the luminescence bands.

This glass-ceramic system has the potential as a laser material in the visible and NIR regions.

Transparent spinel glass ceramics, co-doped with Co^{2+}, Yb^{3+}, and Er^{3+} ions were also studied (Chen et al., 2012).

The absorption and emission spectra of the glass ceramics indicated that $Yb^{3+}$and Er^{3+} remained in the glass matrix, while Co^{2+} occupied tetrahedral sites in $MgAl_2O_4$, spinel nano-crystals, and was able to function as a SA for Er^{3+}.

Figure 2.9 depicts the field emission SEM (FESEM) image of the glass ceramic, containing Co^{2+}, Yb^{3+}, and Er^{3+} ions, heat-treated at 760°C/12 h + 930°C/4 h. It can be seen that $MgAl_2O_4$ nano-crystals were uniformly dispersed in the glass matrix. The sizes of the nano-crystals were in the range of 10–20 nm in diameter, which was in good agreement with the values obtained by X-ray diffraction (XRD) analysis.

The absorption spectra of the precursor glasses and glass-ceramics samples heat-treated according to the above conditions

Figure 2.9 FESEM image of a spinel glass ceramic co-doped with Co^{2+}, Yb^{3+}, and Er^{3+} ions, heat-treated at 760°C/12 h + 930°C/4 h. (After Chen et al., 2012. Reproduced with permission of Elsevier Science Publishing Company.)

were also discussed on the basis of the transitions from the ground state $^4I_{15/2}$ of Er^{3+} ions, as well as the ground states of Yb^{3+} ($^2F_{7/2}$); O_h (octahedral) Co^{2+} ($^4T_{1g}$); and T_d (tetrahedral) Co^{2+} ($^4A_2$) ions to the relevant excited states.

The emission spectra of the aforementioned precursor glass and the glass-ceramic samples were also determined under 980 nm excitation. The samples exhibited IR emission in the 1.5–1.6 μm range. The emission in this wavelength region was assigned to the Er^{3+}: $^4I_{13/2} \rightarrow {}^4I_{15/2}$ transition.

The emission of Er^{3+} ions, as well as the absorption of T_d Co^{2+} ions, makes the transparent Co^{2+}, Yb^{3+}, Er^{3+} co-doped glass-ceramics potential laser materials for compact, efficient, highly stable, self-Q-switched microchip solid-state lasers operating at 1.5–1.6 μm.

More recently, Chen et al. (2013) studied the physical properties and crystallization behavior of $MgO{-}Al_2O_3{-}SiO_2$ glasses with added La_2O_3 (0–10 wt%).

La_2O_3 was found to hardly influence the nucleation and crystallization processes of Co^{2+}-doped $MgO{-}Al_2O_3{-}SiO_2$ glasses in <5 wt%, while the formation of $MgAl_2O_4$ spinel was suppressed in the La_2O_3 content of >5 wt%. Transparent glass ceramics were synthesized by heat treating the glasses with 0–5 wt% La_2O_3 at 760°C/12 h + 930°C/4 h; the changes in absorption spectra indicated that Co^{2+} ions were incorporated into the fourfold coordinated sites of $MgAl_2O_4$ nano-crystals precipitated in the glass ceramics.

Processing of Spinel Glass Ceramics If an efficient nucleation process could be implemented, highly transparent spinel glass ceramics can be made containing crystals of ~10–50 nm in size.

Beall and Duke (1969), who were the first investigators to conduct a comprehensive study in the field of transparent glass ceramics, introduced the zirconium oxide as the most effective nucleating agent for spinel precipitation, particularly in cases where a very small crystallite size is required to afford transparency to the resulting glass ceramic.

According to the above authors, the best transparent spinel glass ceramics can be formed from glasses in the following composition range: SiO_2 65–75%, Al_2O_3 15–25%, ZnO 4–15%, MgO 0–6%, and 5–10% ZrO_2 added as a nucleation agent. Less than 10% of an oxide of a large alkaline cation (e.g., K_2O, Cs_2O) may be added to prevent devitrification of the residual siliceous glass to cristobalite.

They also reported the occurrence of an amorphous phase separation as the first step in the sequence of nucleation and crystallization of spinel glass-ceramic compositions containing ZrO_2 nucleant. This process can be induced at temperatures above the annealing point of the glass, generally around 800°C. The initial crystalline precipitate is invariably tetragonal zirconia, which usually begins to form after a few hours hold at the phase-separation temperature. With an increase in temperature to about 900°C, spinel begins to crystallize, presumably upon the zirconia nuclei. The spinel is generally gahnite ($ZnAl_2O_4$) or a $spinel_{ss}$ $(Zn,Mg)Al_2O_4$, close to the gahnite composition. The spinel crystallites were very fine-grained, initially <500 Å in diameter. Further heating at temperatures considerably above 1,000°C might cause secondary grain growth, leading to light scattering and haze.

Stryjak and McMillan (1978) also showed that zirconia-nucleated glass ceramics derived from the $ZnO-Al_2O_3-SiO_2$ system could be converted to transparent glass ceramics comprising very small crystals of β-quartz if ZrO_2 was <9 wt% and gahnite if its content was >9 wt%. Alkaline earth metal oxides present in these materials were shown not to take part in crystal formation and were incorporated into the residual glass phase. The optimum heat treatment for the development of a fully crystallized glass ceramic was shown to be:

> 800°C/4 h and 950°C/4–6 h for the nucleation and crystallization steps, respectively. The average particle size of the glass ceramics was <30 nm and the volume fraction of crystalline phases was <25%.

The use of TiO_2 or mixtures of TiO_2 and ZrO_2 for nucleation, rather than all ZrO_2, allows for lower melting temperatures of the precursor glass. Although ZrO_2 acts as an actual nucleating agent, with precipitated crystals of tetragonal ZrO_2 serving as the nucleating sites for spinel crystals, no evidence of a crystalline titanate phase is generally observed in TiO_2-nucleated spinel glass ceramics. In this case, nucleation is preceded by a highly uniform, ultrafine-scale phase separation into SiO_2-rich and TiO_2/Al_2O_3-rich areas. Spinel then crystallizes and grows within the latter globular-shaped areas (Beall and Pinckney, 1999).

In summary, the most recommended effective nucleating agents for these glass ceramics are ZrO_2 and/or TiO_2 and primary crystalline phases formed upon the heat treatment of glasses are spinel crystals ranging from gahnite ($ZnAl_2O_4$) to magnesium aluminate spinel ($MgAl_2O_4$). It is interesting to note that in the majority of the papers discussed above, the authors have not discussed the details of nucleation and crystallization mechanisms and the use of the TiO_2 and/or ZrO_2 as effective nucleating agents has been taken for granted!

Table 2.2 summarizes the chemical compositions, structure, and application of some representative transparent spinel glass ceramics.

2.2.2.3 Oxyfluoride Glass Ceramics

Glasses doped with RE ions have recently attracted great interest for applications in the field of up-conversion lasers, fiber amplifiers, and the optical devices for three-dimensional displays. Since most of the transitions of RE ions that play a decisive role in the above applications are initiated from excited levels possessing small energy gaps, materials with low phonon energy are required as a luminescent host to minimize the nonradiative loss and to obtain higher quantum efficiency of the desired fluorescence.

However, as stated previously, most oxide glasses have large phonon energy (1,100 cm^{-1}) due to the stretching vibration of network-forming oxides. Fluoride glasses have an advantage due to their low phonon energy (300–400 cm^{-1}) and higher quantum efficiency of many active transitions, but their relatively poor chemical and mechanical stability compared with oxide-based glasses is still a problem for their application (Tanabe et al., 2002).

The invention of RE ions-doped transparent oxyfluoride glass ceramics containing fluoride nano-crystals has attracted great interest, because of their excellent optical properties owing to the presence of fluoride crystals and good mechanical, chemical properties, resulting from the presence of some oxides such as SiO_2, GeO_2, and Al_2O_3 in the glass structure.

One of the most important advantages of these glass ceramics is the selective incorporation of the RE ions in the **nano-sized fluoride crystals**, which precipitated in the glass matrix after heat treatment. These crystals possess lower

Table 2.2 Composition and Properties of Some Transparent Glass Ceramics Based on Spinel Phases

Sample No.	Composition (wt%)	Nano-crystalline Phase/Size	Dopants	Application	Ref.
1	$x\ La_2O_3$, $(1-\frac{x}{100})$ (14.7MgO, 29.5 Al_2O_3, 47.4SiO_2, 5.3TiO_2, 3.1 ZrO_2) ($x=0$, 2.5, 5, 7.5, 10)	$MgAl_2O_4$ (ss) 8.1–10.2 nm	Co^{2+}	Q-switching	Chen et al. (2013)
2	Yb_2O_3 1.8, Er_2O_3 1.7, MgO 11, Al_2O_3 31.5, SiO_2 47, TiO_2 4.5 ZrO_2 2.5	$MgAl_2O_4$ 10–20 nm	Co^{2+}	Q-switching	Chen et al. (2012)
3	SiO_2 89, Al_2O_3 5.9, ZnO 4.9	$ZnAl_2O_4$ 10–15 nm	Co^{2+}	Laser material in the visible and NIR regions.	Duan et al. (2003)
4	SiO_2 58, Al_2O_3 10, ZnO 21, K_2O 3, Ga_2O_3 3, TiO_2 5	$ZnAl_2O_4$ (ss) 10–15 nm	Ni^{2+}	Broadband optical amplification	Suzuki et al. (2005)
5	SiO_2 39.3, Al_2O_3 26.7, ZnO 21.3, TiO_2 6.3, ZrO_2 6.4	$ZnAl_2O_4$ 11.5 nm	Cr^{3+}/Ni^{2+}	Broadband optical amplifier and tunable lasers	Wu et al. (2008a)
6	$(100-x)(55SiO_2-18Al_2O_3-18MgO-9TiO_2)-x\ Ga_2O_3$ ($x=0$, 2.5, 5, 7.5, 10) mol%	$MgAl_2O_4$ (ss) 7.6 nm	Ni^{2+}	Broadband optical amplifier and tunable lasers	Wu et al. (2008b)
7	$(100-y)$ $(16.7MgO \cdot 16.7Al_2O_3 \cdot 8.3Ga_2O_3 \cdot 50SiO_2 \cdot 8.3TiO_2 \cdot xNiO)$. yBi_2O_3 ($y=0$, 0.25, 0.5, 0.75 and 1.00 when $x=0.3$; $y=0.25$, 0.5, 0.75, and 1.00 when $x=0$) mol%	$MgAl_2O_4$ (ss) ~6 nm	Ni^{2+}/Bi^{3+}	Broadband optical amplifier and tunable lasers	Wu et al. (2009a)

phonon energy, as discussed above, and the glass-ceramic material remains transparent due to the much smaller size of precipitated crystals than the wavelength of visible light.

Processing of Oxyfluoride Glass Ceramics

General Considerations In the great majority of the research studies on oxyfluoride glass ceramics, the main concern and priority of the researchers have been the physical aspects of the subject, i.e., the optical properties and applications of the oxyfluoride glass ceramics. The main issues in the processing of the glass ceramics, such as the effect of glass composition and heat treatment on the nucleation and crystallization mechanisms hence, upon the microstructure of the resulting glass ceramics have been given relatively little attention in these studies.

Considering the key role played by the aforementioned processes in achieving the desired physical properties in nano-glass ceramics, it seems that the study of these topics should be given more attention and priority in the future.

However, many valuable research activities have been carried out in recent years in the field of processing of oxyfluoride glass ceramics, the results of which can be summarized as follows:

The great majority of the studied glass ceramics are based on fluoro-aluminosilicate, and to a lesser extent fluoro-germanate glasses. These glass ceramics usually contained PbF_2 (or solid solution of PbF_2-CdF_2), REF_3, and alkali earth (CaF_2 and BaF_2) nano-crystallites.

Some typical examples concerning the processing details of these glass ceramics are presented here.

Kukkonen et al. (2001) studied the crystallization behavior of transparent oxyfluoride glass ceramics with the composition $30SiO_2-15Al_2O_3-29CdF_2-17PbF_2-5ZnF_2$ (mol%) doped with 1 and 4 mol% ErF_3.

As previously discussed in Section 1.1.2.2.2, HRTEM images of an as-annealed glass prior to heat treatment indicated the absence of phase separation in it. However, HRTEM revealed that after a heat treatment of 6 h at 394°C (very close to T_g), some ordered structures of ~0.5 nm in diameter appeared in the glass specimen containing 4% ErF_3, indicating the onset of nucleation of some fluorite-structured crystallites (see Figure 1.13).

HRTEM/Energy Dispersive Spectroscopy (EDS) proved that Er^{3+} preferentially partitioned to the fluorite-structured crystalline phase and that this phase was PbF_2 and not a complex $Pb_xCd_{1-x}F_2$ solid solution as previously supposed.

The heat treatment of the glass sample at 394°C for 168 h resulted in the formation of a transparent glass ceramic comprising tiny PbF_2 crystallites, with the mean size of ~18 nm, which served as a host to accommodate Er^{3+} ions.

Tanabe et al. (2002) studied the crystallization behavior of a $53SiO_2-11Na_2O-27Al_2O_3-1AlF_3-7LaF_3-0.07ErF_3$ (mol%) glass in order to obtain an Er^{3+}-doped transparent oxyfluoride glass ceramic containing LaF_3 nano-crystals.

After the heat treatment of glass specimens in 550–750°C range, several XRD broad peaks appeared that were attributed to crystalline LaF_3. The peaks became sharper with increasing annealing temperature.

From the obtained peak width of XRD pattern, the crystal size of LaF_3 was also estimated. It was shown that as the heat-treatment temperature increased, crystal size increased from 24 to around 200 nm.

It was proposed that at each crystallization temperature, a highly viscous layer formed around the fluoride crystals during their crystallization, which acted as a diffusion barrier hindering further crystal growth.

The formation of a diffusion barrier was also observed during crystallization of CaF_2 from a glass in the $Na_2O-K_2O-CaO-Al_2O_3-SiO_2-CaF_2$ system (Rüssel, 2005). In this work, it was interestingly observed that while the content of crystalline CaF_2 increased with increasing the crystallization temperature, the mean crystallite size remained constant. It was also noted that the glass-transformation temperature of partially crystallized samples increased with increasing crystallization temperature/time and approached a value equal to the temperature at which the samples were treated. This was explained by the formation of a highly viscous layer enriched in SiO_2 formed during crystallization, which acted as a diffusion barrier, hindering further crystal growth.

A similar phenomenon was also observed during the crystallization of BaF_2 from glasses with the composition: $(100 - x)$ $(2Na_2O \cdot 16K_2O \cdot 8Al_2O_3 \cdot 74SiO_2) \cdot xBaF_2$ (with $x = 0$-6) (Bocker and Rüssel, 2009). The heat treatment of the glass samples with $x = 6$ at temperatures in the range from 500°C to 600°C for 5–160 h led to crystallization of BaF_2 nano-crystals. The quantity of BaF_2 crystals increased with increasing the time of heat treatment, while the mean crystallite size remained constant. The glass-transformation temperature of the samples also increased with increasing crystallization time, similar to the aforementioned glass ceramics containing CaF_2 nano-crystals described above (Rüssel, 2005), approaching the value of heat-treatment temperature of the glass samples. This was attributed to the formation of a highly viscous layer enriched in SiO_2 during the crystallization process, which acted as a diffusion barrier and hindered further crystal growth.

Bhattacharyya et al. (2009), utilizing advanced analytical TEM techniques, experimentally proved the concept of self-limited growth on the nanoscale for a $SiO_2-Al_2O_3-Na_2O/K_2O-BaF_2$ glass in which BaF_2 nano-crystals were formed. It was shown that the growth of these BaF_2 nano-crystals was inherently limited by the formation of a 1-nm wide SiO_2 shell around them.

Pablos-Martín et al. (2011) also investigated the mechanism and kinetics of crystallization of LaF_3 nano-crystals in a

$55SiO_2{-}20Al_2O_3{-}15Na_2O{-}10LaF_3$ (mol%) glass. LaF_3 nano-crystals up to 14 nm in diameter were obtained by heat treatment of the glass at temperatures from T_g+35 to T_g+95°C. The TEM micrograph of the base glass showed the formation of droplets enriched in elements of high atomic number due to the occurrence of a liquid–liquid phase-separation in glass. The droplets were fully amorphous and had an average diameter in the range 30–40 nm (see Figure 1.11A and Section 1.1.2.2.1).

It was proposed that upon the heat treatment of glass specimens, LaF_3 nano-crystals precipitated inside the lanthanum and silicon-rich droplets of phase-separated glass and the silicon redistribution occurring around the growing crystals, forming a shell, which then inhibits further crystal growth.

This was in contrast with the CaF_2 or BaF_2 nano-crystals, which according to the authors, precipitate from homogeneous, nonphase-separated glasses, and the increase in viscosity due to the silica enrichment around the formed crystals limits further crystal growth.

The change in composition in the glass matrix due to demixing and subsequent crystal growth actually resulted in a marked increase in viscosity with temperature and time of treatment at 620°C.

The value of Avrami parameter was also calculated from Differential Scanning Calorimetry (DSC) data, utilizing the Ozawa equation and the average value of $n \approx 1$ was obtained. The value of the m parameter, obtained utilizing various methods, also gave an average value of $m \approx 1$. Such values of m and ($m \approx n \approx 1$) correspond to a bulk crystallization mechanism with a constant number of nuclei (i.e., a well-nucleated sample in which the number of nuclei is independent of the heating rate). Crystal growth occurs in two dimensions, with a rate $\sim\sqrt{t}$ (diffusion controlled), as expected for LaF_3 nano-plates.

Chen et al. (2006) also studied the crystallization behavior of $45SiO_2{-}25Al_2O_3{-}5CaO{-}10NaF{-}15CaF_2{-}xErF_3$ ($x = 0$, 0.1, 0.5, 1, and 2) (mol%) glasses, aiming to prepare transparent oxyfluoride glass ceramics. The as-made samples were transparent and amorphous according to XRD and TEM results.

The DTA traces of Er-free glass samples showed an exothermic peak at 647°C indicative of the CaF_2 crystallization, which decreased with increasing the Er^{3+} content.

The crystallization kinetics studies revealed that CaF_2 crystallization was a diffusion-controlled process, with a continuously decreasing nucleation rate.

After heat treatment at 650°C for 2 h, spherical crystallites of CaF_2 precipitated from all the glass samples, containing differing amounts of Er^{3+}. The size of the crystallites in Er-free glass ceramics was ~20 nm that decreased in the glass ceramics containing 2 mol% Er^{3+}. Correspondingly, the number density of crystallites in the latter sample was more than in Er-free samples, as shown in Figure 2.10A and B. The detailed lattice structure of an individual CaF_2 nano-crystal containing 2 mol% Er^{3+} was presented in the HRTEM image (Figure 2.10C).

On the basis of the obtained experimental results, it was suggested that in the Er^{3+}-doped glass ceramics, a part of Er^{3+} ions that segregated in the interfacial regions probably reduced the crystal growth rate by hindering the atomic diffusion.

The EDS spectra taken from the glass matrix and from a nano-crystal embedded into the glass ceramic doped with 2 mol% Er^{3+} indicated that Er^{3+} ions concentrated mainly in the crystals.

Secu et al. (2009) also prepared transparent oxyfluoride glass ceramics containing CaF_2 nano-crystals of ~65 nm in size by the heat treatment of a $49SiO_2-20Al_2O-30CaF_2-1EuF_2$ (mol%) glass at 760°C for 15 min. It was shown that the addition of Eu^{3+} ions to the glass composition resulted in the decrease of

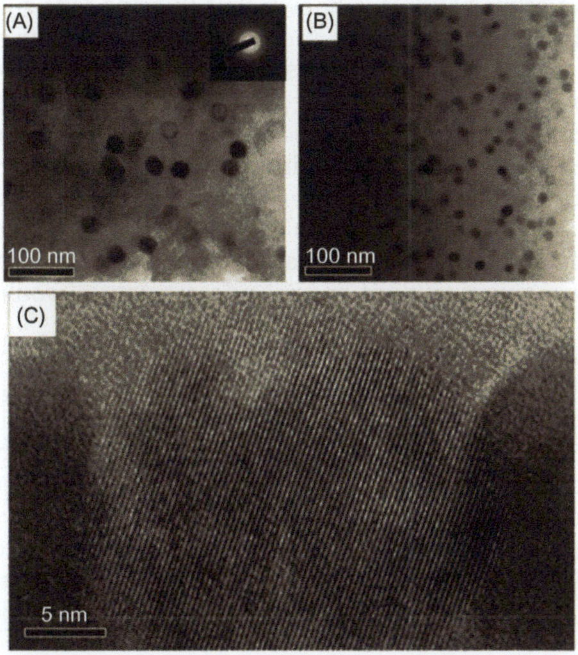

Figure 2.10 TEM bright field images of glass ceramics with different Er doping: (A) the Er-free sample, the inset is the corresponding selected-area electron diffraction (SAED) pattern; (B) sample doped with 2 mol% Er^{3+}; (C) HRTEM image of an individual CaF_2 nano-crystal in (B). (After Chen et al., 2006. Reproduced with permission of Elsevier Science Publishing Company.)

CaF_2 nano-crystal size from ~80 to ~65 nm. This addition also led to the increase of the lattice constant of CaF_2 crystals from 5.457 to 5.497 Å.

This enlargement of the crystal lattice was ascribed to the inclusion of interstitial fluorine ions in the crystal structure of CaF_2 to compensate the excessive positive charge initiated from the substitution of Eu^{3+} ions for Ca^{2+} ions.

More recently Imanieh et al. (2013) investigated the effects of glass composition on the crystallization behavior and optical properties of oxyfluoride glass ceramics in CaF_2–Al_2O_3–SiO_2 system.

According to the authors, the amount of residual fluorine in the glass played a significant role in the precipitation of CaF_2 nano-crystallites and the optimum CaF_2 content in the initial glass composition was determined as 35 mol%.

On the other hand, since the rate of fluorine loss greatly increases with temperature, increasing the melting temperature of the glass batch by the addition of more Al_2O_3 should increase the fluorine loss. However, the effect of Al_2O_3 on glass structure and properties is more complicated.

It was proposed that if the Al_2O_3 content is kept constant, the addition of more CaF_2 to the glass composition would decrease the melting temperature of the glass batch, reducing the loss of F.

On the other hand, it is known that fluorine ions substituted for the bridging oxygens in the glass structure may preferentially be bonded to the Al ions present in the glass structure. When the amount of F in the glass increases, the glass structure becomes more interrupted and the possibility of F loss increases. In this case, the addition of higher amounts of Al_2O_3 may result in the preservation of F ions in the glass structure and reduction of its loss. It was also found that the phase separation occurring by the spinodal decomposition mechanism in the glasses owing to their special microstructure would reduce the crystal growth rate.

On the basis of the above considerations, the glass composition with a SiO_2/Al_2O_3 ratio of 2.18 and the initial CaF_2 content of 35 mol%, owing to its tendency to develop spinodal microstructure of phase separation, as well as its higher viscosity exhibited a limited crystal growth rate and was chosen as the optimum composition.

The glass ceramics resulting from the above glasses owing to the small size of their CaF_2 crystals (20 nm) remained transparent after heat treatment at 740°C.

Properties and Application of Fluorescent Oxyfluoride Glass Ceramics

Up-Conversion Fluorescent Oxyfluoride Glass Ceramics Wang and Ohwaki (1993) were the first investigators who obtained a transparent glass ceramic, after the heat treatment of a $\mathbf{Yb^{3+}-Er^{3+}}$ **co-doped** oxyfluoride glass in the system SiO_2–Al_2O_3–PbF_2–CdF_2–YbF_3–ErF_3 at 470°C.

The glass ceramics contained $Pb_xCd_{1-x}F_2$ nano-crystallites measuring ~20 nm. It was suggested that upon the heat-treatment Yb^{3+} and Er^{3+} ions were preferentially segregated from the precursor glass and dissolved into the $Pb_xCd_{1-x}F_2$ crystallites.

The resonant energy transfer between the aforementioned dissolved ions produced up-conversion fluorescence in the blue-green.

In Figure 2.11A and B, the fluorescence spectra under an excitation of 0.972 μm are presented for the above glass and its glass-ceramic counterpart.

Upon heat treatment, the emission intensities at 545 and 660 nm increased by the factors of ~100 and ~150, respectively. For the comparison, the up-conversion fluorescence spectra of an aluminofluoride glass ($35AlF_3$, $14YbF_3$, $1ErF_3$, $20PbF_2$, $5MgF_2$, $15CaF_2$, $10BaF_2$ mol%) is also presented (Figure 2.11C). This fluoride glass has previously been evaluated as one of the most efficient IR up-conversion materials.

Other investigators also prepared and studied luminescence properties of up-converting oxyfluoride glass ceramics, containing **PbF_2 nano-crystals**, doped with other **RE ions, such as Ho^{3+} and Tm^{3+}**.

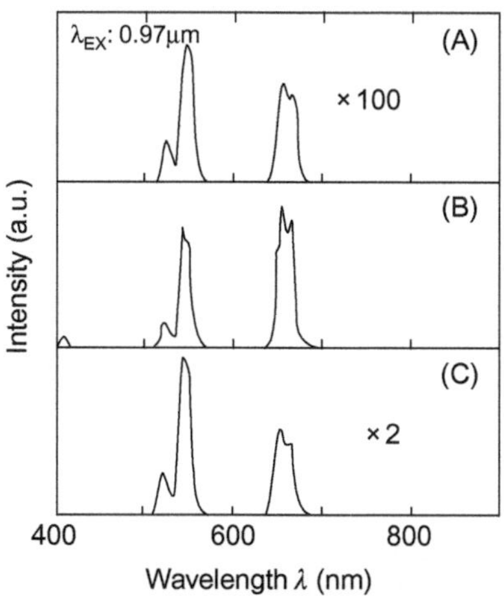

Figure 2.11 Fluorescence spectra under a 0.972 μm excitation for the oxyfluoride glass before (A) and after thermal treatment (B) and for an aluminofluoride glass (C). (After Wang and Ohwaki, 1993. Reproduced with permission of American Institute of Physics.)

For example, Liu and Heo (2006) prepared oxyfluoride glass ceramics containing **β-PbF_2 nano-crystals, doped with Ho^{3+} ions**. The specimens were prepared with the addition of HoF_3 to a glass composition in the SiO_2–GeO_2–Al_2O_3–TiO_2–YF_3–PbF_2 system and their luminescence properties were investigated.

Ho^{3+} ions have many metastable levels which can give rise to several up-converted emissions with wavelengths in the visible spectrum.

HoF_3 concentration in the glasses was controlled from 0.05 mol% to 0.1, 0.2, 0.5 mol% to study the concentration dependence of emission properties and up-conversion mechanisms.

Figure 2.12 shows the emission spectra of the precursor glass and nano-structured glass ceramics (NG) doped with 0.05 mol% and 0.5 mol% of HoF_3 under 898 nm pumping. Four up-conversion emission bands centered at 485 nm ($^5F_3 \rightarrow {}^5I_8$), 540 nm ($^5F_4$, $^5S_2 \rightarrow {}^5I_8$), 650 nm ($^5F_5 \rightarrow {}^5I_8$), and 752 nm ($^5S_2 \rightarrow {}^5I_7$) were observed. Green (540 nm) and red (650 nm) emissions were strongly dependent on HoF_3 concentration. For specimens containing 0.05 mol% HoF_3, green emission was dominant and both the red and green emission intensities increased after thermal treatment. This enhancement in emission intensity was attributed to the partial incorporation of Ho^{3+} ions into the β-PbF_2 nano-crystals, which provided the low-phonon energy environment for RE ions. On the other hand, red emission became dominant with doping of 0.5 mol% HoF_3.

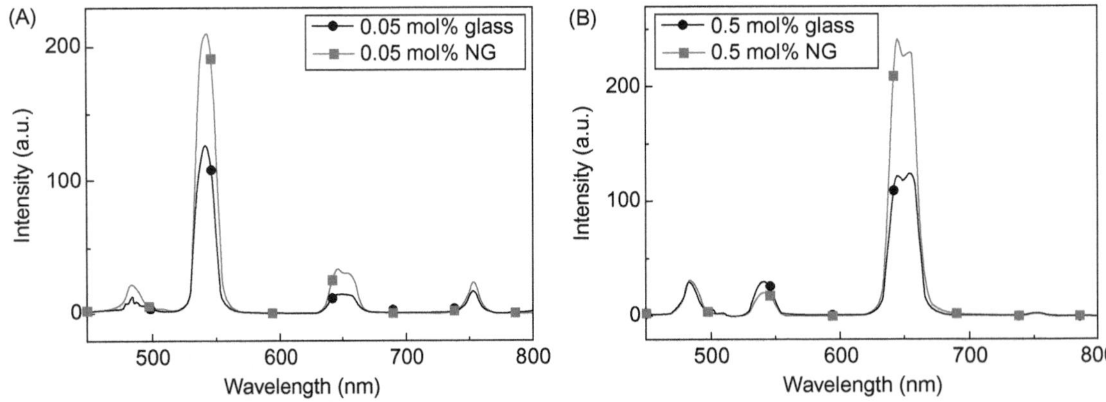

Figure 2.12 Emission spectra of the parent glass and nano-structured glass (glass ceramic) containing (A) 0.05 mol% and (B) 0.5 mol% HoF_3 under 898 nm pumping, measured at room temperature. (After Liu and Heo, 2006. Reproduced with permission of Elsevier Science Publishing Company.)

Hence, it was possible to control the intensities of the green and red emissions, simply by modifying the HoF_3 concentration.

Gouveia-Neto et al. (2004) also observed intense red up-conversion emission around 650 nm in $PbGeO_3$–PbF_2–CdF_2 transparent glass ceramics containing **β-PbF_2: Ho^{3+} nano-crystals**, under the excitation of a 980 nm diode laser source. The 650 nm up-conversion signal (Figure 2.13) was assigned to the $^5F_5 \rightarrow {}^5I_8$ transition of holmium ions. Very low-intensity signals around 490 and 540 nm corresponding to the $^5F_{2,3} \rightarrow {}^5I_8$ and $^4S_2$, $^5F_4 \rightarrow {}^5I_8$ transitions, respectively, were also detected. The up-conversion excitation mechanism was achieved through a combination of stepwise phonon-assisted multiphoton absorption cross-relaxation processes, involving pairs of holmium ions, and ESA.

Transparent glass ceramics in the GeO_2–PbO–PbF_2 system containing **β-PbF_2 nano-crystals doped with Tm^{3+} ions** were also prepared (Hirao et al., 1995).

Up-conversion fluorescence spectra of the glass and glass-ceramic counterpart of this sample are shown in Figure 2.14.

The excitation was carried out at 650 nm for these spectra. The fluorescence peaks at 360, 450, 480, and 800 nm were ascribed to the $^1D_2 \rightarrow {}^3H_6$, $^1D_2 \rightarrow {}^3F_4$, $^1G_4 \rightarrow {}^3H_6$, and $^3H_4 \rightarrow {}^3H_6$ transitions, respectively. The former three transitions correspond to the up-conversion fluorescence.

As stated in the previous section, other host media for accommodation of RE ions have also been utilized, e.g., **alkali earth fluoride nano-crystallites, especially CaF_2.**

In comparison with other fluorides, CaF_2 has high solid solubility to REs, wide transmission range (from ~0.125 to 10 μm),

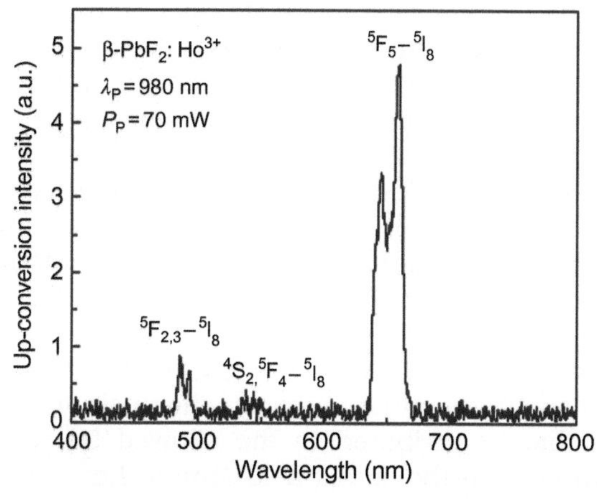

Figure 2.13 Typical visible up-conversion spectrum of radiation emanating from the holmium-doped sample under 70 mW excitation power at 980 nm. (After Gouveia-Neto et al., 2004. Reproduced with permission of Elsevier Science Publishing Company.)

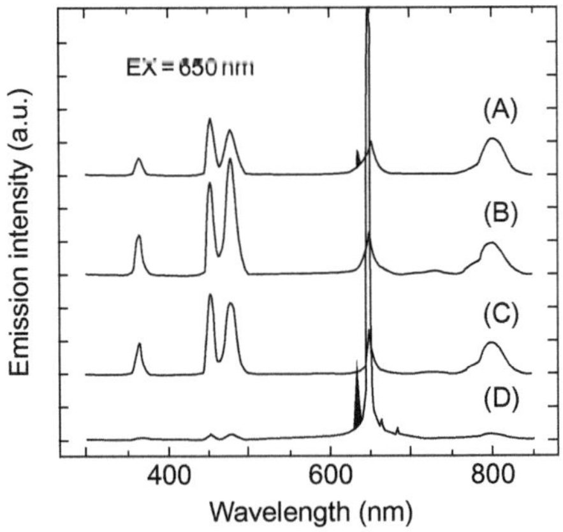

Figure 2.14 Up-conversion fluorescence spectra of glass and transparent glass ceramics. The specimens are (A) $50GeO_2 \cdot 50PbO \cdot 1TmF_3$ glass, (B) $50GeO_2 \cdot 40PbO\text{-}10PbF_2 \cdot 1TmF_3$, glass, (C) transparent glass ceramics heat-treated at 340°C for 10 h, and (D) glass ceramics heat-treated at 360°C for 10 h. The excitation wavelength is 650 nm. The assignments of the fluorescence peaks are as follows: 360 nm; $^1D_2 \rightarrow {}^3H_6$, 450 nm; $^1D_2 \rightarrow {}^3F_4$, 480 nm: $^1G_4 \rightarrow {}^3H_6$, and 800 nm: $^3H_4 \rightarrow {}^3H_6$. (After Hirao et al., 1995. Reproduced with permission of American Institute of Physics.)

lower phonon energy, good chemical stability, and friendliness to environment (Pb and Cd have environmental problems).

For example, Fu et al. (2002) prepared transparent glass ceramics containing CaF_2: Eu^{2+} nano-crystals by heat treating SiO_2—Al_2O_3—CaO—CaF_2 glasses doped with Eu^{2+} ions.

The microstructural observation of the glass-ceramic samples showed that CaF_2 crystallites, 11–18 nm in size, were dispersed in the amorphous phase without clustering. Fluorescence measurements also showed that Eu^{2+} ions incorporated into the CaF_2 crystalline phase and gave a much stronger emission in the glass ceramics than in the corresponding glass.

Figure 2.15 depicts the luminescence spectra for the glasses before and after heat treatment. The emissions from all the samples exhibited a broadband peaking at ~420 nm. No distinct difference in peak position and FWHM is observed, but the intensity of the emission in the glass ceramics is more than twice that in the corresponding glass. The authors stated that Europium in glass may be present as Eu^{3+} and Eu^{2+}. Whereas Eu^{3+} usually shows sharp and strong emission peaks at 590 and 615 nm, due to $^5D_0 \rightarrow {}^7F_1$ and $^5D_0 \rightarrow {}^7F_2$ transitions, respectively, Eu^{2+} exhibits a broad emission band. The position of this emission band, that is attributed to the allowed 5d–4f transition, depends strongly on the ligand field around Eu^{2+} ions, but very

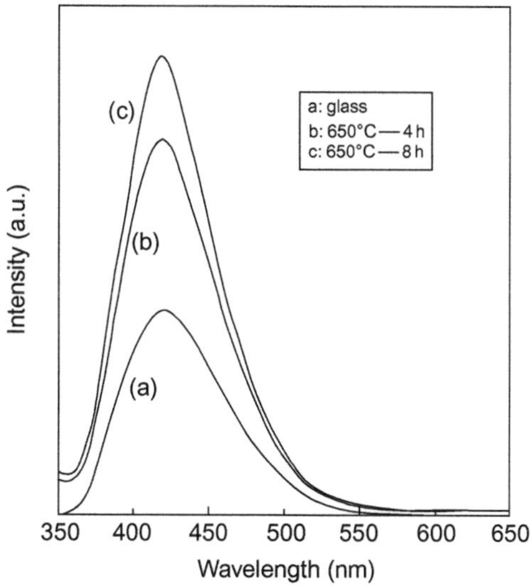

Figure 2.15 Luminescence spectra of the glasses before and after heat treatment. (After Fu et al., 2002. Reproduced with permission of Elsevier Science Publishing Company.)

often appears in the region of <500 nm in most of the oxide hosts.

On the basis of the above explanation, the authors attributed the observed broad emission band to the 5d–4f transition of Eu^{2+} ions.

Kishi et al. (2005) also fabricated **Er^{3+}–Yb^{3+}-doped** transparent oxyfluoride glass ceramics containing **CaF_2 nano-crystals**, by heat treating glasses located in SiO_2–Al_2O_3–CaF_2–LnF_3 (Ln=Er, Yb) systems. Nano-crystals of $(Ca_{1-x}, Ln_x)F_{2+x}$ solid solution precipitated with size of 20–40 nm after heat treatment at 700–750°C. Green up-conversion luminescence at 540 nm and red up-conversion at 660 nm were observed in the glass ceramics at 970 nm pumping. These results were explained by considering that Yb^{3+} and Er^{3+} ions were incorporated into CaF_2 crystals with a phonon energy lower than that of silicate glass matrix.

More recently, Imanieh et al. (2012) investigated the effect of Yb^{3+} content and SiO_2/Al_2O_3 ratio on the crystallization behavior and cooperative emission of up-converting oxyfluoride glass ceramics in CaF_2–Al_2O_3–SiO_2 system. The glass-ceramic specimens were shown to contain CaF_2 nano-crystals doped with Yb^{3+} ions.

The luminescence spectra of the glass-ceramic samples which were heat-treated at 630°C for 4, 48, and 72 h and excited under a 975 nm radiation were recorded.

These emission spectra consisted of two main bands centered in the blue (~487 nm) and in the green region (~550 nm). The blue emission was attributed to the cooperative emission of the Yb^{3+}–Yb^{3+} ion pairs after excitation from the $^2F_{7/2}$ level to the $^2F_{5/2}$ level, whereas the green emission was assigned to the Er^{3+} ions which were present as impurities in the precursor glass. The relative intensities of samples at 487 nm are depicted in Figure 2.16.

The intensity of the cooperative emission depends on the ytterbium concentration and the distance between these ions inside the crystal phase. As can be seen from Figure 2.16, the intensity at 487 nm emission in samples with lower ytterbium concentration (0.5 mol%) is lower than the samples with the concentration of 2 mol%.

The Yb^{3+} ions were considered to be incorporated into the CaF_2 crystals to form (Ca_{1-x}, Yb_x) F_{2+x} solid solution crystals during the heat treatment of the samples.

Increasing the 487 nm emission in the first 48 h of heat treatment was related to raising the ytterbium concentration in the CaF_2 crystals, while the size of the crystals remained almost unchanged.

The decrease in intensity between 48 and 72 h was attributed to the increase in the crystal size and the distance between ytterbium ions in the SA1.8 sample, and the increase in volume fraction of CaF_2 crystals, causing enhanced scattering in the SA2.18 specimen.

The highest and the most efficient cooperative emission intensity was recorded in the SA1.8Yb2 specimen which was heat-treated at 630°C for 48 h. Small crystal size (20 nm), keeping the ytterbium ions close to each other, and the high volume

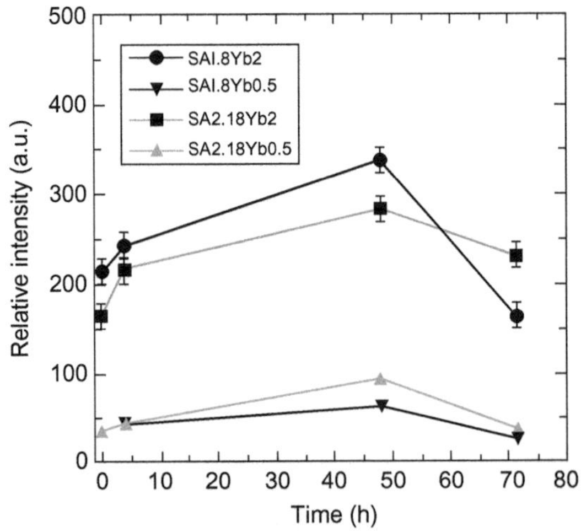

Figure 2.16 Relative intensity of 487 nm of the samples which were heat-treated at 630°C for 0, 4, 48, and 72 h excited by 975 nm. The figures that followed the specimen codes (SA) and Yb denote SiO_2/Al_2O_3 ratio, and the content of YbF_3 in mol% in the glass composition, respectively. (After Imanieh et al., 2012. Reproduced with permission of the American Ceramic Society.)

concentration of CaF_2 phase (30%) were the reasons for this high efficiency.

Some investigators also prepared and studied luminescence properties of up-converting, oxyfluoride glass ceramics, containing **LaF_3 nano-crystals**, doped with different RE ions.

It was suggested that LaF_3 is an ideal host for RE ions, since it has extensive solid solution with all RE ions (owing to the similar ionic radius and the same valence as La^{3+}), a lower phonon energy (350 C/m), and adequate thermal and environmental stability.

For example, Tanabe et al. (2002) prepared transparent oxyfluoride glass ceramics containing Er^{3+} ions-doped LaF_3 nano-crystals and studied their fluorescent characteristic. Through proper heat treatment of the as-made glasses located in the SiO_2–Al_2O_3–Na_2O–Al_2F_6–LaF_3 system, LaF_3 nano-crystals around 30 nm in size precipitated. The glass-ceramic samples, heat-treated at various temperatures, exhibited a quite distinct up-conversion luminescence around 540 nm along with a weaker IR luminescence ~1,000 nm (not shown here) under the excitation wavelength of 970 nm (Figure 2.17).

It was also shown that with increasing the heat-treatment temperature and time, the fluorescence intensity at 540 nm (due to the $^4S_{3/2}$ to $^4I_{15/2}$ transition) increased and the lifetime of the $^4S_{3/2}$ level became longer. These results were attributed to the change of ligand field of Er^{3+} ions and the decrease of effective phonon energy because of the precipitation of LaF_3 nano-crystals and the incorporation of some of Er^{3+} ions into fluoride crystals.

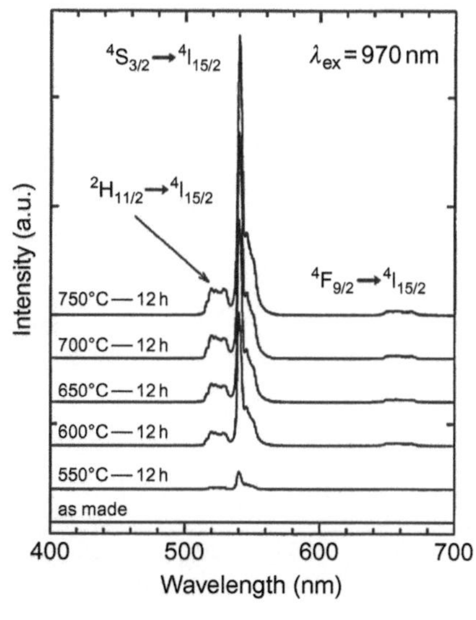

Figure 2.17 Up-conversion fluorescence spectra of Er^{3+}-doped glass-ceramic samples heat-treated at various temperatures. (After Tanabe et al., 2002. Reproduced with permission of Elsevier Science Publishing Company.)

Other Host Nano-Crystals Chen et al. (2009) investigated the crystallization behavior and luminescence properties of Yb^{3+}-doped oxyfluoride glass ceramics containing **TbF_3** nano-crystals. The glass composition comprised SiO_2–Al_2O_3–NaF–TbF_3 and YbF_3. The glass ceramic containing 1.5 mol% YbF_3 (heated at 650°C, 2 h) revealed some intense diffraction peaks, which were assigned to the orthorhombic TbF_3 crystals.

The TEM micrograph demonstrated the homogeneous distribution of TbF_3 nano-crystals with a mean diameter of 22 nm in the glass matrix. According to the authors, Tb^{3+} ions, unlike La^{3+} or Y^{3+} ions, can act as the optically active centers or sensitizers for their partially full 4f electronic configuration.

The incorporation of Yb^{3+} ions into TbF_3 crystals was confirmed by energy-dispersive X-ray spectroscopy. Doping of Yb^{3+} ions resulted in both the visible up-conversion luminescence of Tb^{3+} and the NIR quantum cutting down-conversion emission of Yb^{3+} in these glass ceramics.

Under 976 nm NIR laser excitation, the glass ceramics exhibited intense green Tb^{3+}: $^5D_4 \rightarrow {}^7F_j$ ($j = 3, 4, 5, 6$) up-conversion luminescence due to the cooperative energy transfer from two Yb^{3+} ions to one Tb^{3+} ion (Figure 2.18).

On the other hand, according to the authors, the quantum cutting down-conversion luminescence of Yb^{3+} was realized by cooperative energy transfer from one Tb^{3+} to two Yb^{3+} ions. When excited at 485 nm, Tb^{3+} emissions corresponding to the Tb^{3+}: $^5D_4 \rightarrow {}^7F_j$ ($j = 5, 4, 3$) transitions occur. Meanwhile, two NIR photons originating from Yb^{3+}: $^2F_{5/2} \rightarrow {}^2F_{7/2}$ are also obtained from one absorbed blue photon.

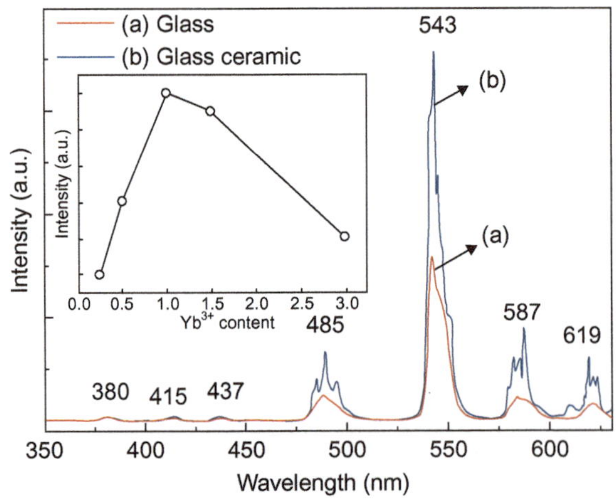

Figure 2.18 Up-conversion emission spectra of the precursor glass and glass ceramic doped with 1.5 mol% Yb^{3+} under 976 nm excitation. The inset shows the dependence of the 543 nm emission intensity on the Yb^{3+} content for the glass-ceramic samples. (After Chen et al., 2009. Reproduced with permission of Elsevier Science Publishing Company.)

Some investigators also prepared and studied luminescence properties of up-converting, oxyfluoride glass ceramics containing Na/KYF_4 nano-crystals doped with Er^{3+}, Yb^{3+} is known as one of the most efficient NIR-to-visible up-converting hosts; its visible emission is at least 2 times greater than that of YF_3: Er^{3+}, Yb^{3+} under optimum excitation; and RE-doped $NaYF_4$ nano-crystals have been found potentially applicable in many fields, such as biology, molecular labeling, and solar cells (Liu et al., 2006).

Liu et al. (2006) fabricated transparent glass ceramics containing Er^{3+}: $NaYF_4$ nano-crystals for the first time and investigated the influences of heat-treatment temperature and Er^{3+} doping level on their up-conversion luminescence.

The prepared glass samples were of the composition of $40SiO_2$, $25Al_2O_3$, $18Na_2CO_3$, $10YF_3$, 7NaF, and $x$$ErF_3$mol% ($x = 0.05$, 0.2, 0.5, 1.0, and 2.0).

After the heat treatment of glass samples at 620°C for 2 h, some nano-crystals of the cubic α-$NaYF_4$ phase with a mean diameter of 22 nm appeared in them which at higher temperature (near 650°C) was transformed to the hexagonal β-$NaYF_4$ phase.

To investigate the influence of heating temperature on up-conversion luminescence, the emission spectra of 1% Er^{3+}-doped transparent glass ceramics heat-treated at different temperatures were measured under 980 nm excitation.

With increasing the heating temperature, an enhancement in the up-conversion intensity was observed that was attributed to the incorporation of more Er^{3+} into the growing $NaYF_4$ crystals. When the heating temperature reached 650°C, an abrupt enhancement in up-conversion emission intensity was observed owing to the occurrence of the phase transition from cubic α-$NaYF_4$ to hexagonal β-NaYF4 phase. For the samples heat-treated at 620°C, the Er^{3+} concentration has significant influence on the red to green emission ratio. For example, when the Er^{3+} doping level was increased from 0.05 to 2.0 mol%, the up-conversion emission was whole-range tunable from monochromatic green to approximately monochromatic red, which could be mainly attributed to the cross-relaxation between Er^{3+} ions.

According to the authors, this interesting characteristic may enable the devices based on Er: $NaYF_4$ glass ceramics to meet the specific color demands, which will have potential applications in the fields of optical display and solid-state laser.

Santana-Alonso et al. (2010) developed transparent nano-glass ceramics containing Eu^{3+}-doped $NaYF_4$ nano-crystals, by heat treatment of precursor sol-gel derived glasses with the composition of $95SiO_2{-}5NaYF_4$ doped with $0.1Eu^{3+}$ (mol%). The dried gel samples (xerogels) finally were heat-treated in air at different temperatures ranging from 550°C to 650°C. The average radii of samples heat-treated at different temperatures, calculated from Scherrer's equation, were in the range 4.3–9.6 nm that was in accord with the value of ~10 nm obtained from results of TEM and HRTEM analysis carried out for the 650°C heat-treated samples.

The nano-structure of the glass ceramics was also studied by HRTEM (Figure 2.19).

Changes in the luminescence spectra with the temperature of heat treatment were also studied and correlated with the crystallinity degree of the samples.

The main emissions were located at ~590 and 613 nm, corresponding to the $^5D_0 \rightarrow {}^7F_1$ and $^5D_0 \rightarrow {}^7F_2$ transitions of the Eu^{3+} ions, respectively.

The examination of the intensity ratios (R) calculated for the emission spectra indicated that it diminished with the heat-treatment temperature from 550°C to 650°C for each excitation wavelength. The intensity ratio of R, which is the ratio of intensities of $^5D_0 \rightarrow {}^7F_2$ to $^5D_0 \rightarrow {}^7F_1$ transitions, and known as asymmetry ratio is used as a probe to assess the local symmetry of the environment of the ions.

The aforementioned diminishing of R factor with the elevation of heat-treatment temperature indicated that the environment

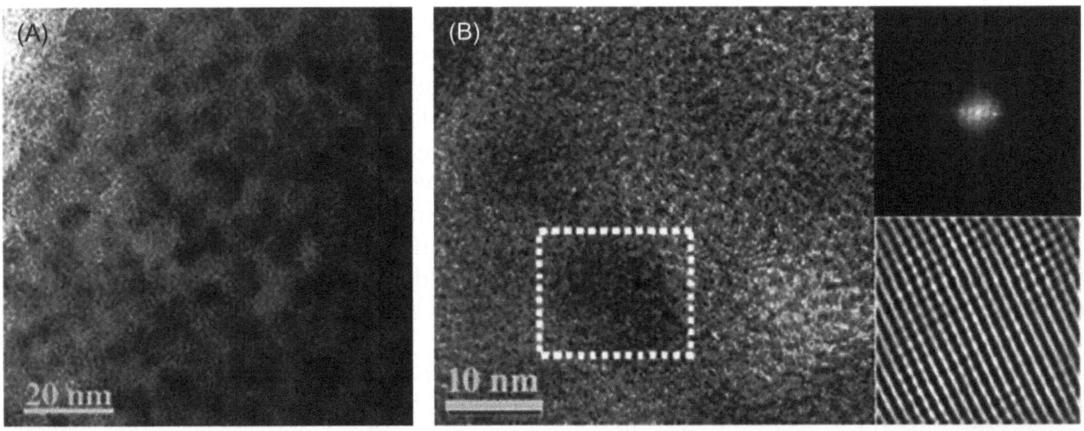

Figure 2.19 (A) TEM bright-field image shows the presence of $NaYF_4$ nano-particles (dark contrast). (B) HRTEM micrograph where $NaYF_4$ nano-crystals are observed. The right-top side insert corresponds to the power spectrum obtained from the region squared in white. The right-bottom inset corresponds to magnified detail of the same region showing the crystalline pattern of a $NaYF_4$ nano-particle. (After Santana-Alonso et al., 2010. Reproduced with permission of Elsevier Science Publishing Company.)

of Eu^{3+} ions got closer to an inversion symmetry site, associated with the partitioning into $NaYF_4$ nano-crystals, which was in agreement with the increasing degree of crystallinity with heat-treatment temperature as proved by XRD measurements.

The site selective spectroscopy also allowed the differentiation between ions remaining in the amorphous silica glassy phase and those preferentially partitioned into the crystalline environment.

Yanes et al. (2013) also prepared transparent nano-glass ceramics containing KYF_4 nano-crystals by heat treatment of precursor sol-gel-derived glasses. The precursor glass has the composition of $95SiO_2{-}5KYF_4$, doped with $0.1Eu^{3+}$ and co-doped with $xYb^{3+}-0.1Tm^{+}$ where $x = 0.3$, 0.6, and 1.2 (mol%).

The dried gel samples were heat-treated in air, at temperatures in the range of 600–700°C. The precipitation of cubic KYF_4 nano-crystals was confirmed by XRD and HRTEM images.

The average radii of nano-crystals in the samples heat-treated at different temperatures, calculated from Scherrer's equation, were in the range ~11–17 nm that was in accord with the value of ~10–15 nm obtained from results of TEM and HRTEM analysis for the samples heat-treated at 700°C.

Utilizing the excitation and emission spectra and using Eu^{3+} as a probe ion, the differentiation between amorphous and the crystal-like environments became possible.

It was also proposed that the unusual high-energy up-conversion emissions observed in $Yb^{3+}-Tm^{3+}$ co-doped samples, that increased with the concentration of Yb^{3+}, as well as the heat-treatment temperature, can be related to the enhancement of the energy transfer from Yb^{3+} to Tm^{3+}, with a subsequent higher efficiency of UV up-conversion emissions associated with five and four photon processes. It was also claimed that the ratio between UV and visible up-conversion emissions could be tuned as a function of pump power excitation. The most efficient UV up-conversion emissions were obtained in the 700°C heat-treated sample with the concentration of $1.2Yb^{3+}-0.1Tm^{3+}$ (mol%), with potential applications in solid-state UV lasers.

Down-Conversion Fluorescent Oxyfluoride Glass Ceramics Tick et al. (1995) were the first investigators who described the optical properties of a **down-converting transparent oxyfluoride glass ceramic**, with efficient luminescence at 1,300 nm, that could be used as amplifier hosts for telecommunications.

According to the authors, the combination of high optical transparency and the long fluorescence lifetimes are essential characteristics of an amplifier host material.

Since the presence of Yb^{3+} ions is undesirable for 1,300 nm amplification (because the quantum efficiency is reduced by an energy back-transfer mechanism), the original Wang–Ohwaki glass composition had to be modified by the substitution of YF_3 or a mixture of YF_3 and ZnF_2 for YbF_3. By heat treatment of the selected glass, consisting of $SiO_2–Al_2O_3–PbF_2–CdF_2–YF_3$ and ZnF_2 at ~470°C, the desired highly transparent glass-ceramic material was developed.

Approximately 20–30 vol% of this material was comprised of cubic, **cadmium–lead fluoride nano-crystals** of the average crystal size ~9–18 nm and the remainder was a predominantly oxide glass.

The fluorescence decay behavior of one example of the aforementioned oxyfluoride glass ceramics, **doped with Pr^{3+}**, is shown in Figure 2.20A. A lifetime of 147±10 μs is shown here, that could be varied by the heat treatment and concentration of Pr^{3+}, between 100 and 160 μs, which is higher in comparison to other materials.

Since the significant fraction of the RE dopant was preferentially partitioned into the fluoride crystal phase, their optical behavior, e.g., long fluorescence lifetimes, is mainly determined by their local fluoride low-phonon energy environment. Hence, the essential conditions for its application as amplifier host material could be met.

Figure 2.20B depicts the down-conversion fluorescence spectrum of the glass ceramic, peaking at 1,310 nm.

Down-converting oxyfluoride glass ceramics, mainly containing **PbF_2 or RE-fluoride nano-crystals**, and doped with other RE ions were also extensively studied in recent decades (Dejneka, 1998; Hayashi et al., 2001; Chen et al., 2008; Tikhomirov et al., 2008; Rodríguez et al., 2010).

Hayashi et al. (2001) fabricated Tm^{3+} ion-doped transparent oxyfluoride glass ceramics containing PbF_2 nano-crystals and

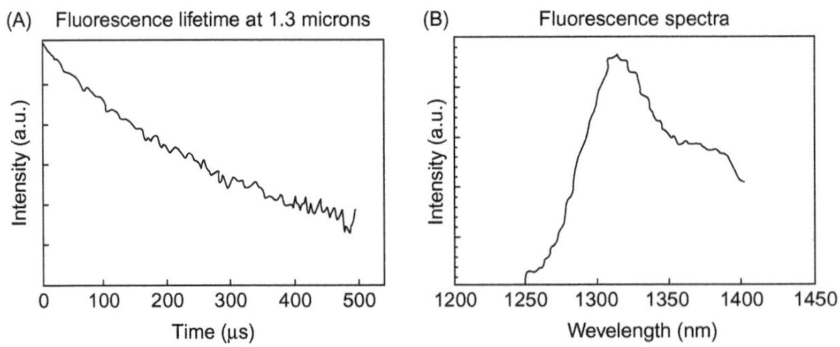

Figure 2.20 (A) Fluorescence lifetime and (B) IR spectra of glass-ceramic sample doped with 125 ppm Pr^{3+}. (After Tick et al., 1995. Reproduced with permission of American Institute of Physics.)

investigated the IR fluorescence properties of them. The glass composition was located in the SiO_2–GeO_2–Al_2O_3–PbF_2–TiO_2 system.

The room-temperature fluorescence spectra of the as-made glass and the glass-ceramic samples heat-treated at 510°C for various times under the excitation wavelength of 792 nm are depicted in Figure 2.21. It can be seen that after cerammization, the emission bands showed sharper structure and inhomogeneous widths decreased. In this process, the relative intensity ratio of the 1,460 nm emission to 1,800 nm increased by 1.5 times with ceramming time. A similar tendency was observed for ceramming temperature.

By luminescence decay measurements, it was revealed that part of the Tm^{3+} ions were incorporated into fluoride crystals by heat treating the as-made glass above glass transition temperature. Temperature dependence of the 1.4 μm band emission properties of Tm^{3+} ions in the transparent glass ceramics was also investigated. With lowering the temperature, the intensity ratio of the 1.4 μm emission band to the 1.8 μm band increased.

At the same time, the center of gravity of the $^3H_4 \rightarrow {}^3F_4$ emission shifted to a longer wavelength and reached ~1,485 nm at 25 K. Under 50 K, the peak of the 1.4 μm band split into two parts and the flat range of the split peaks became flatter.

These properties can be utilized for the 1.5 μm (S-band) optical amplifier applications.

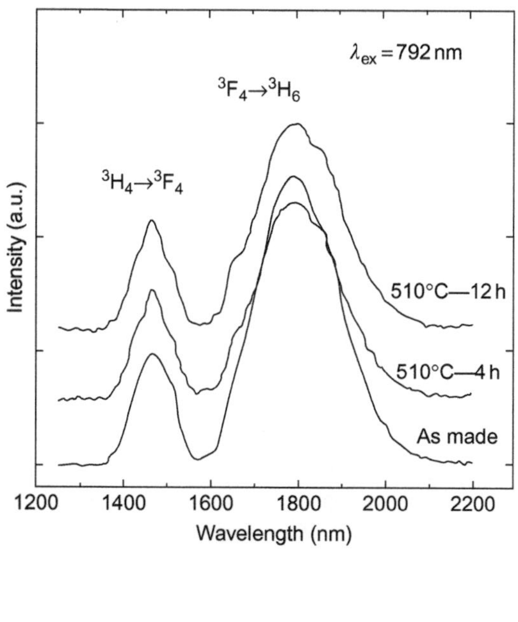

Figure 2.21 IR fluorescence spectra of glass and glass-ceramic samples heat-treated for various times. (After Hayashi et al., 2001. Reproduced with permission of American Institute of Physics.)

Tikhomirov et al. (2008) reported on the preparation and optical properties of a transparent oxyfluoride nano-glass-ceramic host, derived from a SiO_2–Al_2O_3–PbF_2–CdF_2–ZnF_2–REF_3 glass **containing β-PbF_2 nano-crystals.**

In this glass ceramic, heavy doping with RE ions such as Ho^{3+}, Dy^{3+}, Eu^{3+}, Tm^{3+}, Er^{3+} was possible, and the vibration energy affecting the dopant (<250 cm^{-1}) was smaller in comparison with other competitive glassy hosts.

An intense IR emission was observed in these RE-doped transparent glass-ceramic samples.

In these glass ceramics, more than 90% fraction of the RE dopants dissolved in the cubic **β-PbF_2nano-crystals** with a certain diameter of the order of 10 nm, while these nano-crystals were embedded in a robust aluminosilicate glass network. The remarkably low maximal phonon energy coupled to the RE dopants incorporated into the β-PbF_2 nano-crystals permits the efficient IR emission of the dopants from the levels which are nonradiatively quenched in other glassy hosts. Figure 2.22 depicts the normalized emission spectra of the **Ho^{3+}-doped glass ceramic**, at room temperature and 77 K, under the excitation of 640 nm.

Two emission bands of the Ho^{3+} at 1.2 μm ($^5I_6 \rightarrow {^5I_8}$ transition) and 1.47 μm ($^5F_5 \rightarrow {^5I_6}$ transition) fall in the modern extended telecommunication window ranging from 1.2 to 1.7 μm.

Rodríguez et al. (2010) studied a new down-conversion mechanism in Er^{3+} –Yb^{3+} co-doped glass ceramics containing nano-crystals of PbF_2. The aim was to study their potential in increasing the efficiency of photovoltaic solar cells.

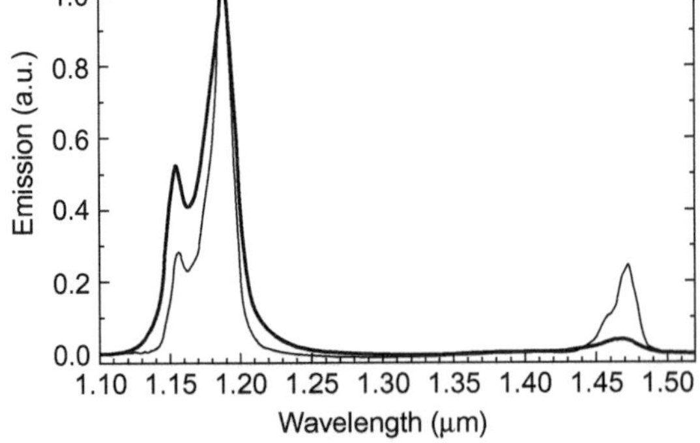

Figure 2.22 Normalized emission spectra of the Ho^{3+}-doped glass ceramic at room (thick solid line) and 77 K (thin solid line) temperatures excited at 640 nm. (After Tikhomirov et al., 2008. Reproduced with permission of Elsevier Science Publishing Company.)

An oxyfluoride nano-glass ceramic was obtained by the heat treatment of a precursor SiO_2–Al_2O_3–PbF_2–CdF_2–ZnF_2–REF_3 glass with the optimized composition.

These materials comprising spherical **nano-crystals of PbF_2** embedded in a silicate-based matrix showed high transparency. A large fraction, ~80% of RE ions, was partitioned into PbF_2 nano-crystals. The concentration of 0.5 mol% Er^{3+} versus 3.0 mol% of Yb^{3+} was selected to obtain efficient transfer from Er^{3+} to Yb^{3+} ions while avoiding energy transfer between Er^{3+} ions.

The emission spectra of the above glass-ceramic sample, shown in Figure 2.23, indicates that the down-conversion mechanisms depend on the excited level of Er^{3+}. The authors analyzed these mechanisms combining the Er^{3+} levels in three groups, and finally concluded that the dominant mechanism is energy transfer to Yb^{3+} ions, with a weight over 89%. After this energy transfer process, 59% of the Er^{3+} ions emit at 660 nm and the rest, 41%, decay to the $^4I_{11/2}$ level. Finally, a high down-conversion efficiency of 199% was obtained for photons absorbed at the violet and near UV. According to the authors, the obtained photons can be efficiently used by tandem or double-junction solar cells. The first layer, made of silicon or gallium arsenide alloy, absorbs the photons in the range 650–1,100 nm and the second germanium layer receives the photons emitted at ~1,540 nm.

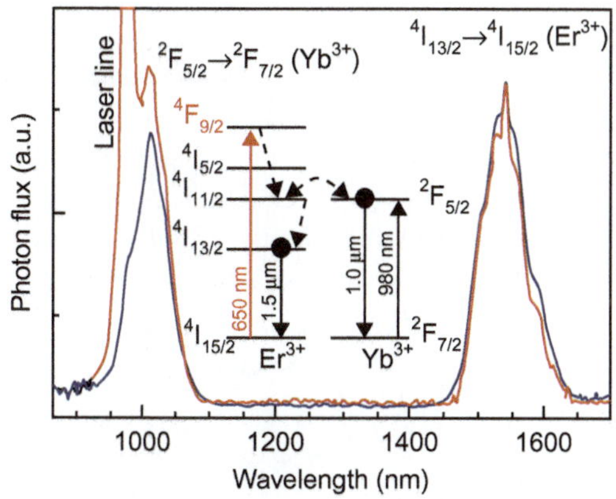

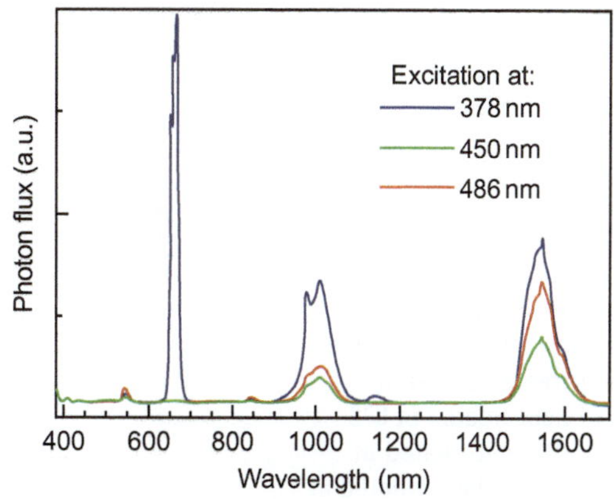

Figure 2.23 Full emission spectra. (A) Spectra excited in to absorption band of Er^{3+} at 650 nm (blue line) and into absorption band of Yb^{3+} at 980 nm (red line). Insert shows the corresponding pump and emission transitions together with population mechanisms. The sharp spike at 980 nm in the red curve corresponds to the laser line. (B) Full emission spectra obtained after excitation of Er^{3+} ions at different wavelengths, as indicated. The spectra have been normalized taking into account the relative quantum efficiencies for the Yb^{3+} excitation. (After Rodríguez et al., 2010. Reproduced with permission of Elsevier Science Publishing Company.)

The preparation of down-converting oxyfluoride glass ceramics containing RE fluoride (mainly LaF_3 and YF_3) nano-crystals was also attempted.

For example, Dejneka (1998) prepared transparent glass ceramics containing **LaF_3 nano-crystals doped with Eu^{3+} ions.**

The as-cast SiO_2–Al_2O_3–Na_2O–AlF_6–LaF_3 glass samples, after annealing for 1 h 20–30°C below the glass transition temperature, exhibited luminescence at 576, 590, and 612 nm but, in the specimens heat-treated above 550°C, Eu^{3+} emission from the $^5D_3$, $^5D_2$, and $^5D_1$ to the ^7F_X manifold were also evident in the blue, green, and yellow portion of the spectrum (Figure 2.24), with fine splitting of the peaks, which is a behavior specific to Eu^{3+} ions in a crystal. Eu^{3+} emission spectroscopy demonstrated that active RE ions could be partitioned into the low phonon energy LaF_3 crystals without clustering.

Chen et al. (2008) investigated the quantum cutting (down-conversion) properties of transparent oxyfluoride glass ceramics, containing **β-YF_3 nano-crystals co-doped with Pr^{3+} and Yb^{3+} ions.** The Pr^{3+} concentration was fixed to 0.1 mol%, while the Yb^{3+} concentrations were 0.1, 0.2, 0.5, 1.0, and 1.5 mol%.

All the resulting glass ceramics, after heat treating the glasses of the SiO_2–Al_2O_3–NaF–YF_3 system for 2 h at 670°C, were highly transparent in the visible region. The TEM image demonstrated the homogeneous distribution of some 18–25 nm particles in the glass matrix, which were identified as β-YF_3 crystals

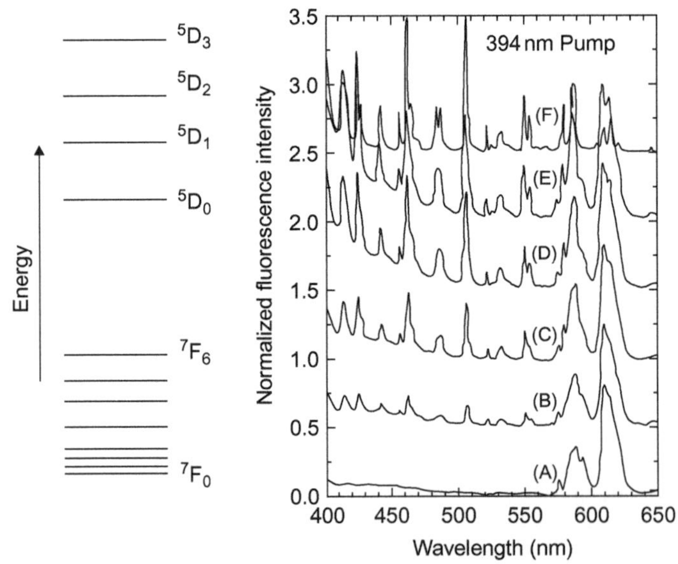

Figure 2.24 Fluorescence of the glass sample (A) as made, (B) heat-treated at 550 for 48 h, (C) heat-treated at 575 for 48 h, (D) heat-treated at 600 for 48 h, (E) heat-treated at 675 for 4 h, (F) LaF_3:Eu crystals. On left, Eu^{3+} energy-level diagram. (After Dejneka, 1998. Reproduced with permission of Elsevier Science Publishing Company.)

by the examination of X-ray and TEM diffraction patterns. In the PL excitation (PLE) spectra shown in Figure 2.25A, intense excitation bands centered at 482, 467, and 441 nm corresponding to Pr^{3+}: $^3H_4 \rightarrow ^6I_1$, ^3P_j ($j = 0,1,2$) transitions were measured by monitoring both the Pr^{3+}: $^3P_0 \rightarrow ^3H_6$ transition at 605 nm and the Yb^{3+}: $^3F_{5/2} \rightarrow ^2F_{7/2}$ transition at 976 nm, verifying the existence of energy transfer from Pr^{3+} to Yb^{3+}. Figure 2.25B shows the PL spectra of the glass ceramics at room temperature with blue light excitation at 482 nm corresponding to the Pr^{3+}: $^3H_4 \rightarrow ^3P_0$ transition. In the visible region of 550–750 nm, a series of sharp lines, ascribed to Pr^{3+}: $^3P_0 \rightarrow ^3H_6$ and $^3P_0 \rightarrow ^3F_{3,4}$ multiplet emissions, and in the 900–1,100 nm region, a broad emission band, attributed to the Yb^{3+}: $^2F_{5/2} \rightarrow ^2F_{7/2}$ transition, are observed; it is noted that, with the increase of Yb^{3+} concentration from 0 to 1.0 mol%, the NIR emission at 976 nm intensifies rapidly, while the red emission at 605 nm weakens monotonically.

Development of such NIR quantum cutting transparent glass ceramics could open a route to enhance the energy efficiency of the silicon solar cells by converting one blue solar photon to two NIR ones.

Chen et al. (2009) investigating the luminescence properties of **Yb^{3+}:TbF_3 nano-crystals-embedded glass ceramics** showed

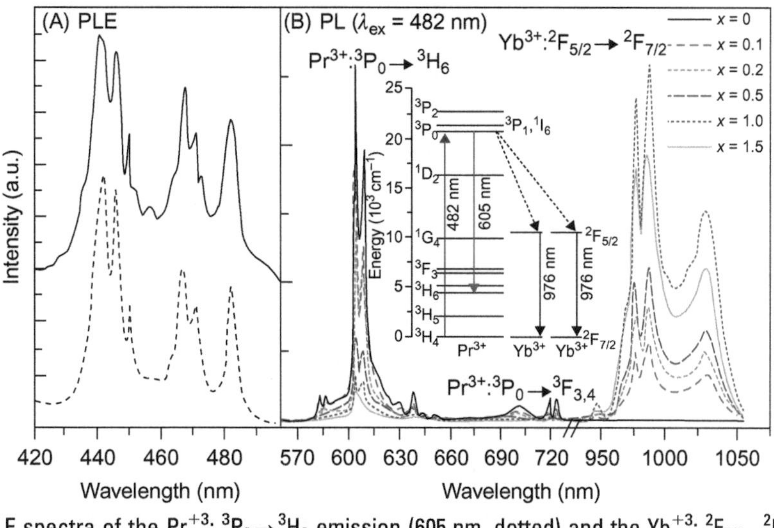

Figure 2.25 (A) PLE spectra of the Pr^{+3}: $^3P_0 \rightarrow ^3H_6$ emission (605 nm, dotted) and the Yb^{+3}: $^2F_{5/2} \rightarrow ^2F_{7/2}$ emission (976 nm, solid) in the glass ceramic. (B) Visible–NIR PL spectra of the glass ceramics with different Yb^{+3} concentrations (x%) on excitation at 482 nm. Inset, schematic energy-level diagram of the Pr^{3+}/Yb^{+3} co-doped glass ceramic, showing the cooperative energy transfer mechanism of NIR QC under visible excitation at 482 nm. (After Chen et al., 2008. Reproduced with permission of Optical Society of America.)

that beside the up-conversion luminescence, that occurred under the 976 nm NIR laser excitation (as discussed before), these glass ceramics also yielded NIR quantum cutting down-conversion emission corresponding to Yb^{3+}: $^2F_{5/2}\rightarrow{}^2F_{7/2}$ transition, under 485 nm excitation. This was ascribed to the cooperative energy transfer from one Tb^{3+} ion to two Yb^{3+} ions.

Velázquez et al. (2012) prepared Eu^{3+}- and Sm^{3+}-doped $95SiO_2$–$5LaF_3$ transparent nano-glass ceramics by sol-gel method and studied the details of their structural and spectroscopic properties. The aim was enhancing the solar cell spectral response by down-conversion of the incident solar spectrum and obtaining a broad excitation range (300–500 nm) for the Eu^{3+} and Sm^{3+} yellow-red emissions.

The amorphous, dried gel samples were heat-treated in air at temperatures in the range of 800–1,000°C that resulted in the formation of hexagonal LaF_3 nano-crystals, 8–19 nm in size as the sole crystalline phase. The luminescence spectra also showed that most of the doped RE ions were incorporated into precipitated LaF_3 nano-crystals.

As a result of the excitation of single Eu^{3+}- or Sm^3-doped specimens, considerable losses, due to the cross-relaxation processes between the dopant ions, were observed, that were appreciably reduced by co-doping of them.

Figure 2.26 depicts the emission spectra of the samples single doped with $0.1Sm^{3+}$ and co-doped with 0.1 or 0.3 mol% of Eu^{3+}, obtained under excitation at 478 nm ($^6H_{5/2}\rightarrow{}^4I_{9/2}$) of Sm^{3+} ions. These results along with other experimental studies indicated that an energy transfer took place from Sm^{3+} to Eu^{3+}

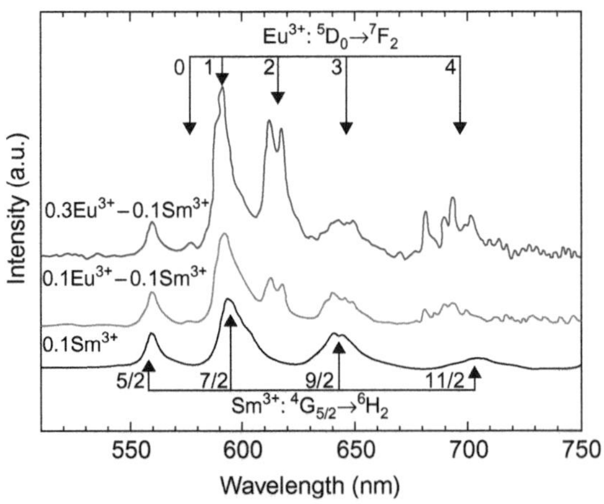

Figure 2.26 Emission spectra of single and co-doped $95SiO_2$–$5LaF_3$ nano-glass ceramics obtained under 478 nm excitation ($^6H5_{5/2}\rightarrow{}^4{}_{19/2}$, Sm^{3+}). Spectra have been normalized to the intensity at 560 nm (Sm^{3+}). (After Velázquez et al., 2012. Reproduced with permission of Springer Science + Business Media, Inc.)

ions in co-doped glass ceramics, the main mechanism of which was summarized as

$$Sm^{3+}(^4G_{5/2}) + Eu^{3+}(^7F_0) \rightarrow Sm^{3+}(^6H_{7/2}) + Eu^{3+}(^5D_0)$$

i.e., the energy transfer from Sm^{3+} to Eu^{3+} ions is produced after the Sm^{3+} ions decay to the $^4G_{5/2}$ emitting level from upper excited levels.

From these emission spectra, the fraction of the total emission coming from Eu^{3+} ions were obtained as 30% and 66% for glass-ceramic samples co-doped with $0.1Eu^{3+}-0.1Sm^{3+}$ and $0.3Eu^{3+}-0.1Sm^{3+}$, respectively, which are indicative of a very efficient energy transfer from Sm^{3+} to Eu^{3+} ions.

More recently, Li et al. (2013) prepared $Eu^{2+}-Yb^{3+}$ co-doped transparent glass ceramics containing nano-crystals of $Ca_5(PO_4)_3F$ (FAP) from glasses of the composition $35.8SiO_2$, $20AlPO_4$, 15CaO, $8Al_2O_3$, $20CaF_2$, $0.2Eu_2O_3$, xYb_2O_3, and $(1.0-x)Gd_2O_3$ ($x = 0$, 0.2, 0.4, 0.6, 0.8, 1.0 mol%). The glass samples were prepared by melting the selected chemicals in alumina crucibles, under reducing atmosphere (5 vol% H_2 + 95 vol% N_2) at 1,450°C for 45 min and quenching the melts.

The as-made glass samples were annealed at 650°C for 2 h and then heat-treated at 730°C for 4 h, under the same reducing atmosphere to prepare the glass-ceramic samples. XRD and TEM analysis indicated the homogeneous distribution of nano-crystals of FAP, with a mean diameter of ~40 nm, within the glass matrix after heat treatment. The spectroscopic measurements indicated that the Eu^{2+} ions mainly partitioned into FAP nano-crystals in the glass ceramics.

More efficient down-conversion luminescence, i.e., NIR emission due to the cooperative energy transfer from Eu^{2+} to Yb^{3+} ions and transition $^2F_{5/2} \rightarrow ^2F_{7/2}$ of Yb^{3+} ions (about 980–1,100 nm) under UV excitation, was observed in glass ceramics.

It was concluded that the $Eu^{2+}-Yb^{3+}$ co-doped glass ceramics containing FAP nano-crystals, owing to their efficient down-converting ability to convert the short wavelength radiation in the broad range of 200–400 nm to ~1 μm IR light, are promising candidate materials for enhancing conversion efficiency of solar cells. The aforementioned short wavelength radiation is not fully and efficiently utilized by the existing solar cells.

Table 2.3 **summarizes the composition, structure, optical properties, and application of some typical luminescent oxyfluoride glass ceramics.**

Table 2.3 Composition and Properties of Some RE-Doped Oxyfluoride Nano-Glass Ceramics

Sample No.	Composition (mol%)	Nano-crystal Phase/Size	Dopants	Luminescence/ Application	Ref.
1	SiO_2 30, Al_2O_3 15, PbF_2 24, CdF_2 20, YbF_3 10, ErF_3 1	$Pb_xcd_{1-x}F_2$ 20 nm	$Er^{3+}-Yb^{3+}$	IR→green up-conversion lasers	Wang and Ohwaki (1993)
2	SiO_2 30, GeO_2 15, Al_2O_3 15, TiO_2 3, YF_3 5, PbF_2 32, HoF_3 0.05–0.5	β-PbF_2 12–17 nm	Ho^{3+}	IR→green up-conversion lasers	Liu and Heo (2006)
3	$PbGeO_3$ 60, PbF_2 10, CdF_2 30	β-PbF_2 5–10 nm	Ho^{3+}	IR→red up-conversion lasers	Gouveia-Neto et al. (2004)
4	GeO_2 50, PbO 40, PbF_2 10, TmF_3 1	β-PbF_2 16–20 nm	Tm^{3+}	Red→UV, blue up-conversion lasers	Hirao et al. (1995)
5	SiO_2 45, Al_2O_3 20, CaO 10, CaF_2 25	CaF_2 11–18 nm	Eu^{2+}	Visible→UV up-conversion scintillators	Fu et al. (2002)
6	SiO_2 53, Al_2O_3 27, Na_2O 11, Al_2F_6 1, LaF_3 7, ErF_3 0.07	LaF_3 30 nm	Er^{3+}	IR→green up-conversion lasers	Tanabe et al. (2002)
7	$SiO_2(44-x)$, Al_2O_3 28, NaF 17, $TbF_3$11, YbF_3 x ($x=0$, 0.25, 0.5, 1.0, 1.5, 3.0)	YbF_3 22 nm ($x=1.5$)	$Tb^{3+}-Yb^{3+}$	NIR→green up-conversion, green-NIR down-conversion	Chen et al. (2009)
8	SiO_2 30, Al_2O_3 15, PbF_2 17, CdF_2 29, YF_3 4, ZnF_2 5, Pr^{3+} 500 ppmw	CdF_2 /PbF_2 9–18 nm	Pr^{3+}	Vis→IR down-conversion amplifiers	Tick et al. (1995)
9	SiO_2 25, GeO_2 10, Al_2O_3 15, PbF_2 36.96, TiO_2 3, TmF_3 0.04, GdF_3 10	PbF_2 20 nm	Tm^{3+}	Red→IR, NIR down-conversion amplifiers	Hayashi et al. (2001)
10	SiO_2 32, Al_2O_3 9, PbF_2 18.5, CdF_2 31.5, ZnF_2 5.5, REF_3 3.5	β-PbF_2 ~10 nm	Ho^{3+}, Dy^{3+}	Red→IR down-conversion waveguides	Tikhomirov et al. (2008)
11	SiO_2 32, Al_2O_3 9, PbF_2 18.5, CdF_2 31.5, ZnF_2 5.5, REF_3 3.5	PbF_2	$Er^{3+}-Yb^{3+}$	Violet, blue, green→red, NIR down-conversion, solar cells	Rodríguez et al. (2010)

(Continued)

Table 2.3 (Continued)

Sample No.	Composition (mol%)	Nano-crystal Phase/Size	Dopants	Luminescence/ Application	Ref.
12	SiO_2 51.70, Al_2O_3 27.62, Na_2O 13.20, AlF_6 0.46, LaF_3 6.97, EuF_3 0.05	LaF_3 15 nm	Eu^{3+}	Violet→blue, green Violet→yellow down-conversion	Dejneka (1998)
13	SiO_2 .44, Al_2O_3 28, NaF 17, YF_3 11	β-YF_3 18–25 nm	Pr^{3+}/Yb^{3+}	Blue→Red, NIR down-conversion solar cells	Chen et al. (2008)

Other Luminescent Glass Ceramics

Transparent YAG Glass Ceramics Optically transparent single-crystal YAG ($Y_3Al_5O_{12}$) is used as an optical host material in various solid-state lasers for its outstanding optical properties when doped with lanthanide or transition elements. YAG-based materials prepared via various routes have found many applications, e.g., in cathode-ray tube field emission displays, phosphors, and electroluminescent materials.

Tarafder et al. (2010a) prepared Eu^{3+}-doped $Y_3Al_5O_{12}$ (YAG) glass ceramics and studied their thermal, structural, and optical properties. The Eu^{3+}-doped glasses, derived from the K_2O–SiO_2–Y_2O_3–Al_2O_3 (KSYA) system, were prepared by melting the batches in a platinum crucible in an electric furnace at 1,680°C for 2 h in air. The glass melts after casting subjected to a controlled crystallization process. The optimum nucleation and crystallization temperatures to obtain transparent nano-glass ceramics were determined as 700°C and 750°C, respectively. It is interesting to note that the above temperatures are much lower than the glass transition (819°C) and softening temperature (867°C) of the prepared glass samples. The optimum heat-treatment conditions to prepare nano-crystalline YAG glass ceramics were determined as 10–50 h at 700°C and 2 h at 750°C. The crystallite size of YAG varied in the range of 35–45 nm with the heat-treatment time at the optimum crystallization conditions.

The FESEM image showed that the parent glass was initially phase-separated prior to the crystallization. The appearance of Al–O and Y–O vibration bands in the Fourier Transform Infrared (Spectroscopy) (FTIR) reflection spectra also evidenced the formation of the YAG phase.

The PL spectra of the Eu^{3+}-doped glass and the glass ceramics exhibited five emission peaks at 579, 592, 613, 653, and

704 nm that was attributed to the transitions of $^5D_0 \rightarrow {}^7F_j$ ($j=0$, 1, 2, 3, and 4) Eu^{3+} ions. The main emission peak near 613 nm was assigned to the induced electric dipole (ED) transition $^5D_0 \rightarrow {}^7F_2$ of Eu^{3+} ions. The absorption and fluorescence spectra also revealed that the Eu^{3+} ions, which entered into the YAG crystal lattice, enhanced the fluorescence performance of the nano-glass ceramics.

Wang et al. (2011) also prepared and studied the optical properties of Ce^{3+}-doped Y_2O_3–Al_2O_3–SiO_2 (YAS) bulk glasses with low SiO_2 content (10–40 mol%) by flame spray-quenching method and hot-press sintering. YAG nano-glass ceramics were then prepared from YAS bulk glass by a controlled crystallization process.

First the well-mixed powders of glass batches were spray-dried into granular spheres and converted into glass microspheres utilizing C_2H_2/O_2 flame spray-quenching method, then the microsphere were sintered into 20 mm thick glass plates by vacuum hot press sintering which was accompanied with their crystallization.

The study of crystallization process showed that glass samples were initially phase-separated prior to the crystallization during the heat-treatment process. The phase-separated grains (100–300 nm) were composed of YAG nano-crystallites, which were uniformly precipitated in the glass matrixes.

The emission spectra of 5 mol% Ce^{3+}-doped four bulk glasses containing 10–40 mol % SiO_2 (denoted by 1–4 respectively) under the excitation of 465 nm are shown in Figure 2.27A. A broad emission band between 500 and 600 nm, which is attributed to the electronic transition from ^2D_j to $^2F_{7/2,\ 5/2}$, is observed in all samples. It implies that Ce^{3+} ions are incorporated in YAS

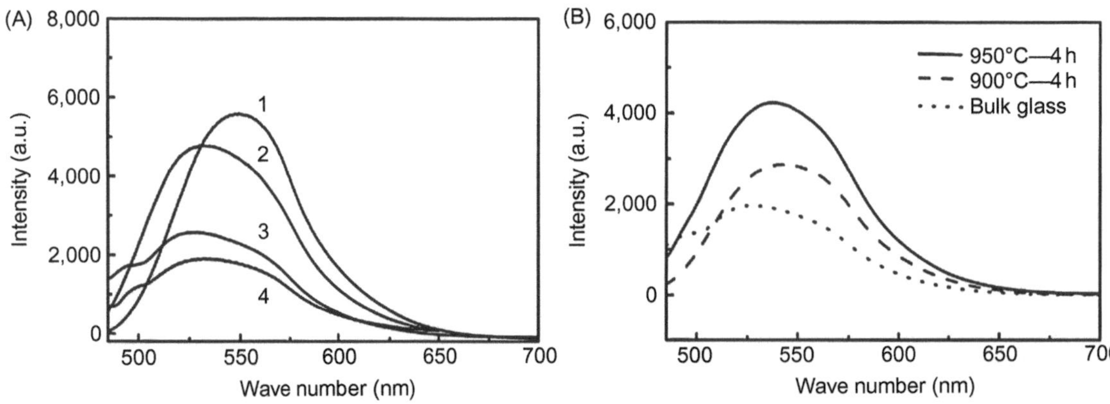

Figure 2.27 (A) Emission spectra of bulk glasses with different contents of silica. (B) Emission spectra of YAS30 (specimen 3) bulk glass and glass ceramics. The numbers 1–4 denote the SiO_2 content of the specimens. (After Wang et al., 2011. Reproduced with permission of the American Ceramic Society.)

bulk glasses. It is also evident that the emission peaks shift to longer-wavelength side and the emission intensity increases with the decreasing content of SiO_2.

This phenomenon was attributed to the possibility of Al_2O_3 and Y_2O_3 acting as network formers in the glasses with the decrease of SiO_2 content, and the consequent decrease in the population of nonbridging oxygen ions. The formation of bridging bonds, like Al–O–Si, results in the change of Ce^{3+} local environment and its luminescent behavior.

Figure 2.27B shows the emission spectra of YAS30 bulk glass and glass ceramic (containing $43.75Al_2O_3$, $26.25Y_2O_3$, and 30 mol% SiO_2). The emission intensities increase and emission peaks become more symmetrical with the increase of the heat-treatment temperatures. It suggests that the transition of ^2D_j to $^2F_{7/2,\ 5/2}$ is dependent on the crystal field of Ce^{3+} ions.

Transparent Willemite Glass Ceramics More recently, Tarafder et al. (2013a) prepared transparent Eu^{3+}-doped willemite glass ceramics by the melt-quenching of $60ZnO{-}5Al_2O_3{-}15B_2O_3{-}20SiO_2{-}0.3Eu_2O_3$ (mol%) glasses and their subsequent controlled crystallization.

The DSC thermogram of the precursor glass exhibited two intense exothermic peaks at 751°C (TP_1) and 842°C (TP_2) which was attributed to the crystallization of alpha and beta willemite, respectively. The T_g and the onset of the first crystallization peak (T_x) were also determined as 588°C and 712°C, respectively.

Interestingly the willemite (Zn_2SiO_4) nano-glass ceramics were obtained by the heat treatment of precursor glasses for 10–30 h at 620°C, which was much lower than the onset of the first crystallization peak temperature, T_x. In fact at the chosen temperature (620°C), owing to the very small size and volume of the crystallizing nano-particles and their sluggish growth, there was little thermal effect to be detected by DSC.

The FESEM micrographs revealed the occurrence of phase-separation in the glassy matrices of the heat-treated samples and the precipitation of willemite crystallites (ZS) within the phase-separated droplets with increase in heat-treatment time.

The average calculated crystallite size of the transparent glass ceramics obtained from XRD was found to be in the range 18–70 nm, whereas the phase-separated droplet size observed in FESEM was 50–250 nm. This confirms the above-mentioned proposition regarding the crystallization of nano-particles within the phase-separated regions.

The emission spectra of the Eu^{3+}-doped glass ceramics under the excitation of a 395 nm light exhibited five emission peaks that can be attributed to the $^5D_0 \rightarrow {}^7F_j$ ($j = 0$, 1, 2, 3, and 4) transitions of Eu^{3+}. The main emission peak near 615 nm was found to be enhanced up to 17-fold with the progress of heat treatment.

Such luminescent nano-glass ceramics are expected to find potential applications in solid-state red lasers, phosphors, and optical display systems.

2.3 Concluding Remarks

The optical properties of nano-glass ceramics can perhaps be considered as the most important properties of these materials. On the other hand, luminescent glass ceramics, including mullite and spinel glass ceramics, doped with transition metal ions, and oxyfluoride glass ceramics, containing RE-doped fluoride nano-crystals, among the optical glass ceramics have attracted considerable attention, in recent years. Moreover, the oxyfluoride glass ceramics have recently gained such an importance that many researchers have included them, as a prominent member, in the list of new materials having made real breakthroughs in the field of optical materials. These glass ceramics, because of their excellent optical properties combined with good mechanical and chemical characteristics, have recently attracted great interest for applications in the field of up-conversion lasers, fiber amplifiers, and the optical devices for three-dimensional displays, as discussed above.

However, in the majority of the aforementioned research studies, the physical aspects of the subject, i.e., the optical properties, and applications of glass ceramics have been the main concern and priority of the researchers. The materials science considerations of the subject, namely, the details of nucleation and crystallization mechanisms and the effect of fabrication parameters, such as glass composition and heat-treatment conditions upon the crystallization process, have been given relatively little attention. Considering the marked influence of the nucleation and crystallization processes on the nano-glass ceramics fabrication and properties, the knowledge of the crystallization mechanism is fundamental in order to design materials with desired properties.

For example, despite some valuable research activities carried out in recent years to understand and explain the crystallization of fluoride crystals in oxyfluoride glass ceramics, it seems that the actual crystallization behavior of these glasses, especially their nucleation process, has not been fully understood and explained yet. In some cases, the actual nucleation mechanism involved in the crystallization of glass, i.e., the occurrence of a homogeneous or heterogeneous nucleation mechanism, as well as the existence or absence of a prior phase-separation has not been identified and explained properly.

Moreover, when the crystallization process is carried out at low temperatures, close to T_g, which is true for most heat treatments discussed above, the possibility of occurrence of some alternative processes, other than the conventional nucleation and growth mechanism, should not be ruled out. In these conditions, the crystallization of glass may take place through the rearrangement of the glass structural elements analogous to the mechanism of disorder–order transition, as described in Section 1.3. Therefore, it seems that more extensive and comprehensive research activities are needed to further clarify the subject of nucleation and crystallization in oxyfluoride nano-glass ceramics.

3

FERROELECTRIC AND ELECTRO-OPTICAL PROPERTIES OF NANO-GLASS CERAMICS

CHAPTER OUTLINE

V.K. Marghussian, Nano-Glass Ceramics, ISBN: 978-0-323-35386-1.

DOI: http://dx.doi.org/10.1016/B978-0-323-35386-1.00003-7

3.1 Theoretical Background

Most ceramic materials are dielectric, i.e., they are electrical insulators. When an electric field is applied to an ideal dielectric material, there is no long-range transport of charge but only a limited rearrangement such that the dielectric acquires a dipole moment and is said to be polarized. In its most elementary form, an electric dipole comprises two equal and opposite point charges Q separated by a distance δx. The dipole moment, p, is defined as

$$p = Q\delta \tag{3.1}$$

where p is a vector with its positive sense directed from the negative to the positive charge.

The polarization in dielectric materials occurs by the following mechanisms.

Atomic polarization, which occurs in all materials, is a small displacement of the electrons in an atom relative to the nucleus; in ionic materials there is, in addition, **ionic polarization** involving the relative displacement of cations and anions. **Orientation polarization** occurs in dipolar materials, which can become polarized because the applied electric field orients the molecules. Finally, **space-charge polarization** is due to the accumulation of charges at the phase interfaces in multiphase dielectrics. This latter mechanism could be of some prominence when considerable resistivity differences exist between various phases comprising the dielectric material.

When a dc voltage V is applied to two parallel plates of area A each, which are separated by a narrow vacuum gap d, the plates are charged with $+q$ and $-q$ charges. In this case, the **electrical flux density** or the **electric displacement vector** D is defined according to Gauss's law. This law states that "the surface integral of the normal component of D over an arbitrary enclosed surface is equal to the net charge enclosed by the surface":

$$\int D \cdot n \, \mathrm{d}S = Aq \tag{3.2}$$

where $\mathbf{n}$ is a unit vector normal to a surface element dS. In the case of the parallel plates above (which, in fact, can be considered as a **capacitor**) D lines are parallel to n; hence the above equation reduces to

$$D_o = q \tag{3.3}$$

where D_o denotes the electric displacement vector in vacuum. The **electrical field strength vector E** between the plates is defined by:

$$D_o = \varepsilon_o E \tag{3.4}$$

where ε_o is defined as the absolute permittivity or dielectric constant of free space.

E is related to the V through:

$$V = \int_{v}^{d} E \cdot J \, \mathrm{d}y = Ed \tag{3.5}$$

where j is a unit vector in the direction of the plate separation which is assumed to be along the y-axis (Varshneya, 1994, pp. 353–355).

The capacitance C_o of the parallel plates is given by

$$C_o = \frac{Aq}{V} \tag{3.6}$$

substituting for q and V from Eqs. (3.3)–(3.5) would result in the following equation:

$$C_o = \frac{A\varepsilon_o}{d} \tag{3.7}$$

If the vacuum between the plates is replaced by a dielectric material, the applied voltage would be reduced. If the voltage was to be maintained constant, then the charges on the plates should increase by a value of P each. This increase in charge P is called the **polarization.** In vectorial representation:

$$p = n \cdot p$$

The capacitance C of the new system is now increased to

$$C = \frac{A(q + p)}{V}$$

and the new electric flux density would be $(q + p)$.

The ratio C/C_o (which would be >1) is called the static relative permittivity ε_r or the **relative dielectric constant**. The net electric displacement D may now be written as

$$D = D_o + P \tag{3.8}$$

Hence, the relative permittivity ε_r would be

$$\varepsilon_r = \frac{C}{C_o} = \frac{q + p}{q} = \frac{D}{D_o} = \frac{D}{\varepsilon_o E} \tag{3.9}$$

From Eqs. (3.8) and (3.9), it is resulted:

$$\varepsilon_r - 1 = \frac{p}{q} = \frac{p}{\varepsilon_o E} = \chi \tag{3.10}$$

where χ is called the **dielectric susceptibility**, and:

$$D = \varepsilon_o E + P = \varepsilon_r \varepsilon_o E = \varepsilon E \tag{3.11}$$

where ε is called the electric permittivity of the dielectric medium.

If the vectorial representation is used, Eq. (3.11) would be represented as

$$D = D_o + P = \varepsilon_o E + P = \varepsilon_r \varepsilon_o E = \varepsilon E \tag{3.12}$$

Also, from Eq. (3.12),

$$P = (\varepsilon_r - 1)\varepsilon_o E = (\varepsilon_r - \varepsilon_o)E = \alpha E \qquad (3.13)$$

where α is the polarizability of the medium (Varshneya, 1994, pp. 353–355).

Frequency and Temperature Dependence of ε_r

The total polarization, P, the total polarizability, α, and the relative permittivity, ε_r, of a dielectric in an alternating field, all depend on the ease with which the dipoles can reverse alignment with each reversal of the field. Some polarizability mechanisms do not permit sufficiently rapid reversal of the dipole alignment. In such a process, the time required to reach the equilibrium orientation is called the relaxation time, and its reciprocal, the relaxation frequency. When the frequency of the applied field exceeds that of the relaxation frequency of a polarization process, the dipoles cannot reorient fast enough and operation of the process ceases. Since the relaxation frequencies of all four polarization processes (as described previously) differ, it is possible to separate the different contributions experimentally. Hence, four frequency ranges can be defined over which the separate polarization mechanisms operate. For example, in covalent solids, only electronic polarization is present; therefore ε_r can be measured optically from the index of refraction. Ionic materials cease to contribute to the total polarization at infrared frequencies. With such materials, it is possible to measure separate ionic and electronic contributions by making both optical and electrical measurements of ε_r. The orientation and space-charge polarization only function at lower frequencies (Rose et al., 1966, pp. 259–260).

The effect of temperature on the dielectric constant of ionic and electronic materials is, in general, small at low temperatures but increases with increasing temperature. At elevated temperatures, ion mobility is appreciable.

The combined effect of temperature and frequency is quite important in ionic materials. This often leads to a sharp rise in ε_r at higher temperatures. The slope of dielectric constant versus temperature also becomes steeper with the frequency decrease in ionic materials. The increase in ε_r may be attributed to an increase in both ion-jump orientation and space-charge polarization. The exponential increase in electrical conductivity with temperature may also contribute.

Energy Loss

The energy losses which occur in dielectrics are due to dc conductivity and dipole relaxation. The loss factor ($\varepsilon_r \tan \delta$) of a dielectric is a useful indication of the energy lost as heat. The maximum dielectric loss for any particular type of polarization process occurs when its relaxation period is the same as the period of the applied field, i.e., when a resonance occurs.

Dielectrics may be divided into low and high loss materials. Typical high loss materials are polar organic materials. Ceramic materials of high dielectric constant like barium titanate are also high loss materials. The major energy losses in ionic crystals and glasses occur at frequencies of $<10^4$ Counts Per Second (cps). They may be attributed to ion-jump relation. Losses due to ion-vibration and deformation are seldom significant at the frequencies used in electronic and power applications. Conduction losses, however, are appreciable; they increase with decreasing frequency at low frequencies (Rose et al., 1966, pp. 261–262).

The Ferroelectric Effect

Some dielectric materials exhibit a spontaneous polarization in the absence of an applied field which can be reversed by an externally applied electric field. This behavior is called the ferroelectric effect.

Ferroelectric behavior is limited to certain materials and to particular temperature ranges for a given material. Above a certain temperature, known as Curie point, T_c, the spontaneous polarization falls to zero and the properties change to those of a paraelectric (Moulson and Herbert, 2003, pp. 59–60).

Many ferroelectrics possess very high-permittivity values making them quite useful for capacitors. The permittivity that varies considerably with both applied field strength and temperature usually reaches a peak at the Curie point and falls off at higher temperatures (Moulson and Herbert, 2003, pp. 59–60).

The reason for coining the term "ferroelectric" is that the relationship between field and polarization for a ferroelectric material bearing electrodes takes the form of a hysteresis loop similar to that relating magnetization and magnetic field for a ferromagnetic body.

Various models have been suggested to explain why some materials are ferroelectric. The most recent and successful model involves a consideration of the vibrational states of the crystal lattice (Moulson and Herbert, 2003, pp. 59–60).

Some dielectrics can generate a potential difference when subjected to mechanical stress or change physical shape if an external voltage is applied across the material. This property is called **piezoelectricity**. Piezoelectric materials are another class of very useful dielectrics.

Polar dielectric materials also exhibit **pyroelectric** effect. True pyroelectricity results from the temperature dependence of the spontaneous polarization P_s and is therefore shown by ferroelectric materials whether they are single-domain single crystals or poled ceramics. Because a change in polarization in a solid is accompanied by a change in surface charges, it can be detected by an induced current in an external circuit. If the pyroelectric material is perfectly electrically insulated from its surroundings, the surface charges are eventually neutralized by charge flow occurring because of the intrinsic electrical conductivity of the material.

Since pyroelectric materials are polar, they are also piezoelectric, and the strain resulting from thermal expansion will result in the development of a surface charge. However, this is a small effect that seldom exceeds 10% of the primary pyroelectric effect. Because P_s falls to zero at the Curie point, ferroelectric materials are likely to exhibit high pyroelectric coefficients just below their transition temperatures.

Electro-Optic Effect

An electro-optic effect is a change in some optical properties of a material (such as the absorption and refractive index) in response to an electric field that varies slowly compared with the frequency of light.

Nonlinear Optics

Nonlinear optics (NLO) is the branch of optics that describes the behavior of light in nonlinear media, i.e., media in which the dielectric polarization P responds nonlinearly to the electric field E of the light. This nonlinearity is typically only observed at very high light intensities (values of the electric field comparable to interatomic electric fields, typically 10^8 V/m) such as those provided by pulsed lasers. In NLO, the superposition principle no longer holds.

Second Harmonic Generation

Second harmonic generation (SHG) or frequency doubling is the generation of light with a doubled frequency (half the wavelength); two photons are destroyed creating a single photon at two times the frequency.

Third Harmonic Generation

Third harmonic generation is the generation of light with a tripled frequency (one-third the wavelength). Three photons are destroyed creating a single photon at three times the frequency.

3.2 Structure, Properties, and Application of Ferroelectric Nano-Glass Ceramics

Barium titanate ($BaTiO_3$) the first ceramic material in which **ferroelectric behavior** was observed is isostructural with the mineral perovskite ($CaTiO_3$).

The perovskite structure ABO_3 can be visualized as based on a cubic close-packed (Face-Centered Cubic) assembly of O^{-2} anions located at the center of cube feces and A and B cations at the cube corners and the center of cube, respectively. A consideration of the geometry shows that for a perfect fit, the following relationship between the ionic radii holds (Moulson and Herbert, 2003, p. 71):

$$R_A + R_o = 2(R_B + R_o)$$

For many compounds having the perovskite structure, the relationship will not hold exactly because of small variations in the sizes of the A and B ions. Therefore, to allow for this, the above equation should be written as

$$R_A + R_o = t2(R_B + R_o)$$

in which "t" is termed the "tolerance factor" with a value typically in the range

$$0.95 < t < 1.060$$

In the case of $SrTiO_3$, $t = 1$. When $t \neq 1$, then small lattice distortions (the octahedra tilt) occur in order to minimize lattice energy. These distortions have a significant effect on dielectric properties.

Above its Curie point (~130°C), the unit cell of barium titanate is cubic with the ions arranged in the way as explained above. Below the Curie point, the structure is slightly distorted to the tetragonal form with a dipole moment along the c direction. Other transformations occur at temperatures close to 0 and −80°C.

Below 0°C, the unit cell is orthorhombic with the polar axis parallel to a face diagonal and below −80°C, it is rhombohedral with the polar axis along a body diagonal (Moulson and Herbert, 2003, p. 71).

Soon after the discovery of the high permittivity of ferroelectric barium titanate in 1943, the barium titanate ceramics, fabricated by the conventional methods, have been used as capacitors, thermistors, and so on in various applications.

Since then, the application of these materials have been greatly expanded and diversified, and many other ferroelectric ceramics including lead titanate ($PbTiO_3$), lead zirconate titanate, lead lanthanum zirconate titanate, strontium titanate ($SrTiO_3$), lithium niobate ($LiNbO_3$), and lead magnesium niobate have also been developed. They are now widely utilized in many electronic devices such as thermal switches, thermal imaging, multilayer capacitors, electro-optic devices for data storage and displays, piezoelectric sonar and ultrasonic transducers, radio communication filters, sensors, and ferroelectric thin film memories.

Among the various methods adopted to prepare ferroelectric materials, **the glass-ceramic route** has also attracted considerable attention in recent decades.

There are a number of advantages in preparing the materials by the glass-ceramics route in respect to other fabrication methods.

These include low-cost and high-speed fabrication processes offered by glass technology in comparison to other methods; the absence of pores that imparts these materials higher mechanical and dielectric breakdown strengths; the possibility of changing the properties by varying the volume fraction of the active phase dispersed in the glass matrix and its nano/microstructure; the relative ease and flexibility of this route in fabricating articles of intricate sizes and shapes, and close tolerance in dimensions (Prasad and Varma, 2005).

On the other hand, the volume fraction of the ferroelectric phase and the grain size, which are critical parameters dictating the properties of the material, can be controlled by the composition and heat-treatment processes. This class of materials is of great importance because it offers a combination of physical properties, which are otherwise difficult to achieve.

Moreover, the careful control of the glass crystallization process and intelligent choice of the glass composition can lead to the fabrication of **nano-structured transparent ferroelectric glass ceramics** with quite interesting electro-optic properties.

The great majority of ferroelectric glass ceramics can broadly be classified into **titanate**- and **niobate**-based compositional groups.

3.2.1 Titanate-Based Glass Ceramics

In the past few decades, many researchers have studied the fabrication methods and properties of the glass-ceramic materials containing various **perovskite titanates** as their main ferroelectric phase, such as:

$BaTiO_3$ (Herczog, 1964, 1984; Parkash et al., 1986; Bhargava et al., 1988; Sadhukhan et al., 1999; Kim et al., 2007; Al-Assiri et al., 2009; Al-Assiri and El-Desoky, 2011, 2012, 2013; Ferreira et al., 2013).

$PbTiO_3$ (Bergeron and Russel, 1965; Russel and Bergeron, 1965; Grossman and Isard, 1969a,b; Kokubo and Tashiro, 1974; Lynch and Shelby, 1984; Shyu and Yang, 1995; Golezardi et al., 2010; Shankar and Deshpande, 2010, 2012; Yao et al., 1996).

$SrTiO_3$ (Swartz et al., 1988a,b; Thakur et al., 1995, 1996, 2002, 2003).

Solid solution of the above perovskites or other phases (Ruiz-Valdés et al., 2004; Sahu et al., 2006a,b; Gautam, 2013).

General Processing of Titanate Glass Ceramics

In principle, the ferroelectric perovskites can be produced in the glassy form in the pure state without addition of a network former, but owing to the relatively low viscosity of the melt this requires rapid quenching of the melt, of the order of 10^6 degrees/s or more, producing only very small thin platelets. On the other hand, crystallization of bodies made of glass powder may produce undesirable porous structures. The addition of network formers, such as SiO_2, B_2O_3, and Al_2O_3, increases the viscosity of the melt. This helps producing a few millimeter thick disks or plates by cooling of the melt at a moderate rate of a few hundred degrees/s (Parkash et al., 1986). Moreover, due to the high crystal nucleation rates of these glasses, usually no nucleating agents were used in their crystallization process.

During ceramization of various glasses containing a combination of perovskite-forming oxides such as PbO, BaO, SrO, and TiO_2 in some instances, the formation of pure perovskite titanates have been observed. In these cases, the relative amount of various oxides would usually determine which perovskite phase would be formed.

For example, in the case of Pb,Ba(TiO_3) perovskites, the ratio of PbO/BaO and (PbO+BaO)/TiO_2 determines whether $PbTiO_3$ or $BaTiO_3$ would be precipitated out during heat treatment. Other substitutions such as Sr for Ba and Zr for Ti usually result in the precipitation of other phases, rather than solid solutions (Parkash et al., 1986).

However, since the solid solution ferroelectric ceramics usually exhibit more desirable properties, in comparison with the pure ferroelectric materials, many attempts have also been made to precipitate solid solution crystallites in the glass ceramics.

Titanate glass ceramics, as stated above, can be classified according to their main perovskite phases in four groups, namely, $BaTiO_3$, $PbTiO_3$, $SrTiO_3$, and solid solution perovskite glass ceramics. Table 3.1 summarizes the chemical composition, major crystalline phases, and some properties of titanate glass ceramics. The processing details and properties of some of these glass ceramics are discussed here.

Processing, Properties, and Application of Major Titanate Glass Ceramics

3.2.1.1 $BaTiO_3$ Glass Ceramics

The first perovskite-type barium titanate ($BaTiO_3$) glass ceramic was prepared and reported by Herczog (1964). The author described the properties and compositions of glasses suitable for crystallization of $BaTiO_3$. The crystallization of some glasses resulted in the formation of $BaTiO_3$, as the major and barium feldspar ($BaAl_2SiO_3$) as the minor phases, respectively. It was found that within the range 30–60% by volume of $BaTiO_3$ at about 1 μm grain size, the measured dielectric constant increased from 100 to 1,200. As the grain size approached 0.1 μm, the dielectric constant decreased and became nearly independent of the temperature because of the predominance of surface states.

Later on, Sadhukhan et al. (1999) prepared the first nanostructured $BaTiO_3$ glass ceramics in the PbO–BaO–V_2O_5 system (sample 1, Table 3.1) by a fast quenching technique followed by a controlled heat-treatment process.

While, the X-ray diffraction (XRD) and scanning electron microscopic (SEM) studies indicated the existence of homogeneous glasses in the wide compositional range of $x < 30$ wt% $BaTiO_3$, the transmission electron microscopic (TEM) study showed the presence of some nano-crystalline particles.

The dielectric constants of these glasses were found to be very high and comparable to that of bulk $BaTiO_3$. This was about two orders of magnitude higher than those of $BaTiO_3$-free

Table 3.1 Composition and Properties of Some Titanate Nano-Glass Ceramics

Sample No.	Chemical Composition	Major Crystalline Phase/Size	Heat-Treatment Temp. (°C)/Time (h)	Room Temp. Dielectric Constant/Dissipation Factor (at 100 KHz)	Ref.
1	$(80\ V_2O_5 - 20\ PbO) + xBaTiO_3$ ($x = 5$–30) (wt%)	$BaTiO_3$/ 10–50 nm ($x = 30$)	300/9	~1500	Shadhukhan et al. (1999)
2	$4BaTiO_3 \cdot SiO_2$	$BaTiO_3$/32 nm	900	–	Kim et al. (2007)
3	$xBaTiO_3$, $(100 - x)$ V_2O_5 ($x = 35$–50)	$BaTiO_3$/ 20–35 nm ($x = 45$)	450/1	$\sim 10^4$	Al-Assiri et al. (2009)
4	39PbO, 1BaO, $25TiO_2$, $1B_2O_3$, $9.7A1_2O_3$, $24.3SiO_2$ (mol%)	$PbTiO_3$/ ~200 nm	830/3–12	70–80	Shyu and Yang (1995)
5	$(50 - x)$ PbO, xSrO, $25TiO_2$, $25B_2O_3$ ($x = 0,5,10,15$) mol%	$PbTiO_3$/~100 nm ($x = 10$)	733 K/14 + 793 K/14	~125 (at 1 KHz)	Shankar and Deshpande (2012)
6	46.67PbO, $23.33TiO_2$, $20B_2O_3$, $10SiO_2$, (0.5, $1Bi_2O_3$) mol%	$PbTiO_3$/20–25 nm	455–485/ (1–3) + 575/1	(140–400)/0.04–0.4	Golezardi et al. (2010)
7	37.9SrO, $26.6TiO_2$, $23.1SiO_2$, $12.4Al_2O_3$ (wt%)	$SrTiO_3$/30 nm	1000/1	~35/0.0005	Swartz et al. (1988a,b)
8	$65(SrO \cdot TiO_2)$, $35(2SiO_2 \cdot B_2O_3)$ (wt%)	$Sr_2B_2O_5$/10 nm	860/1	~55/0.002	Thakur et al. (1995)
9	$64(SrO \cdot TiO_2)$, $35(2SiO_2 \cdot B_2O_3)$, 1CoO (wt%)	$Sr_2B_2O_5$	900/3	~12/0.15	Thakur et al. (1996)

(*Continued*)

Table 3.1 (Continued)

Sample No.	Chemical Composition	Major Crystalline Phase/Size	Heat-Treatment Temp. (°C)/Time (h)	Room Temp. Dielectric Constant/Dissipation Factor (at 100 KHz)	Ref.
10	$65(SrO \cdot TiO_2)$, $35(2SiO_2 \cdot B_2O_3) + xBi_2O_3$ ($x = 0, 1, 5, 10$) wt%	$SrTiO_3 + TiO_2$ ($x = 10$)	950/6 + 1000/3	~72/0.22	Thakur et al. (2003)
11	$64(SrO \cdot TiO_2)$, $35(2SiO_2 \cdot B_2O_5)$, $1La_2O_3$ (wt%)	$Sr_2B_2O_5$, $Sr_3Ti_2O_7$, TiO_2	950/3	~250/~0.26	Thakur et al. (2002)
12	$29BaTiO_3$, $17PbTiO_3$, $38TiO_2$ $13Al_2O_3$, $3B_2O_3$	$(Pb_{1-x})Ba_xTiO_3$/ 90–200 nm	900/2	640/0.03 (at 800 KHz)	Ruiz-Valdés et al. (2004)
13	65((Pb1-xSrx)O · TiO_2), $25(2SiO_2 \cdot B_2O_3)$, $5K_2O$, 5BaO ($x = 0.2-0.9$) mol%	$(Pb_{1-x})Sr_xTiO_3$	750/3	22–129/0.0005–0.010 (at 1 MHz)	Sahu et al. (2006a,b)
14	$(100 - x)(K_2O \cdot 2SiO_2)$, $x(2Bi_2O_3 \cdot 3TiO_2)$, ($x = 10-50$) mol%	$Bi_4Ti_3O_{12}$/ 85–90 nm ($x = 45$)	580/10	39/0.014 (at 1 MHz)	Molla et al. (2011)
15	$(100 - x)(K_2O \cdot 2SiO_2)$, $x(BaO \cdot Bi_2O_3 \cdot 4TiO_2)$ ($x = 10-60$) mol%	$BaBi_4Ti_4O_{15}$/ 10–20 nm ($x = 50$)	520/10 + 610/100	~55/0.32 (at 3 MHz)	Molla et al. (2013)

($80V_2O_5$–20PbO) glasses. The authors attributed this character to the presence of nano-crystalline particles of $BaTiO_3$ in the glass matrix.

The glass-ceramic samples heat-treated at 300°C for 9 h in air also showed nano-crystalline particles ~10–50 nm along with some larger clusters ~50–90 nm in size identified as the major $BaTiO_3$ phase along with traces of PbV_2O_6, $PbTiO_3$, TiO_2, and $Pb_2V_2O_7$ phases.

Figure 3.1 represents the frequency-dependent dielectric constants of a typical glass-ceramic sample (with $x = 30$ wt% $BaTiO_3$) indicating a broad peak, around 390 K. This transition resembles the ferroelectric Curie temperature of bulk $BaTiO_3$ crystal.

The broadness of peak in the glass-ceramic sample was attributed to the presence of glassy and impurity phases in the samples.

The authors also stated that though the dielectric constants of these glasses and glass ceramics are extremely high, they are not very suitable candidates as capacitor materials because of their high losses (which have not been given). However, these materials could be used where only high values of dielectric constant are desired.

Yao and Zhu (2002) successfully deposited Ba–Ti–B glass-ceramic thin films with a substantial number of perovskite barium titanate ($BaTiO_3$) grains on Pt-coated silicon substrates by sol-gel processing. A Pt-coated Si wafer was used as the

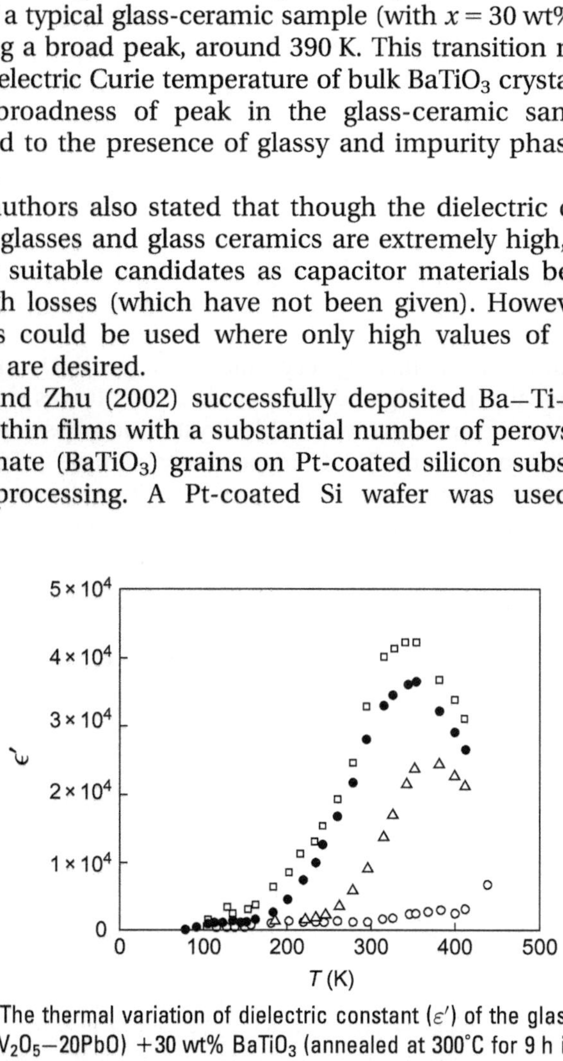

Figure 3.1 The thermal variation of dielectric constant (ε') of the glass-ceramic sample ($80V_2O_5$–20PbO) +30 wt% $BaTiO_3$ (annealed at 300°C for 9 h in air) showing one broad peak around 390 K at four different frequencies 100 Hz (□), 1 kHz (•), 10 kHz (Δ), and 100 kHz (○), corresponds to ferroelectric transition of $BaTiO_3$ at 393 K. (After Sadhukhan et al., 1999. Reproduced with permission of American Institute of Physics.)

substrate. The Ba–Ti–B precursor film was spin-coated onto the substrate at a speed of 3,000 rev/min. After the deposition, the films were heated in a furnace at 475°C for 18 min in a controlled oxygen atmosphere to remove the residual organics. The films were then annealed at various higher temperatures for 18 min in oxygen and $BaTiO_3$ (BT) grains were crystallized from the glass matrix. Field emission scanning electron microscopy (FESEM) revealed that the $BaTiO_3$ grains were ~40 nm in diameter when annealed at 625–675°C. The grains were bonded with each other by the glass phase in the glass–ceramic thin films.

An interesting phenomenon observed in the glass-ceramic thin films was the self-healing of microcracks provided that the thickness of the film was below 200 nm.

Electrical characterization results showed that the dielectric constant and dielectric loss were, respectively, in the range of 120–350 and 0.009–0.025 at 1 kHz, depending on the heat-treatment conditions.

The experimental results indicated the potential value of $BaTiO_3$ glass–ceramic thin films for integrated high-dielectric media.

The kinetics of $BaTiO_3$ crystallization in $BaO–TiO_2–SiO_2$ glasses was also investigated (Kim et al., 2007). It was shown that the glass transforms into two stable crystalline phases of $BaTiO_3$ and $BaTiSiO_5$ at 672 and 711°C, respectively (sample 2, Table 3.1).

The temperature dependence of the transformed crystal volume fraction of $BaTiO_3$ was also determined (Figure 3.2). The values of obtained volume fractions were used in the

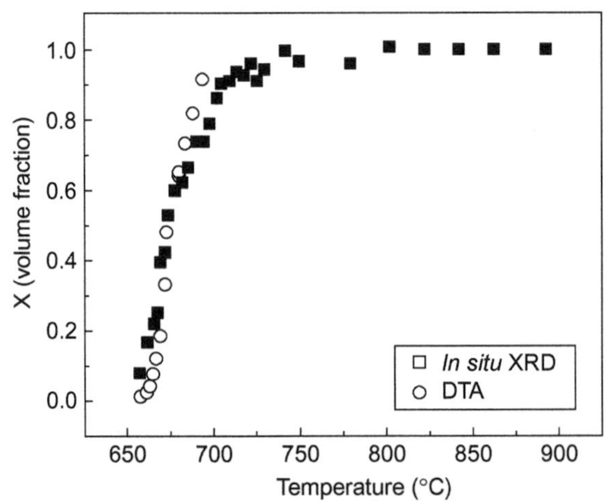

Figure 3.2 Temperature dependence of the transformed $BaTiO_3$ crystal volume fraction measured by DTA (○) and *in situ* XRD (▪). (After Kim et al., 2007. Reproduced with permission of Elsevier Science Publishing Company.)

calculation of activation energies for $BaTiO_3$ crystallization, utilizing the modified Ozawa equation.

The Avrami exponent was also determined from the Johnson–Mehl–Avrami equation, utilizing the transformed volume fractions of $BaTiO_3$ as a function of time at 655°C.

The calculated value of Avrami constant $n \approx 1$ indicated a mechanism of diffusion-controlled crystallization with zero nucleation rate.

The $BaTiO_3$ particle size as a function of temperature was also determined and an average size of ~32 nm was obtained for the specimen heated up to 900°C, with a constant heating rate of 2 K/min.

In recent years, a great resurgence in research interest and activity in the field of nano-structured $BaTiO_3$ glass ceramics is observed (Al-Assiri et al., 2009; El-Desoky et al., 2009; Al-Assiri and El-Desoky, 2011, 2012, 2013; Ferreira et al., 2013; Harizanova et al., 2013; Szreder et al., 2014).

This can perhaps be related to the possibility of achieving superior NLO and electro-optical properties in ferroelectric nano-glass ceramics, as well as the rapid development of multilayer ceramic capacitors and the miniaturization technology in recent years that have created an increasing demand for high dielectric materials, such as nano-structured barium titanate glass ceramics.

Al-Assiri et al. (2009) studied the nano-structural details and electrical properties of $BaTiO_3$ glass ceramics (sample 3, Table 3.1). The glass samples were prepared by melting and quenching the mixture of pure $BaTiO_3$ and V_2O_5 raw materials. The high-resolution transmission electron micrograph (HRTEM) along with the selected-area electron diffraction (SAED) of $45BaTiO_3$–$55V_2O_5$ glass ceramics, prepared by heat treating the glass samples around 450°C for 1 h in air, indicated the presence of nano-crystals with a particle size of 20–35 nm in the heat-treated samples (Figure 3.3). The lattice spacing, a, obtained by SAED pattern (inset of Figure 3.3) proved that they were barium titanate crystallites.

The ferroelectric properties of various glass and glass-ceramic samples were also determined. It was shown that the dielectric permittivity (ε) values of these glasses were very high as compared to the known vanadium-containing glasses, e.g., a glass-ceramic sample of the composition $40BaTiO_3$–$60V_2O_5$ exhibited a very high dielectric constant of ~10,000 at room temperature and a fixed frequency of 100 kHz. The high dielectric permittivity of the glasses attributed to the presence of nano-crystalline $BaTiO_3$, V_2O_5, and TiO_2 grains or clusters within the glass matrix.

These glass ceramics exhibited ferroelectric behavior and their average Curie temperature was around 352 K which is very

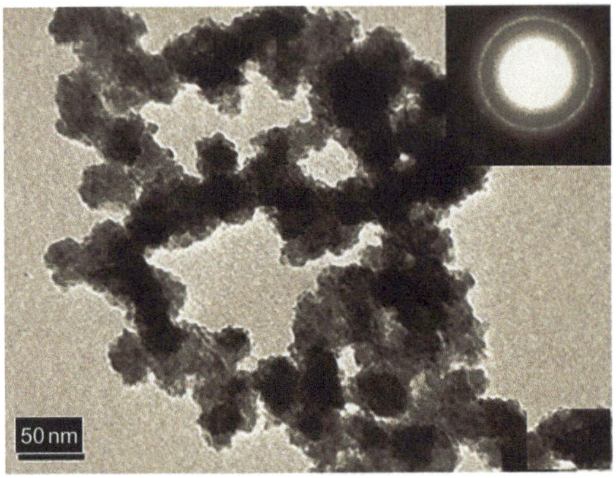

Figure 3.3 TEM and SAED for $45BaTiO_3{-}55V_2O_5$ nano-crystalline glass ceramic (inset). (After Al-Assiri et al., 2009. Reproduced with permission of Elsevier Science Publishing Company.)

close to the Curie temperature of pure $BaTiO_3$ with grain size <100 nm. This result indicated the existence of dipolar orientation and occurrence of phase transition in the nano-structured glass-ceramic samples.

The electrical conductivity of the nano-crystalline glass-ceramic samples were also measured and found to be higher than that of the corresponding glass samples. This was attributed to the nonadiabatic small polaron hopping (SPH) of electrons between vanadium ions.

Al-Assiri and El-Desoky (2011) also gave a more detailed account of the electrical conductivity of barium titanate and vanadium oxide containing glasses and glass ceramics. The glass samples were prepared by melting and quenching the batch of raw materials by adding Bi_2O_3 to the $BaTiO_3{-}V_2O_5$ binary system. The resulting glass samples were then heat-treated at the crystallization temperature, which was in the temperature range of 347–357°C for the glass sample with the composition of $15BaTiO_3{-}65V_2O_5{-}20Bi_2O_3$ mol%.

After the heat treatment, samples consisted of small nano-crystals of $Ba_3TiV_4O_{15}$, $BaTiO_3$, V_2O_5, and Bi_2O_3 phases, with an average grain size of 25 nm, that were embedded in the glassy matrix.

The resulting materials exhibited much higher electrical conductivity than the initial conduction in the glasses of the $BaTiO_3{-}V_2O_5{-}Bi_2O_3$ system. This was explained by the development of interfacial regions between crystalline and amorphous phases, in which the concentration of $V^{4+}{-}V^{5+}$ pairs, responsible for electron hopping, had higher values than inside

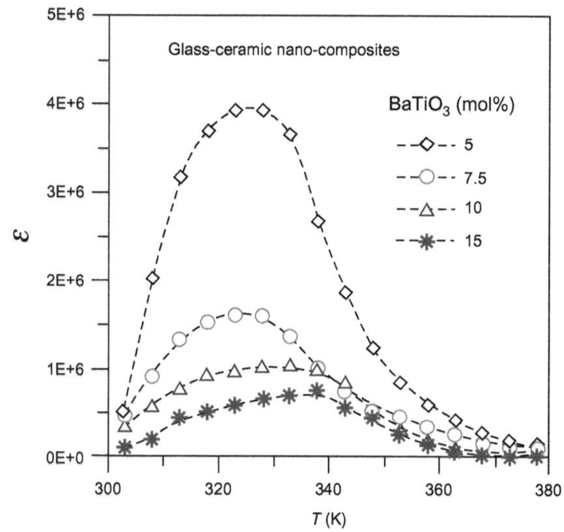

Figure 3.4 Temperature dependence of dielectric constant at fixed frequency (100 kHZ) for $BaTiO_3$—V_2O_5—Bi_2O_3 glass-ceramic nano-composites. (After Al-Assiri and El-Desoky, 2012. Reproduced with permission of Elsevier Science Publishing Company.)

the glassy matrix. This led to smaller average distance between the hopping centers and resulted in higher electronic conductivity and lower activation energy, since all the previous investigations showed, when the distance R between hopping centers decreases, regions of higher than average conductivity form a kind of "easy conduction paths" for electrons.

According to the authors, a "core–shell" concept can explain the conductivity improvement caused by nano-crystallization. The overlapping and intersecting defective shells around crystalline cores formed a complicated system of paths for facilitated electron transport that took place by SPH of electrons between vanadium ions.

Al-Assiri and El-Desoky (2012) also concentrated on the study of the dielectric and ferroelectric properties of the aforementioned nano-glass ceramics of $BaTiO_3$—V_2O_5—Bi_2O_3 system, prepared via a similar procedure as explained above.

The temperature dependence of dielectric constant was measured for different compositions in the temperature range 293–523 K at 100 kHz. Most of the glass ceramics exhibited very high values of dielectric constant and all of them showed a transition from ferroelectric to paraelectric phase at ~324 K which was the Curie temperature (T_C) of these glass ceramics (Figure 3.4).

The variation of dielectric constant as a function of temperature at different frequencies for a selected $10BaTiO_3$–$70V_2O_5$–$20Bi_2O_3$ nano-glass ceramic showed that the dielectric constant of the glass-ceramic sample was higher than the value obtained for the as-received glass for all the frequencies (1 kHz–1 MHz). These curves exhibited clear broad peaks at about 324 K at all the frequencies which, as explained above for the frequency of 100 kHz, was assigned to the average Curie temperature of the prepared glass ceramics. This temperature was very close to the Curie temperature of pure $BaTiO_3$ with grain size <100 nm. Therefore, it was inferred that the isolated barium titanate particles with the grain size of 20–30 nm, uniformly distributed in the glass matrix, demonstrated a ferroelectric response. The authors concluded that the size of $BaTiO_3$ nano-crystals can determine their critical behavior, i.e., ferroelectric–paraelectric transition, a characteristic that can be utilized to control T_c by proper adjustment of $BaTiO_3$ content and crystal size.

Very recently, Szreder et al. (2014) studied the ac and dc electrical conductivity of barium titanate-doped bismuth-vanadate glass ceramics. Samples of the composition $x[BaO,TiO_2]$–$(80 - x)V_2O_5$–$20\ Bi_2O_3$, $x = 5$, 10, and 15 (mol%) were prepared by a conventional melt-quenching technique and heat-treated at the temperature of 633 K for 6 h.

The XRD patterns of 10 $(BaO \cdot TiO_2)$–$70V_2O_5$–$20Bi_2O_3$ glass and glass-ceramic samples, both indicated the presence of the crystalline phases of $BiVO_4$, Bi_2O_3, V_2O_5, TiO_2, and $BaTiO_3$.

The ac and dc electrical conductivity of glasses and glass ceramics was measured in the frequency range from 1 to 1 MHz and in the temperature range from 153 to 423 K.

The plot of imaginary versus real part of impedance (Nyquist plot) for 10 $(BaO \cdot TiO_2)$–$70V_2O_5$–$20Bi_2O_3$ glass samples indicated that only one relaxation process was taking place in as-quenched glasses, which corresponded to one conduction mechanism, namely, SPH. Whereas, in the heat-treated glasses, two relaxation processes was taking place.

It was suggested that during heat treatment, possibly a number of vanadium ions took part in the formation of crystallites and the rest of them remained in the glass matrix. Hence, in these samples, two different relaxation processes may be operative, caused by two different conduction mechanisms, i.e., within the crystallites and in the interfacial regions.

It was also shown that the heat-treated samples had significantly higher conductivity (by about two orders of magnitude) and lower activation energy than as-quenched samples, which is in accord with the results of other investigators and the core–shell model.

Ferreira et al. (2013) investigated the crystallization behavior of a $70TeO_2$–15BaO–$15TiO_2$ glass, doped with 0.1% Eu^{3+} and Er^{3+}, in order to assess the possibility of obtaining transparent glass ceramics containing rare earth-doped $BaTiO_3$ nano-crystals. The glass samples prepared by a conventional melt-quenching technique were heat-treated either by one-step heat treatment at 410°C or two-step heat treatment at 410°C and 430°C.

The XRD patterns of all samples, including the base glass and the doped glass samples, after the heat treatment for differing times at 410°C (10°C above T_g) exhibited intense diffraction halos, indicating the presence of large amounts of residual glass phase, as well as two diffraction peaks. The peaks were attributed to barium titanate, $BaTiO_3$, and an unidentified phase, *X*, which was isostructure with lanthanum tellurate ($La_{10}Te_6O_{27.2}$). However, crystalline phases appeared only after very long holding times (20–70 h). Since one-step heat treatment resulted in the impaired transparency and a low $BaTiO_3$ volume fraction, all samples were subjected to a two-step heat treatment at 410°C for nucleation and at 430°C for crystal growth.

Though the heat-treated samples exhibited increased crystalline volume fractions, especially the Eu-doped samples, some of the aforementioned problems persisted. Finally, the glass sample containing Eu additive was doped with 0.017 mol% of gold (Au^{3+}), with the aim of promoting the heterogeneous nucleation and volume crystallization of the glass sample.

The Au-doped sample after being subjected to a heat-treatment process at 410°C for 5 h exhibited more enhanced volume crystallization, preferential precipitation of $BaTiO_3$, and higher transparency. This suggested the effectiveness of gold as a nucleating agent for heterogeneous nucleation and volume crystallization in tellurite glasses.

3.2.1.2 $PbTiO_3$ Glass Ceramics

Lead titanate is a ferroelectric material having a structure similar to $BaTiO_3$ with a high Curie point (490°C). The phase transition from the paraelectric cubic phase to the ferroelectric tetragonal phase during cooling of $PbTiO_3$ below the Curie point accompanies with a large strain of >6%. This is the result of a phase transformation from a cubic ($c/a = 1.00$) to tetragonal ($c/a = 1.064$) crystal structure. Because of this strain, the fabrication of the pure $PbTiO_3$ articles in the bulk form via the conventional ceramic route is very difficult.

$PbTiO_3$ has great promise as a stable pyroelectric and piezoelectric material for high-temperature or high-frequency applications, owing to its properties, such as high transition temperature, a low ratio for the planar-to-thickness coupling factor a low value of dielectric constant and a low aging rate of the dielectric constant (Gautam et al., 2013).

The controlled crystallization of lead titanate ($PbTiO_3$) from a $PbO-B_2O_3-TiO_2$ precursor glass was first reported by Bergeron and Russell (1965). Later on Grossman and Isard (1969a) studied the precipitation of the ferroelectric $PbTiO_3$ crystals from a base glass of the composition 23.8PbO, 20BaO, 23.8TiO_2, 25B_2O_3, 3SiO_2, 3Al_2O_3, 1.5TiO_2 (wt%) with progressive substitution of ZnO for PbO, which effectively increased the Ba/Pb ratio in the glass samples. All the glass samples were heat-treated at 520°C for nucleation and 610°C for crystal growth for a total of 36 h, producing fine-grained materials with crystal clusters of 2–3 μm in diameter, comprised of tiny, ~0.5 μm crystallites.

The measurement of the axial ratio for the tetragonal lead titanate crystals precipitated in the base glass ($c/a = 1.038$) showed a marked decrease in comparison with the pure stress-free $PbTO_3$ at room temperature (1.063). According to the authors, this might be a result of either strain, initiated due to the suppression of the cubic to tetragonal transition in $PbTO_3$ crystals, or solid solution formation.

Considering the alteration in unit cell volume of the pure $PbTiO_3$ crystals upon the formation of a solid solution with $BaTiO_3$, as well as the difference in ionic radii of Ba^{2+} and Pb^{2+}, the authors attributed the observed reduction in axial ratio for the glass-ceramic materials to a "**clamping effect**" imparted by the glass matrix, rather than by a substitution of Ba for Pb.

The role of ZnO in clamping effect was explained by the rise of viscosity and the annealing point of glass upon the addition of ZnO. Since the annealing point of the glass was close to the Curie temperature of the lead titanate crystals, in some conditions, the annealing temperature of glass might surpass the Curie temperature and the crystals underwent a limited spontaneous strain during the cooling process. In this condition, the resulting crystals could exhibit reduced axial ratios.

Shyu and Yang (1995) also studied the crystallization of the perovskite $PbTiO_3$ crystals from a $PbO-BaO-TiO_2-Al_2O_3-SiO_2$ base glass (sample 4, Table 3.1).

They showed that the primary and secondary crystalline phases formed in the resultant glass ceramics were the bulk-crystallized $PbTiO_3$ perovskite and transient $PbTi_3O_7$ phases, respectively. A surface-crystallized $PbO \cdot Al_2O_3 \cdot 2SiO_2$ phase was also present.

It was also shown that when the heating temperature/time increased, the structure of the precipitated perovskite crystals was transferred from cubic to tetragonal, and the measured c/a ratio increased to approach a constant value which was lower than that for the strain-free crystal.

The dielectric constant of the glass ceramics heat-treated at 830°C for 3–12 h was determined as ~70–80 in the frequency range of 200–10^5 Hz. It was shown that this dielectric constant value was mainly determined by the amount of the perovskite phase.

Yao et al. (1996) fabricated lead titanate glass ceramics in Pb–Ti–B, Pb–Ti–Si, Pb–Ti–B–Si, and Pb–Ti–A1–Si systems by sol-gel processing. Gels with high Pb and Ti content were successfully prepared and lead titanate crystallites were obtained in these gel-derived glass-ceramic systems through the *in situ* growth process.

The crystallization behavior of the amorphous dried gels was studied utilizing differential thermal analysis (DTA) and XRD analyses. For example, the DTA results of the Pb-Ti-Si (PTS) samples with the composition ratios of Pb:Ti:Si of 1.05:1:0.5 exhibited four exothermic peaks in the DTA curve, the first two peaks were attributed to the thermal decomposition of organic components, while the third peak at 601°C was attributed to the crystallization of the $Pb_2Ti_2O_{6+x}$ pyrochlore phase, and the last peak at 644°C to the formation of the perovskite $PbTiO_3$ phase from its precursor pyrochlore phase. This transformation process was described by the following formula:

$$Pb_2Ti_2O_{6+x} \rightarrow 2PbTiO_3 + \frac{x}{2}O_2$$

The XRD patterns for the PTS samples confirmed the above deductions. The grain sizes in these four glass-ceramic systems varied with composition. SEM micrographs showed the small grain sizes of 200 nm for the PTS sample treated at 700°C and of ~100 nm for the Pb-Ti_Al-Si (PTAS) sample with the composition ratios of Pb:Ti:Al:Si of 1.05:1:0.25:0.25, treated at 800°C.

The presence of small crystallites in the latter system were attributed to the addition of Al ions and the competition of Ti^{4+} and Al^{3+} ions for oxygen ions in order to improve their coordination numbers, or the presence of very fine Al precipitates in the precursor PTAS gel.

As in the field of $BaTiO_3$ glass ceramics, an increased research interest is observed in the field of nano-structured $PbTiO_3$ glass ceramics, in recent years.

Shankar and Deshpande studied the effect of BaO (2010) and SrO (2012) additions on the crystallization behavior and electrical properties of lead titanate glass ceramics. The glass samples with composition of $(50-X)PbO-X(BaO/SrO)-25TiO_2-25B_2O_3$ (where $X=0$, 5, 10, and 15 mol%) were prepared using conventional quenching technique. It was found that the addition of BaO or SrO both resulted in the increase and decrease of the glass transition temperature, T_g and crystallization temperature T_c, respectively. The glass samples were converted into glass ceramics by a two-stage heat-treatment process. The XRD results revealed the formation of ferroelectric lead titanate phase as a major crystalline phase, in all glass-ceramic samples belonging to both glass-ceramic series. In the glass ceramics containing BaO, the additional BaB_2O_4 and barium–titanium–borate phases as well as some other minor phases were also formed. Whereas, in the glass ceramics containing SrO, the secondary crystalline phases $Sr_2B_2O_5$ and PbB_2O_4 were observed.

The SEM microstructures of glass ceramics containing BaO or SrO both exhibited the uniform distribution of fine crystallites in rounded shape corresponding to the major phase of perovskite, $PbTiO_3$ for samples with $X=0$ and the presence of the latter phase as well as the formation of needle-like crystals containing higher concentration of BaO in the heat-treated samples containing barium oxide. Whereas, in the SrO-containing glass ceramics, SEM micrographs showed the presence of a dense population of $PbTiO_3$ crystallites in the range of 100 nm, grown in the glass matrix.

The dielectric constants of the glass and glass-ceramic samples indicated higher values for all the glass-ceramic samples containing either oxide, in comparison with the corresponding glass samples. This was attributed to the formation of lead titanate phase in the glass ceramics. Further, it was shown that with addition of BaO or SrO, the dielectric constant for glass-ceramic samples increased upon addition of up to 10 mol% of BaO/SrO and beyond this concentration it decreased. The relatively high value of ε (~125) at room temperature observed for the glass ceramics containing 10 mol% SrO (sample 5, Table 3.1).

The observed hysteresis (P versus E) loop at room temperature confirmed the ferroelectric behavior of both series of glass ceramics.

Shankar and Deshpande (2011) also studied the effect of PbO/TiO_2 ratio on the crystallization behavior and electrical properties of lead titanate glass ceramics.

The glass samples of the composition $(50-X)PbO-(25+X)TiO_2-25B_2O_3$ (where $X=0$, 5, 10, and 12.5 mol%) were prepared by the conventional quenching technique and then were subjected to a fixed two-stage heat treatment at 733 and 793 K for 14 h at each stage.

The XRD results revealed the formation of tetragonal lead titanate as a major crystalline phase in all the glass-ceramic samples. A lead borate phase also appeared as a minor phase in some samples.

It was also found that with increase in X (mol%), the amount of lead titanate phase increased.

The SEM micrograph of glass-ceramic sample $X=0$ revealed the presence of discrete crystallites of $PbTiO_3$ ~100 nm in size, that were uniformly distributed in the glass phase. The other glass-ceramic samples showed larger and well-developed grains.

The variations of dielectric constant as a function of temperature at 1 KHz exhibited that the dielectric constants of all glass-ceramic samples were higher than that of corresponding glass samples. The variation of dielectric constant as a function of X (mol%) at room temperature and 673 K along with volume fraction of the crystalline $PbTiO_3$ phase is depicted in Figure 3.5.

It is evident from this figure that ε increases with content of X (mol%) at room temperature and 673 K. This may be attributed to the increase in the volume fraction of lead titanate phase.

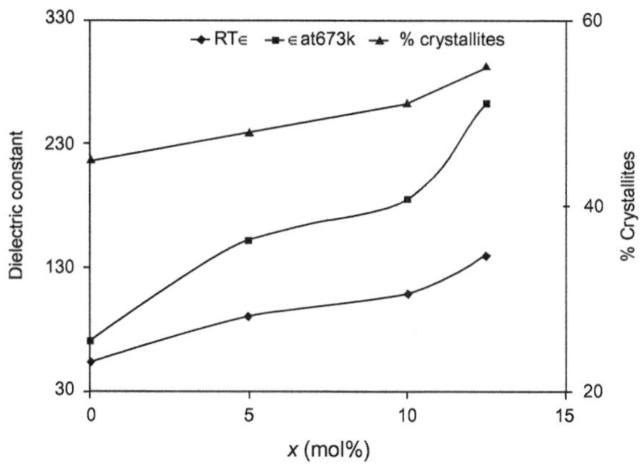

Figure 3.5 Variation of dielectric constant at RT and 673 K and volume fraction of crystallites with X (mol%). (After Shankar and Deshpande, 2011. Reproduced with permission of Elsevier Science Publishing Company.)

The glass-ceramic samples also exhibited moderately good hysteresis loops between polarization and electric field right from the room temperature, confirming the ferroelectric nature of the glass-ceramic samples.

Golezardi et al. (2010) studied the crystallization behavior and properties of $PbO-TiO_2-B_2O_3-SiO_2$ glasses in the presence of Bi_2O_3 as a nucleating agent (sample 6, Table 3.1).

The authors have reported that by choosing the total amount of glass-forming oxides as $(SiO_2 + B_2O_3) = 30$ mol%, the stoichiometric ratio of $PbO/TiO_2 = 1$ (in $PbTiO_3$) had to be increased to $PbO/TiO_2 = 2$ in order to prepare a stable glass, with no sign of crystallization during cooling. To the above base glass, 0.5 and 1.0 mol% Bi_2O_3 were added as the nucleating agent.

The glass samples were heat-treated in the range of 455–485°C for 1–3 h and 575°C for 1 h respectively, for nucleation and growth. It was found that the addition of Bi_2O_3 nucleant resulted in the formation of homogeneous, nano-structured glass ceramics with a mean crystallite size of 20–25 nm and $PbTiO_3$ as the major crystalline phase. The dielectric constant and dissipation factors for the prepared glass ceramics were in 140–400 and 0.04–0.4 ranges, respectively. The measured dielectric constant for the glass ceramic containing 0.5 mol% Bi_2O_3, nucleated for 3 h at 485°C and 1 h at 575°C was ~400. This value is among the highest values reported for this type of glass ceramics. According to the authors, the resistivity difference existing between crystals and glassy matrix was responsible for the large space-charge polarization due to the charge buildup in the interfacial region, which led to the elevation of dielectric constant values.

It was deduced that the reduction in the crystallite size of the lead titanate phase, brought about by the addition of Bi_2O_3 nucleant, extended the interfacial region and resulted in higher polarization effects and larger dielectric constant values. Figure 3.6 depicts the microstructure of the aforementioned specimen.

3.2.1.3 $SrTiO_3$ Glass Ceramics

$SrTiO_3$ glass ceramics have found application as cryogenic capacitive temperature sensors and have been proposed for several other cryogenic applications. They have also been proposed for ambient temperature applications requiring temperature compensation of the dielectric constant, which is achieved

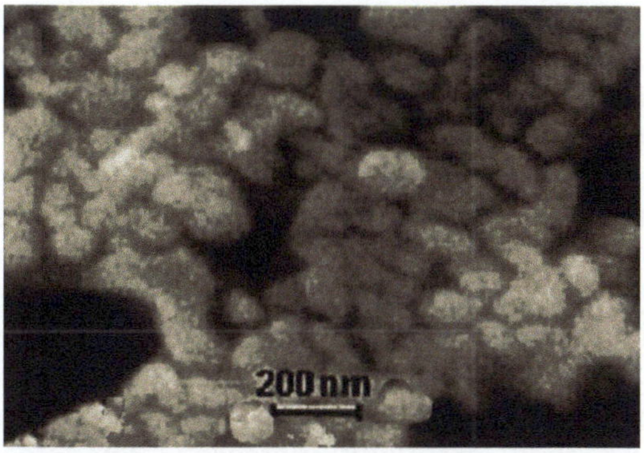

Figure 3.6 SEM micrograph of specimen S0.5 with 0.5% Bi_2O_3, after nucleation at 485°C for 1 h and crystallization at 575°C for 1 h. (After Golezardi et al., 2010. Reproduced with permission of Elsevier Science Publishing Company.)

through an appropriate balance between the negative temperature coefficient of $SrTiO_3$ and the positive temperature coefficient of the parent glass.

Swartz et al. (1988a) studied the crystallization behavior and microstructure of the glasses in the SrO is–TiO_2–SiO_2–Al_2O_3 system (sample 7, Table 3.1). They found that at low crystallization temperatures (800–900°C), the initial crystallization of perovskite $SrTiO_3$ occurred via a spherulitic crystal growth mechanism, resulting in dendritic $SrTiO_3$ crystal clusters, which were composed of 20 nm crystallites. As the crystallization time was increased, the secondary phase, hexacelsian ($SrAl_2Si_2O_8$), precipitated interstitially, interrupting the dendritic growth of $SrTiO_3$. At still higher crystallization temperatures (900–1,000°C), the $SrTiO_3$ and $SrAl_2Si_2O_8$ (hexacelsian) phases compete with each other for strontium, and the microstructure consisted of dispersed $SrTiO_3$ crystallites (50 nm) and larger (200 nm) hexacelsian crystals. At a higher crystallization temperature of 1,100°C, the $SrTiO_3$ phase consisted of larger faceted crystals (500 nm) and hexacelsian gradually transformed to its anorthite polymorph. At this temperature, titania (TiO_2) was also crystallized, first as small (300 nm) crystals, which gradually transformed into larger acicular rutile crystals which grew after 16 h to a length of ~80 μm.

The dielectric properties of the above glass ceramics have also been determined over the frequency range of 10–1,000 kHz, and the temperature range of −170°C to 200°C (Swartz et al., 1988b).

It was found that the dielectric properties were strongly dependent on the crystallization conditions, which determined the amounts of $SrTiO_3$ and secondary crystalline phases, and the microstructure of the glass ceramics.

For example, the plots of the dielectric constant against crystallization time at temperatures of 800°C, 850°C, and 900°C are shown in Figure 3.7. These data were consistent with the amount of $SrTiO_3$ versus crystallization time, as determined by the XRD data. For example, for the specimens crystallized at 800°C and 850°C after 4 h and 20 min respectively, the dielectric constant remained almost constant until $SrTiO_3$ was detected by XRD and then gradually increased. The steep increase observed after 16 h in the dielectric constant, for the specimen crystallized at 850°C, was related to the crystallization of hexacelsian, $SrAl_2Si_2O_8$.

Thakur et al. (1995) studied the crystallization behavior of a strontium titanate borosilicate glass-ceramic system (sample 8, Table 3.1).

An attempt was made to crystallize the strontium titanate phase in this glass ceramic. Different phases precipitated out during heat treatment in the 800–950°C range for 1–48 h, but due to high reactivity of SrO with B_2O_3, the formation of $SrTiO_3$ proved to be impossible. Strontium borate ($Sr_2B_2O_5$) crystallized instead, as the major phase, followed by TiO_2 (rutile) and $Sr_3Ti_2O_7$ phases. During the initial stage of crystallization, particles of 50–100 nm in diameter appeared in the glass. These particles were agglomerates of very fine crystallites

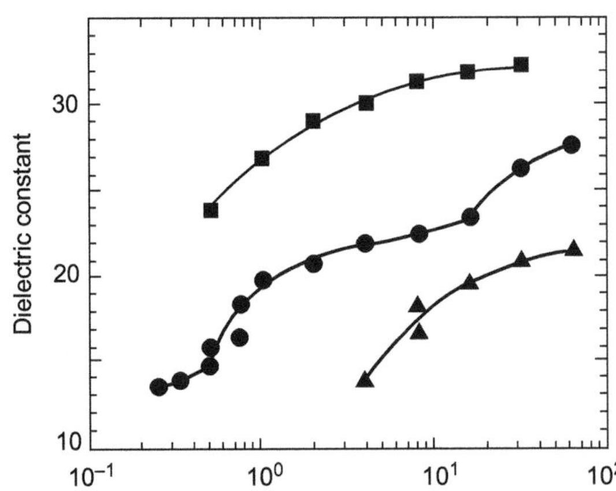

Figure 3.7 Dielectric constant plotted against crystallization time for glass ceramics crystallized at 800 (▲), 850 (●), and 900 (■) °C. (After Swartz et al., 1988b. Reproduced with permission of Springer Science + Business Media, Inc.)

of 10 nm or less. These fine crystallites served as a nucleus for the growth of star-shaped crystals. As the crystallization temperature was raised to 950°C, no major change was observed in morphology of this glass-ceramic sample, except a little increase in crystallite size.

All micrographs showed uniform distribution of precipitated crystals in the glass matrix.

The dielectric constant of these glass ceramics was relatively low and almost temperature independent over a wide range of temperatures and the values of the dissipation factor were also small.

The crystallization behavior and properties of strontium titanate borosilicate glass ceramics, in the presence of some additives such as CoO, Bi_2O_3, and La_2O_3 were also studied (Thakur et al., 1996, 2002, 2003).

Thakur et al. (1996) studied the effect of addition of 1 wt% CoO nucleant on the crystallization behavior of strontium titanate borosilicate glass ceramics. The glass samples were subjected to heat treatment at 900°C and 950°C. The phase formation sequence in these glass ceramics consisted of the formation of $Sr_2B_2O_5$ as the primary crystalline phase followed by rutile (TiO_2), $Sr_3Ti_2O_7$, $SrB_2Si_2O_8$, and $Sr_3B_2SiO_8$ as secondary phases. The microstructural studies showed that $Sr_2B_2O_5$ exhibited relatively larger grain size, while the other phases such as TiO_2 and $Sr_3Ti_2O_7$ appeared in smaller sizes. The authors concluded that the crystallization of $SrTiO_3$ phase in these glasses is difficult.

The resulting glass ceramics exhibited relatively low dielectric constant and high dissipation factor (sample 9, Table 3.1), but as the authors have stated, they possessed higher thermal stability and may be used in the same applications as other strontium titanate glass ceramics.

Thakur et al. (1997) also studied the crystallization behavior of strontium titanate borosilicate glasses, with the composition of $65(SrO \cdot TiO_2)$ and $35(2SiO_2 \cdot B_2O_3) + xBi_2O_3$, where $x = 1$, 5, and 10 (wt%).

This study confirmed the earlier reports concerning the difficulty of crystallization of $SrTiO_3$ phase in the glass ceramics, due to the higher reactivity of SrO with some oxides other than TiO_2. For most of the heat treatments, $Sr_2B_2O_5$ appeared as the primary crystalline phase followed by TiO_2, $Sr_3Ti_2O_7$, and some unidentified phases.

But it was stated that in some glass compositions or by adoption of some special heat treatment schedules, the crystallization of $SrTiO_3$ could be possible.

For example, by the direct insertion of glass ceramics containing 1 wt% Bi_2O_3 into the furnace at higher temperatures (≈1,000°C), the crystallization of $SrTiO_3$ became possible.

The $SrTiO_3$ phase also appeared in the case of glass ceramics containing higher amounts of Bi_2O_3 (10 wt%) if subjected to a special heat-treatment schedule. This was explained by some structural changes occurring in the glasses having fourfold coordinated boron which resulted in redissolution of earlier formed phases, such as $Sr_2B_2O_5$, and $Sr_3Ti_2O_7$ to form $SrTiO_3$.

It was also concluded that the glass ceramics containing 5 wt% Bi_2O_3 possessed fine grains of submicron size evidencing its higher nucleation rate and minimal crystal growth.

The electrical properties of the aforementioned strontium titanate borosilicate glass ceramics containing Bi_2O_3 additive were also studied (Thakur et al., 2003).

In this interesting study the ac electrical data, measured in the frequency range 0.1 kHz–1 MHz by impedance spectroscopy technique, were used to study the electrical response of the glass-ceramic system.

Different regions of the glass-ceramic sample are assumed to have an equivalent circuit comprising the resistance and capacitance parameters in parallel. The complex plane plots of various quantities, associated with the data of dielectric properties for various glass-ceramic samples reveal contributions from simultaneously operating polarization mechanisms to overall dielectric behavior.

The authors explained the dielectric behavior of various glass ceramics in terms of contributions arising from different polarization processes occurring at glassy matrix, crystalline phases, glass–crystal interface, and blocking electrodes.

The interfacial or space-charge polarization, which is of great importance in a glass-ceramic material, was attributed to the differences existing between the conductivities of various phases present in the material. In this case, the charge buildup at the interface may result in a large polarization and high value of effective dielectric constant.

For example, in the case of the variation of dielectric constant (ε') and dissipation factor ($\tan \delta$) of the glass-ceramic sample with 5 wt% Bi_2O_3, crystallized at 950°C, 3 h (Figure 3.8), the relatively small increase in ε' and $\tan \delta$ with temperature was explained by the uniform distribution of fine crystallites of different phases within glassy matrix and relatively little variation in their amount and structure values over a wide temperature range. Table 3.1 depicts some of the electrical properties of a glass-ceramic sample containing 10 wt% Bi_2O_3 (sample 10).

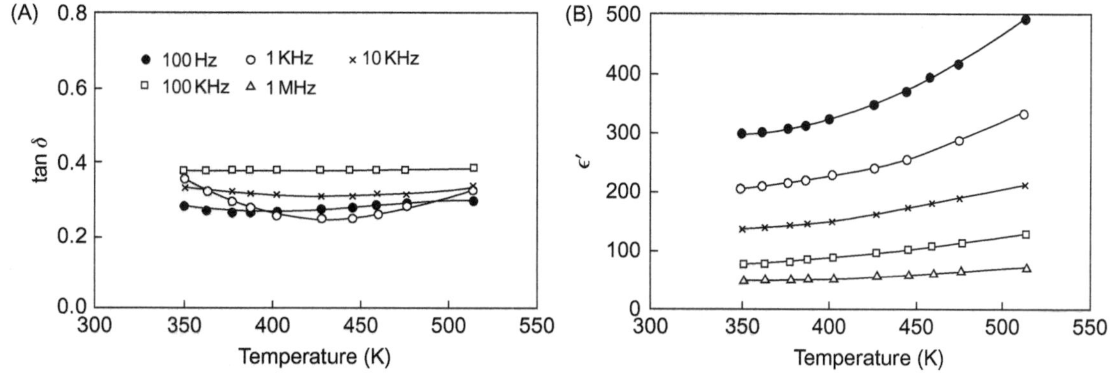

Figure 3.8 Temperature variation of (A) dissipation factor and (B) dielectric constant for a strontium titanate borosilicate glass ceramic containing 5 wt% Bi_2O_3. (After Thakur et al., 2003. Reproduced with permission of Elsevier Science Publishing Company.)

The crystallization behavior, microstructure development, and dielectric properties of various strontium titanate borosilicate glass ceramics, with addition of 1, 5, and 10 wt% La_2O_3 were also investigated (Thakur et al., 2002).

The glass compositions were $(65-x)[SrO\ TiO_2]$ (35) $[2SiO_2 \cdot B_2O_3] - (x)La_2O_3$, where $x = 1$, 5, and 10 (wt%).

It was shown that the glass-ceramic samples containing different concentrations of La_2O_3 exhibited different crystallization behavior and different crystalline phases, such as $Sr_2B_2O_5$, $Sr_3Ti_2O_7$, and rutile (TiO_2) as the major phases, hence, resulting in different dielectric properties.

For example, in the glasses containing 10 wt% La_2O_3, the XRD peak intensity of $Sr_3Ti_2O_7$ phase was first increased (up to 950°C) and then started decreasing at higher temperature (1,000°C) treatments. This was attributed to the decomposition of the latter phase at high temperatures.

Some of the studied glass ceramics exhibited quite high values of dielectric constants. For example, the glasses containing 10 wt% La_2O_3, heat-treated at 900°C for 3 h, showed a very high value of dielectric constant. This was attributed to the space-charge polarization effect, initiated due to the conductivity difference between the formed crystallites and glassy matrix.

Table 3.1 depicts some of the properties of the glass ceramic containing 1 wt% La_2O_3 (specimen 11).

3.2.1.4 Solid Solution Perovskites

Ruiz-Valdés et al. (2004) investigated the vitrification, crystallization, and dielectric properties of glasses in the

$BaO-TiO_2-PbO-Al_2O_3$ (B_2O_3) system. According to the authors, in order to suppress the marked tendency of spontaneous crystallization in the glasses of this system, the contents of the added glass formers (Al_2O_3 and B_2O_3) must be regulated. In this way, stable glasses with relatively low fusion temperature, acceptable temperature range of regulated crystallization, and potentially controllable crystallization properties could be prepared in the certain composition range (sample 12, Table 3.1). The phase composition of such glass ceramics included different $Pb_xBa_{(1-x)}TiO_3$ solid solutions depending on the starting chemical composition and the heat-treatment conditions. The treatment at 710–850°C resulted in typical solid solutions, $Pb_{(1-x)}Ba_xTiO_3$ ($x = 0.1-0.3$) as well as $PbTiO_3$. At higher temperatures (900–950°C), stronger reflections of $Pb_{(1-x)}Ba_xTiO_3$ ($x = 0.4-0.9$) appeared. Though the exact value of x has not been specified, the formation of higher amounts of $Pb_{(1-x)}Ba_xTiO_3$ solid solutions, containing more BaO (and less free $PbTiO_3$), were proposed at these high-temperature ranges. According to the authors, since the presence of $PbTiO_3$ crystals, due to their large tetragonality, leads to higher dielectric losses, the crystallization at 900–950°C is preferable. The crystal sizes reported for the aforementioned glass ceramics were 90–200 nm which, according to the authors, is quite acceptable for ferroelectric properties. By controlling the crystallization time at 900°C, it was possible to obtain glass-ceramic samples with relatively low dielectric losses and variable permittivity in the range 30–650 (Figure 3.9).

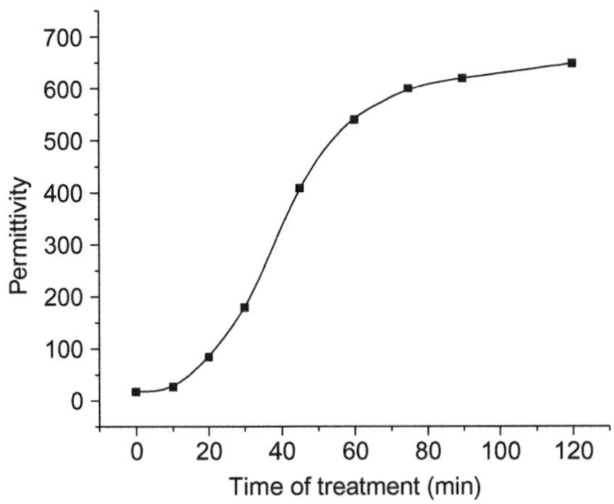

Figure 3.9 Permittivity values versus time of crystallization at 900°C for the glass-ceramic sample No. 12 Table 3.1 ($f = 800$ kHz, $T = 25$°C). (After Ruiz-Valdés et al., 2004. Reproduced with permission of Elsevier Science Publishing Company.)

Sahu et al. (2006a) studied the crystallization behavior and microstructural characteristics of glass samples prepared in the $65[(Pb_{1-x}Sr\ x)O \cdot TiO_2)] \cdot 25[2SiO_2 \cdot B_2O_3] \cdot 5[K_2O] \cdot 5[BaO]$ system, where x varied in 0.2–1.0 range, with stepwise increments of 0.1 (sample 13, Table 3.1).

All the bulk glass-ceramic samples prepared from the above glasses contained perovskite titanate as their major phase.

Interestingly, the titanate phase precipitated in all the glass-ceramic samples with $x \leq 0.5$ had tetragonal structure, whereas for $x > 0.5$, cubic structure was observed. It was also found that the lattice parameter(s) and axial ratio were less influenced by the crystallization temperature and time for a fixed content of SrO in the glass samples. But, with increasing SrO content in the glass-ceramic samples, a systematic variation in c, a, and c/a was observed.

It was also shown that the composition of the crystallites which appeared in all the glass-ceramic samples were $(Pb_{1-x}Sr_x)TiO_3$ solid solution, and the calculated intensity of $(Pb_{1-x}Sr_x)TiO_3$ crystals varied systematically with changes in Pb^{2+}/Sr^{2+} ratio. The microstructure of some glass ceramics consisted of fine interconnected crystallites, exhibiting a uniform distribution which is more desirable for better performance as capacitor material.

The electrical properties of the above-mentioned glass ceramics were also determined (Sahu et al., 2006b).

The dielectric constants and dissipation factors of various glass-ceramic samples against temperature were plotted at different frequencies and a comparative study of various glass-ceramic samples has also been made by plotting ε' against temperature (Figure 3.10).

Variation of the Curie temperature of these samples with composition was also studied. A systematic shift in Curie temperature to lower temperatures was observed by the substitution of Sr for Pb in the glass composition.

According to the authors, since strain due to crystal clamping and particle size has a negligible effect on T_c, the variation in T_c was solely due to $(Pb_{1-x}Sr_x)TiO_3$ solid solution formation.

It was also shown that glass-ceramic samples rich in SrO (e.g., $x = 0.8$ and 0.9) exhibited the lowest dielectric constant and dissipation factors due to the formation of $(Pb_{1-x}Sr_x)TiO_3$ solid solutions very rich in strontium. The latter samples are also expected to show the lowest Curie temperatures that in some cases could be negative.

Mady (2011) investigated the crystallization behavior and electrical properties of ferroelectric lead barium titanate–based

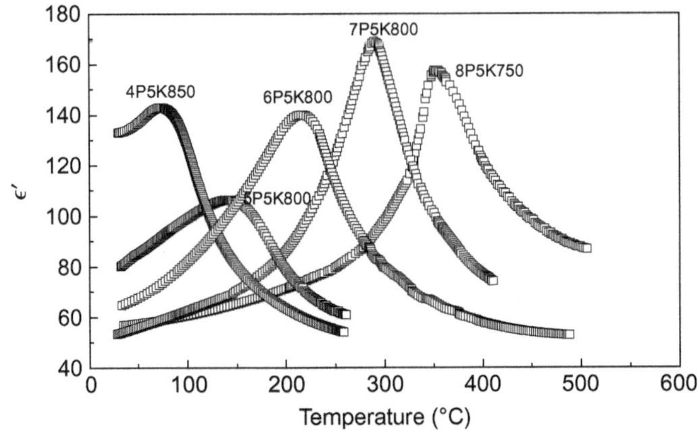

Figure 3.10 Variation of dielectric constant (at 1 MHz) with temperature for different glass-ceramic samples. The glass codes 4P, 5P, 6P, 7P, and 8P denote the *x* values of 0.6, 0.5, 0.4, 0.3, and 0.2, respectively, and the last numbers in the codes (e.g., 850) indicate the crystallization holding temperature of the glass samples. The holding time was 3 h for all the samples. (After Sahu et al., 2006b. Reproduced with permission of Springer Science + Business Media, Inc.)

glass ceramics by the heat treatment of melt-quenched glasses of the composition $0.5PbO{-}0.5BaO{-}1.0TiO_2{-}0.5Na_2B_4O_7$. The heat-treatment process was performed in two steps, at 650°C, 2 h, and 750°C, 8 h. XRD patterns revealed the formation of tetragonal perovskite $Ba{-}PbTiO_3$ phase with particle size of 21.3 nm, as the primary phase after the heat treatment. The secondary phases Pb_3O_4 and $PbTiO_3$ were also observed.

The electrical properties were investigated for both the as-quenched glasses and the glass ceramics. It was found that the electrical conductivity considerably increased after the heat treatment at 650°C as a result of the growth of tiny crystallites in the glass matrix. Ferroelectric Curie temperature was determined as 310°C. Heat treatment at higher temperature of 750°C and longer time 8 h led to the complete crystallization of the glass sample and decreased electrical conductivity. Such decrease in conductivity was attributed to the formation of larger crystals and reduction of the extent of interfacial region which apparently acted as easy electronic conduction paths. The SPH was suggested as the dominant mechanism of conductivity. Gautam et al. (2013) prepared various glass samples in the system:

$$[(Pb_xSr_{1-x})O \cdot T_iO_2] - [2SiO_2 \cdot B_2O_3] - [K_2O] - [BaO] - [Nb_2O_5](0.5 \leq x \leq 1)$$

by melt-quench method.

DTA traces of the prepared and annealed glass samples (400°C, 3 h) showed that glass transition temperature T_g was in the temperature range 520–695°C and increased with increasing the value of x. The increase in T_g was attributed to the increase in the viscosity of the melt with x. Three exothermic peaks were also observed in the DTA patterns in the temperature range 600–800°C. On the basis of DTA results, various glass-ceramic samples were prepared by heat treating the glass samples in the temperature range 605–1,060°C for 3 and 6. Lead titanate ($PbTiO_3$) or solid solution $(Pb,Sr)TiO_3$ crystallized as major phases followed by secondary phases such as $PbTi_3O_7$, PbB_2O_4, and TiO_2 (rutile) in the heat-treated samples. The observed crystallized phases developed in all the glass-ceramic samples had tetragonal crystal structure. The lattice parameter c of the perovskite titanate phase was found to decrease with decreasing Pb^{2+}/Sr^{2+} ratio.

SEM showed the uniform distribution and highly interconnected morphology of the fine crystallites throughout the glassy matrix in the microstructure of most glass-ceramic samples. The average grain and crystallite size were in range of 0.40–2 μm and 17–40 nm, respectively, for all the glass-ceramic samples. With increasing concentration of Sr, the crystallite and grains size were found to decrease.

3.2.1.5 Other Titanate Glass Ceramics

Bismuth Titanate Glass Ceramics Ferroelectric bismuth titanate ($Bi_4Ti_3O_{12}$) (BiT) ceramics are one of the most promising classes of materials for optical memory, piezoelectric, and electro-optic device applications that have attracted great research interest and activity in recent decades.

They are usually fabricated using various techniques, such as solid-state reaction process, coprecipitation, and molten salt synthesis. These methods all have disadvantages which result in unacceptable microstructure and poor properties of the fabricated ceramics (Molla et al., 2011, 2012).

Fabrication of bismuth titanates via the glass-ceramics route, in addition to the common advantages of glass ceramics, such as the possibility of obtaining pore-free, fine-grained, nanostructured transparent materials, the obtained nano-structure comprising BiT crystallites embedded in a low-permittivity, high-resistivity glass matrix result in higher resistivity, which is more desirable for some of their proposed applications. Despite the extensive research carried out in the field of rare-earth-doped $Bi_4Ti_3O_{12}$ thin films as one of the most potential materials for nonvolatile ferroelectric random access memories because of their high fatigue resistance and high remanent polarization value, fast switching speed, and excellent optical properties (such as large optical nonlinearity), very limited work has been reported on preparation of bulk BiT glass ceramics (Molla et al., 2012).

Molla et al. (2011) prepared melt quenched bismuth titanate glasses of the composition $(100-x)KS_2-x$ BiT, where KS_2 is potassium disilicate ($K_2O \cdot 2SiO_2$) with $x=10$, 20, 30, 40, 45 and 50 and BiT is $Bi_2O_3 \cdot 3TiO_2$. By controlled heat treatment of the prepared glasses at various temperatures and duration, transparent $Bi_4Ti_3O_{12}$ (BiT) glass ceramics were obtained.

Various thermal, optical, and electrical properties were also evaluated and correlated with their composition.

The systematic increase observed in dielectric constants (ε_r) as well as refractive index (n) and third-order susceptibility (χ^3) with increase in BiT content of glasses was attributed to the combined effects of high polarization and ionic refraction of bismuth and titanium ions.

Microstructural studies showed the development of granular polycrystals of 70–90 nm and nano-rods of average diameter 85–90 nm in the samples heat-treated at 580°C for 1 and 10 h, respectively. The sharp increase in the value of (ε_r) demonstrated upon the aforementioned crystallization of glass was explained by the formation of noncentrosymmetric and ferroelectric BiT nano-crystals in the glass ceramics. The dielectric loss values of the glass ceramics were also determined and found to be considerably low which is important with respect to their device application point of view (sample 14, Table 3.1).

The emission spectra of the glass-ceramic samples, with identical composition to the above glass ceramics, and doped with Eu^{3+} ions were also recorded by exciting the samples at 467 nm (Molla et al., 2012). A red emission at 612 nm corresponding to the ${}^5D_0 \rightarrow {}^7F_2$ transition was observed, that with increase in heat-treatment time showed an eightfold increase in intensity. This was attributed to the occupancy of Eu^{3+} ions in the low phonon energy crystal sites of BiT, since the ${}^5D_0 \rightarrow {}^7F_2$ (612 nm) transition is electric dipole allowed and very much sensitive to the ligand environment.

Recently, Molla et al. (2013) also prepared the first barium bismuth titanate ($BaBi_4Ti_4O_{15}$) (BBiT) glass ceramics and studied their properties.

By controlled heat treatment of the melt-quenched glasses of the composition $(100-x)KS_2-x$BBiT, (where KS_2 is potassium disilicate [$K_2O \cdot 2SiO_2$] and $x=10$, 20, 30, 40, 50, and 60 and BBiT is $BaO \cdot 2Bi_2O_3 \cdot 4TiO_2$), with and without Eu_2O_3 doping, nanocrystalline $BaBi_4Ti_4O_{15}$ glass ceramics were obtained.

FESEM micrographs of Eu^{3+}-doped BBiT glass ceramics (with the composition of $x = 50$) heat-treated at 520°C for 10 h and 610°C, for a varying time of 1, 10, 20, 40, 60, and 100, showed the formation of randomly oriented hexagonal rod-shaped crystals of 200–400 nm, whereas the TEM images exhibited 10–20 nm crystallites (Figure 3.11). The marked difference observed in the size of crystallites through TEM and FESEM studies was explained by the fact that TEM studies showed the size of the individual crystallites, whereas FESEM studies only

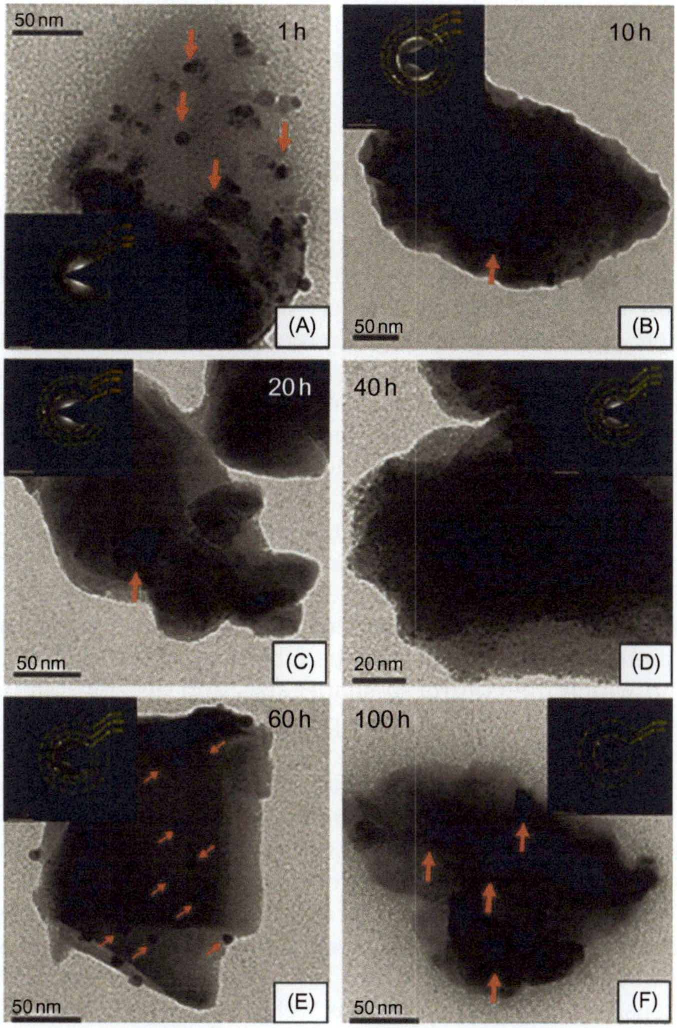

Figure 3.11 TEM and SAED images of Eu^{3+}-doped BBiT glass-ceramics heat-treated at 520°C for 10 h for nucleation and 610°C for (A) 1; (B) 10; (B) 20; (D) 40; (E) 60; and (F) 100 h for growth of crystals (crystals are indicated by arrow). (After Molla et al., 2013. Reproduced with permission of the American Ceramic Society.)

revealed the size of clusters of polycrystals which consisted of a large numbers of crystallites.

The heat-treated glass ceramics showed higher dielectric constant (ε_r) values (25–55), in comparison with the precursor glasses (20–37). For glass ceramics, these values increased with an increase in frequency up to 3.0 MHz and then decreased up to 5.0 MHz. Dielectric losses for all the samples increased from 0.005 to 1.0 with an increase in frequency from 100 Hz to 5.0 MHz (see sample 15, Table 3.1).

Emission spectra were also recorded by exciting the glass-ceramic samples at 466 nm. Longer heat-treatment times led to a 15-fold increase in the intensity of the main red emission at 612 nm, assigned to the induced ED transition $^5D_0 \rightarrow {}^7F_2$ of Eu^{3+}. This increase was attributed to the segregation of Eu^{3+} ions into the low phonon energy BBiT crystallites.

3.2.2 Niobate-Based Glass Ceramics

Lithium niobate is a ferroelectric material with a distorted perovskite structure. It has a high Curie temperature (1,140°C) and large second-order NLO response. The single crystals of lithium niobate ($LiNbO_3$) are very attractive materials because of their combination of large ferroelectric, pyroelectric, piezoelectric, photoelastic, acousto-optical, and NLO properties.

Because of the above-mentioned promising properties, $LiNbO_3$ has found many applications in the field of optical waveguides, modulators, switches, and optical circuits fabrication (Abouellell and Leonberger, 1989).

Because the preparation of the $LiNbO_3$ crystals, using conventional routes (Czochralski process) is difficult and involves high costs, the possibility of preparing glass ceramics containing ferroelectric crystals has received considerable attention. In this method, the materials can be fabricated easily and at lower cost in which the control of shape, size, and distribution of the crystalline phase and thus, the properties of the end product is possible.

Since $LiNbO_3$ (and other alkali niobates, such as $NaNbO_3$ and $KNbO_3$) have poor glass-forming ability, the addition of network formers, such as SiO_2, B_2O_3, and Al_2O_3, to the selected glass-ceramic compositions is necessary.

The preparation of glass ceramics with high NLO response requires the right balance between SHG efficiency and transparency of the glass ceramic. The SHG efficiency is connected to the density number of crystals as well as their average size.

On the other hand, it is well known that glass does not exhibit second-order optical nonlinearity due to its inversion symmetry. Nevertheless, some techniques were developed to induce NLO responses in glass-based materials. They include electrical poling, alternating electric field induction, thermal poling, or facilitating oriented crystallization in glass. The latter mechanism of crystallization may be achieved via plastic deformation processes (e.g., hot extrusion) (Habelitz et al., 1997) or kinetically (Keding and Rüssel, 1996) and thermodynamically (Rüssel, 1997) controlled processes.

For the initiation of optical nonlinearity in the glass-based materials, the fabrication of glass-ceramic materials comprising oriented nano-crystals, with the desired electro-optical properties, would be a more straightforward and economical route in comparison with other methods such as thermal or electrical poling processes. In this way, if careful and intelligent selection of appropriate glass compositions was realized and suitable heat-treatment procedure was adopted, it is possible to produce glass-ceramic materials with higher NLO response and SHG efficiency as promising candidates for the nano-structured functional optics.

The niobate glass ceramics according to the glass-forming oxides used in their composition can be classified in the following groups.

3.2.2.1 *Niobate Glass Ceramics with TeO_2-Based Glasses*

These glass ceramics are mainly located in the $R_2O{-}Nb_2O_5{-}TeO_2$ system. In the great majority of these glass ceramics, Li_2O is used as the alkali oxide (R_2O) component hence, the main niobate nano-crystal in the prepared glass ceramics is $LiNbO_3$, though K_2O was also used by some researchers to a lesser extent.

The reason of using TeO_2-based glasses as matrix in the aforementioned glass ceramics are their promising electrical and nonlinear optical properties. They exhibit large refractive indices, high dielectric constants, large third-order NLO susceptibilities, and wide infrared transmittance (Nasu et al., 1989, 1990; Kim et al., 1993). The close matching of the refractive index of TeO_2-based glasses ($n = 2.0{-}2.2$) and lithium niobate, $LiNbO_3$ crystal ($n = 2.296$), has also motivated intense research, aimed at the fabrication of optical quality TeO_2-based glasses and glass ceramics containing microcrystalline $LiNbO_3$ (Shankar and Varma, 1999).

Komatsu et al. (1991) fabricated transparent glass ceramics with the composition of $(100 - x)TeO_2$, $xLiNibO_3$ ($x = 10 - 60$)

containing ferroelectric $LiNbO_3$ crystals. Glasses were prepared by using a conventional melt-quenching method and their crystallization behaviors were studied.

It was found that $LiNbO_3$ was formed in the glasses with $x \geq 40$, through the transformation of a metastable pyrochlore-type compound at temperatures above 500°C. The authors claimed that the fabrication of transparent TeO_2-based glass ceramics, containing ferroelectric $LiNbO_3$ microcrystallites, is possible via the crystallization of the above glasses.

The glass-formation ability and crystallization behavior of $(100 - x)TeO_2 - xLiNbO_3$ $(10 \leq x \leq 50)$ glasses and the dielectric and optical properties of the resulting glass ceramics were also studied (Shankar and Varma, 1999).

It was found that the most thermally stable glass, in the above system, was obtained around the composition with $x = 20$ and fabrication of homogeneous (uncrystallized) glasses with $LiNbO_3$ content above $x = 50$ was prevented by rapid reduction in stability against crystallization.

This behavior was explained by the structural model for TeO_2–Li_2O–Nb_2O_5 glasses consisting of NbO_6 octahedra homogenously distributed in the network structure composed of TeO_4 trigonal bipyramids (tbp). According to the authors, the interaction of the above structural units plays a deciding role in the stability of the glass and its crystallization tendency.

It was demonstrated that $LiNbO_3$ crystallites were directly precipitated on the surface of the glass, with the composition $50TeO_2$–$50LiNbO_3$, by a single-step heat treatment at 200°C for 12 h.

The surface-crystallized $50TeO_2$–$50LiNbO_3$ transparent glass-ceramic samples exhibit a strong pyroelectric response, ferroelectric hysteresis, and intense SHG of 1,064 nm wavelength radiation.

Kim et al. (1993) reported the direct precipitation of $LiNbO_3$ crystals, without passing through the metastable pyrochlore-type compound. They used $50TeO_2$–$(50 - x)Li_2O$–xNb_2O_5 glasses with high Li^+ content, and a two-step heat-treatment procedure.

The optical SHG in electrically poled Li_2O–NbO_5–TeO_2 glasses was also studied (Tanaka et al., 1995). Electrical poling was performed at 250°C for 30 min using an applied dc potential of 3–4 kV. The SHG was observed in binary Li_2O–TeO_2 and Nb_2O_5–TeO_2 glasses as well as ternary $10Li_2O \cdot 10Nb_2O_5 \cdot 80TeO_2$ and $15Li_2O \cdot 15Nb_2O_5 \cdot 70TeO_2$ glasses. The observation of SHG in tellurite glasses lacking Li^+ ions was explained by the presumable existence of long-range orientation of electric dipoles due to asymmetrical structural units, i.e., TeO_4 (trigonal bipyramids) and TeO_3 (trigonal pyramids) in the glass structure rather than the Li^+ ion migration.

The fabrication and electrical properties of some transparent glass ceramics, located in the K_2O–Nb_2O_5–TeO_2 system, were also investigated.

For example, Kim et al. (1996) prepared some glass ceramics in the above system, containing nano-crystals of an apparently face-centered cubic crystal structure.

The glass ceramics exhibited an SHG behavior, which was attributed to the possibility of existence of a very small distortion from the cubic structure in the precipitated crystals.

Shioya et al. (1995) also fabricated transparent glass ceramics with the composition $15K_2O$–$15Nb_2O_5$–$70TeO_2$ (mol%) containing a crystalline phase of cubic structure, which was not identified. This metastable phase, consisting of tiny crystallites of 20–40 nm, was formed by the heat treatment at ~390°C for 1 h and transformed into a stable phase at temperatures >450°C. The glass ceramics which consisted of the stable crystalline phase were opaque. The transparency of the glass ceramics was attributed to the small particle size of the metastable cubic crystalline phase.

The optical and dielectric properties of the transparent glass ceramics obtained by 1 h heat treatment at 425°C were also reported. The values were $\varepsilon_r = 44 \pm 1$ (dielectric constant) and $\chi^{(3)} = 3.3 \times 10^{-13}$ esu (third-order nonlinear optical susceptibility). These values were larger than those for the original base glass. SHG was also clearly observed in the transparent glass ceramics.

The crystallization behavior and SHG properties of the mixed K_2O/Na_2O niobate glass ceramics were also studied (Jeong et al., 2006). The glasses located in the xK_2O–$(14-x)Na_2O$–$14Nb_2O_5$–$72TeO_2$ ($x = 0$–14 mol%) system were prepared by a conventional melt-quenching method.

While the crystalline phases present in the glass-ceramic samples with $x = 0$–8 mol% showed face-centered cubic structures, the samples with $x = 10$–12 mol% exhibited increased lattice constants and slightly distorted structures as a result of the substitution of K^+ for Na^+.

It is well known that a cubic crystalline phase is isotropic and has inversion symmetry which does not allow second-order NLO processes. Therefore, the samples with $x = 0$–8 mol% exhibiting face-centered cubic structure, had no SHG properties, but the samples with $x = 10$–12 mol% with distorted cubic structures exhibited SHG (Figure 3.12). The latter glass ceramics were transparent and consisted of ~35 nm nano-particles.

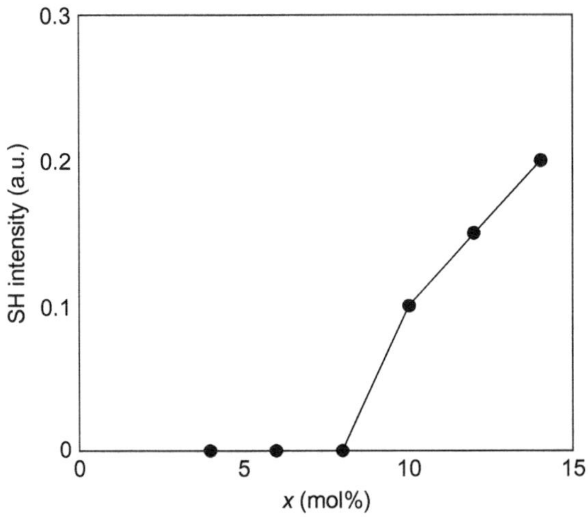

Figure 3.12 Variation of SH intensity with *x* for transparent $xK_2O-(14-x)Na_2O-14Nb_2O_5-72TeO_2$ glass ceramic. (After Jeong et al., 2006. Reproduced with permission of Elsevier Science Publishing Company.)

3.2.2.2 Niobate Glass Ceramics with Silicate and Aluminosilicate-Based Glasses

Niobate Glass Ceramics Containing $NaNbO_3$ Nano-Crystals As early as in 1965, Borrelli et al., as pioneers of this field, prepared transparent glass ceramics composed of $NaNbO_3$ crystals in a high silica glassy matrix, exhibiting a marked electro-optic effect. The dielectric constant of this material was nearly independent of temperature. Hence, it was expected that electro-optical devices made from such materials would be insensitive to temperature changes.

Transparent, nano-structured glass ceramics containing $NaNbO_3$ crystals were also obtained by the controlled crystallization of $NaNbO_3-SiO_2-Al_2O_3$ glasses (Radonjiċ et al., 2004). The authors indicated that in order to obtain clear, transparent parent glasses in the investigated system, a very high cooling rate from the liquid to the glassy state is necessary to prevent crystallization, and these glasses could be prepared only in a very narrow composition range.

It was proved that the glass compositions with an atomic ratio of $B/N > 1$ resulted in opaque, crystallized samples, indicating that Layton's criterion $B/N > 1$ for glass transparency, did not hold in the investigated system. In order to get transparent, homogeneous glass samples, it was necessary to increase the SiO_2 content at the expense of $NaNbO_3$ content, resulting in samples with the ratio $B/N < 1$. Also, alumina was added to the glass composition in order to decrease the liquidus temperature and the temperature of phase separation.

According to the authors, in order to prepare high transparency sodium niobate nano-structured glass ceramics, a high nucleation rate of the niobate crystals should be realized at relatively low temperatures (e.g., 700°C). Therefore, the temperature of phase separation of the glass samples, as the precursor step of the nucleation, should be decreased.

The addition of alumina to the parent silica-based glass reduced the temperature of phase separation to ~660°C. The boundaries of phase-separated regions in glass served as heterogeneous nucleation sites for the sodium niobate, leading to faster nucleation rates, hence finer crystallites.

For example, for a typical glass sample (sample 1, Table 3.2), it was observed that after heat treatment at 660°C, phase separation appeared on a scale 10–15 nm and at 700°C, the specimen exhibited a homogenous crystalline microstructure on a rather fine scale (~40–60 nm).

Radonjic et al. (2005) also prepared transparent, nano-structured glass ceramics based on ferroelectric solid solutions of the type $Na_{1-x}Li_xNbO_3$ (for $x = 0.12$ and 0.93) and $LiNb_{1-y}Ta_yO_3$ ($y = 0.5$) by controlled crystallization of $Li_2O-Na_2O-Nb_2O_5-(Ta_2O_5)-Al_2O_3-SiO_2$ glasses. The parent glass samples were prepared by conventional melt-quenching techniques.

It was proposed that crystallization in all of the investigated glasses was preceded by phase separation. Apparently, the phase-separated regions served as nucleation sites for the subsequent crystallization, as explained in the previous sections. In this way, the crystallized solid solutions $Na_{1-x}Li_xNbO$ and $LiNb_{1-y}Ta_yO_3$ in all the investigated glass samples had crystal size on nanoscale (<100 nm), and the resulting glass ceramics were transparent.

It was shown that the addition of a small amount of lithium to sodium niobate (e.g., $x = 0.12$, sample 2, Table 3.2) caused the dielectric constant to be decreased at room temperature (at 1 kHz) and dielectric loss factor increased with respect to pure sodium niobate. (The dielectric constant and loss factors were 260/0.012 for pure sodium niobate). On the other hand, in the system with $x = 0.93$ (sample 3, Table 3.2) by addition of sodium to lithium niobate, the dielectric constant increased at room temperature (at 1 kHz) and tan δ decreased with respect to pure lithium niobate.

Du et al. (2005) fabricated glass ceramics in the $Na_2O-PbO-Nb_2O_5-SiO_2$ system via roll-quenching of

Table 3.2 Composition and Properties of Some Niobate and Tantalate Nano-Glass Ceramics

Sample No.	Chemical Composition	Major Crystalline Phase/Size	Heat-Treatment Temp. (°C)/Time	Room Temp. Dielectric Constant/Dissipation Factor (at 1 KHz) (*SHG* Intensity, Relative to α-Quartz	Ref.
1	$12.3Na_2O$, $52.7Nb_2O_5$, $30SiO_2$, $5Al_2O_3$ (mol%)	$NaNbO_3$/32 nm	700/10 min	220–230 (at 1 KHz)	Radonjic et al. (2004)
2	$3Li_2O$, $18Na_2O$, $21Nb_2O_5$, $53SiO_2$, $5Al_2O_3$ (mol%)	$Na_{088}Li_{0.12}NbO_3$/ 60–100 nm	700/15 min	220/0.028(at 1 KHz)	Radonjic et al. (2005)
3	$27Li_2O$, $2Na_2O$, $29Nb_2O_5 \cdot 37SiO_2$, $5Al_2O_3$ (mol%)	$Na_{0.07}Li_{0.93}$ NbO_3/ 60–100 nm	660/15 min	220/0.028 (at 1 KHz)	Radonjic et al. (2005)
4	$22.6Na_2O$–$11.3PbO$–$45.1Nb_2O_5$–$21SiO_2$ (cation mol%)	$NaNbO_3$,$PbNb_2O_6$ and $NaNbO_3$–$PbNb_2O_6$ss/ 10–40 nm	850/3 h	~650/0.02	Du et al. (2005)
5	$29.2Li_2O$, $28.2Nb_2O_5$, $37.2SiO_2$, $5.4Al_2O_3$ (mol%)	$LiNbO_3$/150–350 nm	650/10 min	~155–185 (at 1 KHz)	Todorovic, Radonjic (1997)
6	$(100 - x)$ $LiNbO_3$, $xSiO_2$ $(20 \leq x \leq 35)$ mol%	$LiNbO_3$/42 nm $(x = 30)$	700/1 h	~140/0.063 (at 100 kHz)	Prapitpong-wanich et al. (2009)
7	$25.0K_2O$, $25.0Nb_2O_5$, $50.0SiO_2$ (mol%)	$KNbO_3$	900/4 h	(3.0, SHG intensity)	Pernice et al. (1999)
8	$40.0K_2O$, $20.0Nb_2O_5$, $40.0SiO_2$ (mol%)	$KNbO_3$	820/3 h	(54, SHG intensity)	Tanaka et al. (2003)
9	$16.6K_2O$, $16.6Nb_2O_5$, $66.7SiO_2$ (mol%)	$KNbSi_2O_7$	1100/3 h	(300, SHG intensity)	Tanaka et al. (2003)
10	$xSrO$, $(32 - x)BaO$, $32Nb_2O_5$, $36B_2O_3$ $(x = 16, 19.5, 24)$ mol%	$Sr_{0.61}Ba_{0.39}Nb_2O_6$/ 20–40 nm $(X = 19.5)$	658/10 h	129	Yamazawa et al. (2009)
11	$25.53Li_2O$, $21.53Ta_2O_5$, $35.29SiO_2$, $17.65Al_2O_3$ mol% ($0.5Nd_2O_3$ wt%)	$LiTaO_3$/16 nm	650/2 h + 680/5 h	~30.5	Tarafder et al. (2010b)

$22.6Na_2O-11.3PbO-45.1Nb_2O_5-21SiO_2$ (cation mol%) glasses, followed by controlled crystallization at 750–900°C for 10–180 min.

The authors aimed to prepare some glass ceramics with two ferroelectric phases, $NaNbO_3$ and $PbNb_2O_6$, since their solid solution anticipated to be a promising dielectric material for the capacitor applications. X-RD studies indicated that $Pb_2Nb_2O_7$ crystallized at 750°C and disappeared at 850°C, $NaNbO_3$ was the primary phase at 850°C, along with the secondary $PbNb_2O_6$ phase, the amount of latter phase increased at 900°C. The dielectric constant continuously increased with the crystallization temperature from 650°C to 800°C. A marked increase occurred in the 800–850°C range, that was attributed to the high crystallization rate of high-ε_r $NaNbO_3$. The highest dielectric constant >600 observed in the sample treated at 850°C for 3 h (sample 4, Table 3.2). Figure 3.13A and B depicts the bright-field TEM micrographs of the samples after heat treatments at 850°C for 10 and 180 min. Figure 3.13A shows the onset of crystallization of nanometer-sized crystallites within the phase-separated glass phase. With longer treatment, the particles tend to grow and coalesce to ~10–40 nm in size (Figure 3.13B). The selected-area diffraction pattern inserted in the image of Figure 3.13B shows thin diffraction rings, indicating the polycrystalline nature of the particles. In HRTEM image of the glass-ceramic sample treated at 850°C for 180 min (Figure 3.13C), the residual glass is seen located between adjacent ceramic particles and at triple junctions, which is important for high dielectric constant. Results of microchemistry of

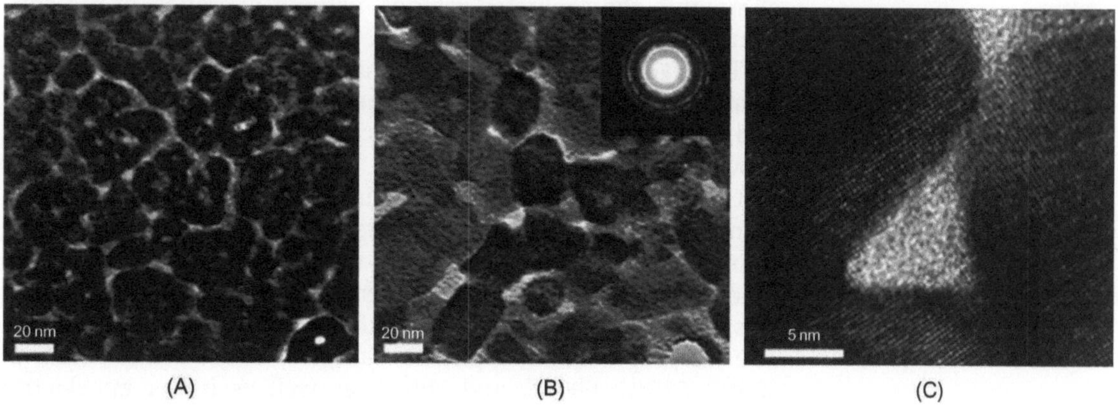

Figure 3.13 (A) and (B) Bright-field TEM of the glass ceramics treated at 850°C for 10 and 180 min. (C) HRTEM image obtained from the glass ceramic heat-treated at 850°C for 180 min, showing the primary location of the residual glass.

the glass-ceramic samples indicated the presence of $NaNbO_3$ and $PbNb_2O_6$. crystals along with their solid solution in the microstructure of samples.

Niobate Glass Ceramics Containing $LiNbO_3$ Nano-Crystals Todorovic and Radonjic (1997) investigated the crystallization of glasses in the system $LiNbO_3$–SiO_2–Al_2O_3. It was shown that the heat treatment of glass samples at 610°C led to the phase separation on the scale of 20–50 nm, followed by the appearance of $LiNbO_3$ nuclei at 620°C. At 650°C, $LiNbO_3$ crystallized intensively on a larger scale (150–350 nm). By heat treating the glass samples at higher temperatures, e.g., 700–900°C for 10 min, or at the lower crystallization temperature of 650°C for the longer periods of 20–60 min, opaque glass ceramics with $LiNbO_3$ crystals ~800–1,000 nm in size, were obtained. The authors explained the mechanism of nucleation in the aforementioned glass system using B/N criterion as suggested by Layton and Herzog. Here B is the content of Nb (ions of the ferroelectric oxide Nb_2O_5) and N is the total content of Si and Al (cations of glass-forming SiO_2 and Al_2O_3 oxides). Layton and Herzog (1967) have shown that the transparent glass ceramics can be obtained in the above system only in the cases where $B/N > 1$. This criterion has been related to the occurrence of phase separation in these glasses which is essential for achieving high nucleation rate without substantial grain growth in the glass specimens. It was suggested that the nuclei possibly can be originated from micro-heterogeneous regions rich in Nb_2O_5. The possibility of occurrence and the extent of phase separation would obviously be increased with increasing B/N ratio. Further grain growth can then take place by the coalescence of crystalline particles within the separated regions, without alteration of the concentration of the respective phases.

The values of dielectric constant and refractive index, at a frequency of 1 KHz at room temperature, were also determined and reported for various samples. One of the results has been provided in Table 3.2 (sample 5).

Graça et al. (2008) by omitting Al_2O_3 from the common composition of glasses prepared $34SiO_2$–$33Li_2O$–$33Nb_2O_5$ (mol%) glasses and investigated their crystallization behavior. In order to suppress the early crystallization of the prepared glasses, they were made by the melt-quenching method. The as-prepared, transparent samples were heat-treated at 550°C, 575°C, 600°C, and 700°C, as well as under an electric field at 575°C (thermoelectric treatment).

$LiNbO_3$ crystal phase was detected in the samples treated without electric field applied, at temperatures above 575°C.

The dc and ac conductivity and the dielectric constant decreased with the rise of the treatment temperature.

The presence of an electric field, during the heat-treatment process, promoted the growing of $LiNbO_3$ particles in a preferential crystal orientation.

Prapitpongwanich et al. (2009) prepared transparent glasses of the composition $(100-x)LiNbO_3-xSiO_2$ (mol%) where $20 \leq x \leq 35$, by conventional melt-quenching technique.

Heat treatment of the quenched glass samples at temperatures ranging from 580°C to 975°C resulted in the precipitation of lithium niobate nano-crystals. SEM micrographs showed that the glass samples heat-treated at 600°C, 700°C, and 975°C developed randomly oriented crystallites of $LiNbO_3$ with the average size of 30, 50, and 200–300 nm, respectively.

The ε_r values were generally in the range of 80–180 and increased with increasing $LiNbO_3$ concentration.

The variation of the relative dielectric constant (ε_r) as a function of frequency (f) for the samples with $x=30$ for various temperatures revealed that the ε_r and $\tan\delta$ ($\delta=$ loss angle) values of the aforementioned sample at 100 kHz were ~140 and 0.063, respectively, after heat treatment at 700°C (sample 6, Table 3.2). The quenched sample and samples heat-treated at 580°C and 600°C were transparent, whereas the samples heated at 700°C and 975°C were opaque due to the presence of larger crystals. It is interesting to note that the effect of temperature on the mean size of the $LiNbO_3$ crystals in the sample with 20 mol% SiO_2 was surprisingly small, namely 15 and 42 nm after heating at 600°C and 975°C, respectively. This opened up the possibility for preparation of transparent glass ceramics even after heating at 975°C.

During crystallization of $LiNbO_3$, the residual glassy phase was enriched in SiO_2, hence its viscosity increased during the course of the crystallization. In analogy to previous reports on the crystallization of fluoride crystals from silicate glasses, this led to a diffusion barrier around the crystals, which hindered crystal growth as well as Ostwald ripening.

Niobate Glass Ceramics Containing $KNbO_3$ Nano-Crystals The crystallization behavior of potassium niobium silicate (KNS) glasses was also investigated. For example, Pernice et al. (1999) studied the glasses of general formula $xK_2O \cdot xNb_2O_5 \cdot (1-2x)SiO_2$, with $x=0.167$, 0.182, 0.200, 0.220, and 0.250.

According to the authors, the KNS system is more profitable than the sodium niobium silicate system because the SHG signal of the $KNbO_3$ perovskite exceeds by about 700 times the one for the $NaNbO_3$ crystals, hence transparent glass ceramics with higher optical nonlinearity could be prepared in the KNS system. Moreover, the possibility of precipitation of $KNbSi_2O_7$ phase with favorable ferroelectric, ion-conductive, or NLO properties exists in this system.

It was shown that by heat treatment of glasses with the lowest silica content, at temperatures just above T_g, transparent glass ceramics based on potassium niobates, with remarkable SHG signal values, could be obtained, whereas, at higher temperatures, the $KNbSi_2O_7$ ferroelectric phase was crystallized, as the main phase, from all of the glasses.

Grain-oriented crystallization was induced in the above glasses resulting in the development of significantly anisotropic $KNbSi_2O_7$ crystallites in the presence of a dc electric field.

The SHG properties of the two silica rich and low silica glasses were also determined. At temperatures up to ~950–1,000°C, slow growth of SHG signals with the increase of heat-treatment temperature was observed that were connected predominantly to the bulk precipitation of potassium niobate in low silica glass and the beginning stage of surface crystallization of $KNbSi_2O_7$ in high silica glasses. The range of SHG intensity values were 1–3 and >5 (relative to α-quartz crystals), respectively, for the low silica glasses and for both glass compositions after the crystallization of the $KNbSi_2O_7$ phase at higher temperatures. The latter glass ceramics, in the case of intensive crystallization of $KNbSi_2O_7$, were nontransparent. The composition and properties of a transparent glass ceramic containing potassium niobate crystallites have been given in Table 3.2 (sample 7).

Aronne et al. (2004) also studied the crystallization behavior of KNS glasses the composition of which was characterized by the K_2O/Nb_2O_5 molar ratio of 0.85–1.2 and SiO_2 content of 50–54 mol%.

It was found that all glasses crystallized in three stages as determined by DTA exothermic effects. At the first two DTA exothermic peak temperatures, some unidentified phases were formed, while at higher temperatures (1,000–1,100°C), the crystallization of $K_3Nb_3O_6Si_2O_7$ ($K_3Nb_3Si_2O_{13}$) took place.

The details of crystallization behavior of the sample containing 50 mol% SiO_2 and 27 mol% Nb_2O_5 show that the as-quenched glass samples were phase-separated, droplets of ~50–60 nm, were observed uniformly dispersed in a matrix. After heating the glass sample for 2 h at the temperature of $T_g = 680°C$, only some changes in the amorphous phase-separated regions, regarding their extension and microstructural details, took place. After 10 h, some ordering in glass structure was observed that became well evident after 24 h indicating the precipitation in the amorphous matrix of crystallites extremely small in size <5nm. It was suggested that the crystallization started at the interface between the droplets and the matrix and that the growth of the crystallites took place inside these droplets. This phenomenon caused a change of the matrix composition that prevented the further growth of crystallites, regardless of the heat-treatment time. This resulted in the formation of a stable transparent biphasic structure. It was shown that only for the glasses in which $K_2O/Nb_2O_5 = 0.85$ and $SiO_2 = 50$ mol%, was it possible to control the above processes by isothermal heat treatments at 680°C in order to obtain transparent nano-structured samples.

The crystalline phases and SHG intensities in the crystallized glasses of $K_2O-Nb_2O_5-SiO_2$ system were also studied (Tanaka et al., 2003). It was shown that the crystallization behaviors depended strongly on SiO_2 contents, and three stable crystalline phases, $KNbSi_2O_7$, $K_3Nb_3Si_2O_{13}$, and $KNbO_3$, were formed in fully crystallized glasses.

$KNbO_3$ crystal is one of the most excellent NLO materials and shows an extremely large Second Harmonic (SH) intensity. Therefore, it is expected that the SH intensity of crystallized glasses increases steeply with increasing the amount of $KNbO_3$ crystals. The crystallized glasses which consisted of only perovskite-type, $KNbO_3$ phase, were also fabricated. The authors indicated that the precipitation of ferroelectric $KNbO_3$ phase was realized through crystallization of $K_2O-Nb_2O_5-SiO_2$ glasses with a large amount of K_2O (40–45 mol%) and Nb_2O_5 (20–25 mol%) or a small amount of SiO_2 (30–40 mol%). It was also confirmed that the crystallized glasses with a perovskite-type $KNbO_3$ phase exhibited a clear SHG with an intensity of 20–54 (relative to α-quartz powder).

Figure 3.14 depicts the SH intensities (relative to α-quartz powder) of fully crystallized glasses containing $KNbSi_2O_7$, $K_3Nb_3Si_2O_{13}$, or $KNbO_3$ crystalline phases in $K_2O-Nb_2O_5-nSiO_2$ glass series and Table 3.2 provides the composition and properties of one of these glass ceramics (sample 8).

The crystallized glasses containing $KNbSi_2O_7$ phase also showed large SH intensities (sample 9, Table 3.2). The formation of $K_3Nb_3Si_2O_{13}$ phase was also reported, and it was demonstrated that this phase exhibited a second-order optical nonlinearity, although its SH intensity is small in comparison with $KNbSi_2O_7$ or $KNbO_3$ phases.

More recently, Chaliha et al. (2011) studied the crystallization behavior and dielectric properties of potassium niobate

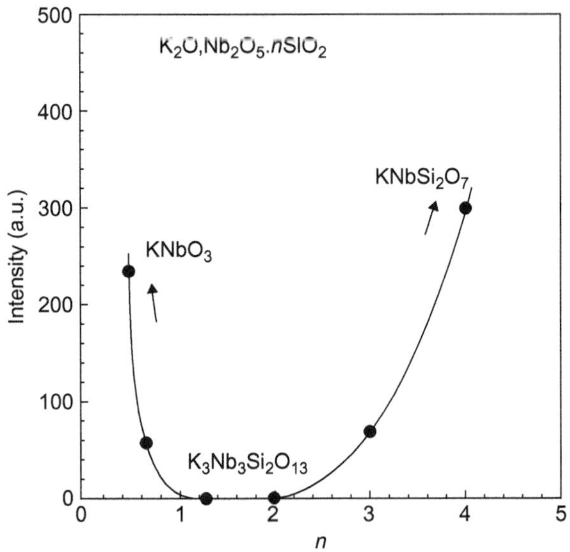

Figure 3.14 SH intensities (relative to α-quartz powder) of fully crystallized glasses containing $KNbSi_2O_7$, $K_3Nb_3Si_2O_{13}$, or $KNbO_3$ crystalline phases in $K_2O-Nb_2O_5-nSiO_2$ glass series. (After Tanaka et al., 2003. Reproduced with permission of Elsevier Science B.V.)

nano-glass ceramics by the crystallization of $25K_2O-25Nb_2O_5-50SiO_2$ (mol%) melt-quenched glasses, and isothermally heat treating them at 800°C for 0–200 h.

The examination of Flourier Transform Infrared (FTIR) reflectance spectra of the as-prepared samples revealed the existence of a broad reflection band as a result of wide distribution of the structural units, which was attributed to the existence of (SiO_4) tetrahedra and (NbO_6) octahedra in the amorphous network.

On the basis of the information obtained by systematic monitoring of the variation of FTIR spectra during the heat-treatment process, it was deduced that the crystallization of $KNbO_3$ crystals took place through an initial phase separation followed by advancement of crystal growth in the glass matrix.

The TEM and HRTEM images of the sample heat-treated for 1 and 200 h (Figure 3.15A–D) clearly shows the crystallites and their atomic or lattice fringes. The distance between any two planes was found to be 2.46 and 7.85 Å for samples heat-treated for 1 and 200 h, respectively, which was consistent with the XRD data for potassium niobate crystals.

The size of the crystallites estimated from the XRD patterns and TEM (6–12 nm) were less than the droplets observed in

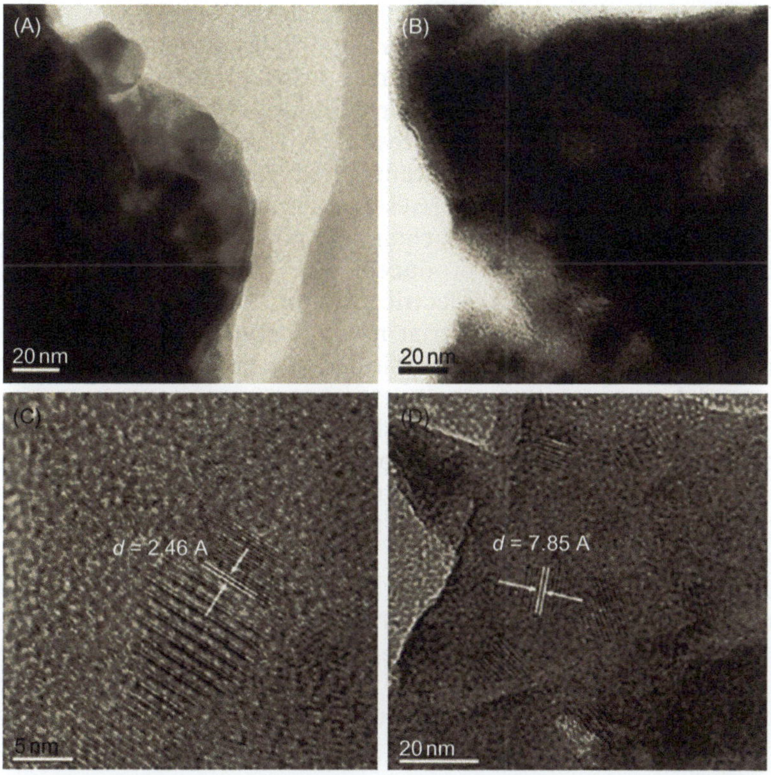

Figure 3.15 TEM images of heat-treated glasses at 800°C for (A) 1 h and (B) 200 h. HRTEM images of lattice fringe of heat-treated glasses at 800°C for (C) 1 h and (D) 200 h. (After Chaliha et al., 2011. Reproduced with permission of Springer Science + Business Media, Inc.)

FESEM (20–40 nm), confirming the initiation of crystallization from the droplets/matrix interface and the subsequent growth of the formed crystallites inside these droplets. This led to a change of matrix composition, which restricted the further growth of the crystallites, regardless of the increase in the heat-treatment time, and a stable transparent biphasic structure was obtained.

The room temperature dielectric constant and loss factor were found to be independent of the heat-treatment time and was determined for all the samples as ~34 and 0.05, respectively.

3.2.2.3 Niobate Glass Ceramics with Borate-Based Glasses

Prasad and Varma (2005) discussed the results of the glass formation and evolution of the nano-crystalline $LiNbO_3$ phase in the glasses of the system $(100-x)LiBO_2$–xNb_2O_5($5 \leq x \leq 25$ mol%).

The authors tried to crystallize the desired phase by *in situ* chemical reaction. For this purpose, they chose $LiBO_2$ as a matrix, richer in Li (in comparison with the other borate glasses known in the literature). According to the authors, it was essential to overcome the formation of impurity phase, $LiNb_3O_8$. They were successful in preparing $LiBO_2$–Nb_2O_5 glasses in a binary system with different compositions, from which $LiNbO_3$ crystallites were grown as a result of the reaction between Li and Nb_2O_5. This resulted in transparent glass ceramics.

By heat treating the glasses containing 15 mol% Nb_2O_5 at 500°C/3 h ($x = 15$), glass ceramics containing 100 nm size $LiNbO_3$ crystallites were produced.

The observed pyroelectric response and ferroelectric (*P* versus *E*) hysteresis loop at room temperature of the aforementioned glass ceramics confirmed the polar nature of these materials.

The heat-treated samples exhibited moderately good ferroelectric behavior, as evidenced by the hysteresis loops of polarization versus electric field. Figure 3.16A and B shows the loops recorded at 25°C and 150°C for the samples of the composition $x = 15$ heat-treated at 500°C. Very interestingly, the values obtained for E_c (coercive field) were very much less than that reported for single crystalline $LiNbO_3$ at 200°C.

Malakho et al. (2006) also reported the preparation of glass ceramics by crystallization of $NaNbO_3$ in sodium niobium borate glasses. A homogeneous bulk crystallization of the

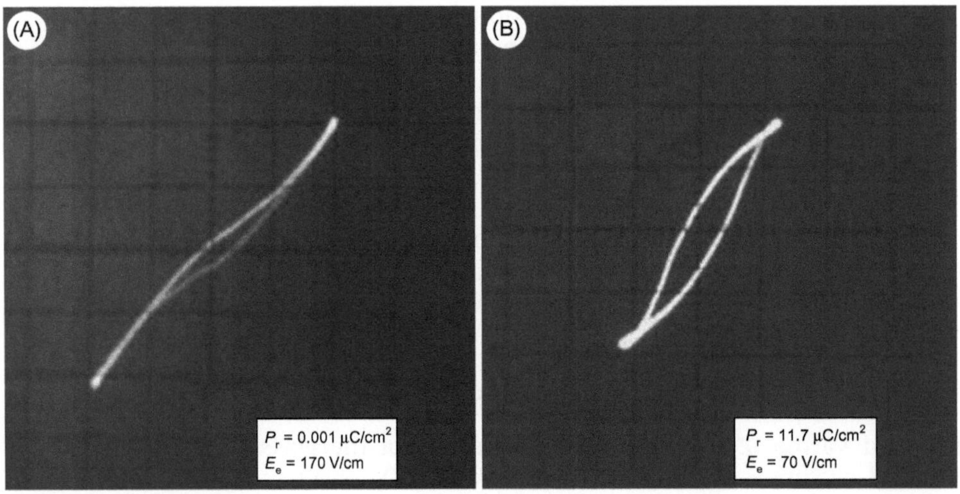

Figure 3.16 *P–E* hysteresis loops recorded on the glass nanocomposite (heat-treated at 500°C/3 h) of the composition $x = 15$ at two different temperatures (A) 25°C and (B) 150°C. (After Prasad et al., 2005. Reproduced with permission of Elsevier Science B.V.)

$NaNbO_3$ phase was observed during heat treatments that produced visible–near infrared (NIR) transparent materials, containing ~30 nm $NaNbO_3$ nano-crystallites. Upon thermal poling, a strong Na^+-depleted NLO thin layer was observed at the anode side that should induce a large internal static electric field. In thermal poling, a dc electric field is applied to the glass sample at an elevated temperature, and then the sample is cooled before removing the electric field.

As a result, a nonlinear layer with a thickness of 1–10 μm is created at the anode surface of the glass due to a charge migration. A model of built-in internal electric field created in a thin anode layer is proposed to explain the origin of NLO responses in thermally poled glasses.

The authors reported an increase of the second-order nonlinear optical susceptibility $\chi^{(2)}$ response of the poled glass-ceramic samples from 0.2 up to 1.9 pm/V with the rate of crystallization. The origin of this behavior was also explained.

Yamazawa et al. (2009) synthesized glass ceramics consisting of $Sr_xBa_{1-x}Nb_2O_6$ (SBN) nano-crystals ($x = 0.5$, 0.61, and 0.62) with a tetragonal tungsten–bronze structure. The glass samples of general formula xSrO–$(32 - x)$ BaO–$32Nb_2O_5$–$36B_2O_3$ with $x = 16$, 19.5, and 24 were prepared using the conventional melt-quenching method. The prepared glass samples were then heat-treated at various temperatures in the range of 600–660°C for 10 h. The XRD patterns for the bulk glass samples after the heat treatment indicated the formation of $Sr_{0.5}Ba_{0.5}Nb_2O_6$, $Sr_{0.61}Ba_{0.39}Nb_2O_6$, and $Sr_{0.62}Ba_{0.38}Nb_2O_6$ crystals with a tetragonal tungsten–bronze structure in the glasses containing 16%, 19.5%, and 24% SrO, respectively. All crystals had rod-like shapes with an average diameter of 20–40 nm as determined from XRD and TEM examinations. The polarization (P) as a function of electric field (E), P–E curves at room temperature exhibited a clear hysteresis for all the crystallized samples. Hence, it was proposed that the nano-crystals of $Sr_xBa_{1-x}Nb_2O_6$ formed in the prepared glass ceramics had ferroelectric behavior. The dielectric constants, ε_r, of the glass and glass-ceramic samples containing 19.5% SrO were also determined (at the frequency of 100 kHz, at room temperature) and were found to be 66 for the sample crystallized at 608°C, and 129 for the sample crystallized at 658°C (sample 10, Table 3.2).

Electro-optic activity was also observed for the aforementioned crystallized glasses containing SBN nano-crystals, indicating their possible suitability for the active optical device applications.

3.2.3 Other Ferroelectric Nano-Glass Ceramics

3.2.3.1 Tantalate Nano-Glass ceramics

Lithium tantalate ($LiTaO_3$, LT) single crystal is one of the most important lead-free ferroelectric materials in the $A^{1+}B^{5+}O_3$-type perovskite family, which, owing to its large nonlinear constant and SHG coefficient, exhibits unique piezoelectric, acousto-optic, electro-optic, and NLO properties. These characteristics, combined with its good mechanical and chemical stability, have made this material very attractive for both scientific and industrial circles. Hence, many researchers explored the processing parameters and properties of LT single crystals, powders, thin films, glass ceramics, and so on in recent years.

Rare-earth-doped transparent $LiTaO_3$ nano-glass ceramics also offer excellent luminescent properties, because $LiTaO_3$ nano-crystallites provide a low phonon energy environment for luminescent ions, as well as being due to the good mechanical and chemical properties of the oxide glassy matrix. This ability, combined with the inherent NLO properties of ferroelectric crystals, offers a possibility to design self-frequency-doubling laser sources (Tarafder et al., 2010b).

3.2.3.1.1 Processing and Properties of Nano-Structured $LaTaO_3$

As early as 1978, Ito et al. studied the stability range and the crystallization behavior of glasses with various compositions in the $LiTaO_3$–SiO_2–Al_2O_3 system. The glasses were melted in platinum crucibles at 1,550°C for 3 h and the melts were poured onto a steel plate and pressed into plates ~1 mm thick. The glass plates were heated from room temperature to temperatures ranging from 750°C to 1,050°C at a rate of 5°C/min. From the glasses in the $LiTaO_3$–SiO_2 system, no transparent glass ceramic was obtained even when their $LiTaO_3/SiO_2$ mole ratios were as high as 2.33. The diameters of the $LiTaO_3$ crystal grains precipitated in these glasses were 5–15 μm, whereas the diameters of the $LiTaO_3$ crystal grains precipitated in the glasses containing Al_2O_3 were as small as 10–20 nm. Hence, transparent glass ceramics with a very broad compositional region located in the $LiTaO_3$–SiO_2–Al_2O_3 system were made from the latter glasses. The above compositional region encompassed the compositions for which the $LiTaO_3/SiO_2 + AlO_{1.5}$ molar ratio was as low as 0.25.

High nucleation rates of the $LiTaO_3$ crystals in the Al_2O_3-containing glasses were explained in terms of structural inflexibility induced in the glass network by the addition of Al_2O_3 to the $LiTaO_3$–SiO_2 system.

Hase et al. (1996) observed SHG in *c*-axis orientated surface crystallized 35Li_2O, 30Ta_2O_5, 35SiO_2 glasses, poled at room temperature for 24 h under a 0.7 kV dc voltage that was attributed to the residual polarization in $LiTaO_3$ crystals.

The value of d_{33} obtained in this experimentation was much lower than the $LiTaO_3$ single crystals. The authors interpreted this by stating two possibilities: the achievement of no perfect *c*-axis orientation and/or the initiation of small residual polarization. To improve this, a greater degree *c*-axis orientation and poling at higher voltage and temperature were suggested.

Recently, Tarafder et al. (2010b) reported the processing and the structural, dielectric, and optical properties of neodymium ion (Nd^{3+})-doped transparent glass ceramics containing $LiTaO_3$ nano-crystals in the $Li_2O-Ta_2O_5-SiO_2-Al_2O_3$ system in the presence of Nd_2O_3. Nd^{3+} owing to its highly efficient emission in NIR region or up-conversion/frequency doubling in the green–red wavelength has found many applications as a laser active ion in various hosts. Most of the Nd^{3+} lasers operate in the NIR region on two lasing transitions $^4F_{3/2} \rightarrow {}^4I_{11/2,\ 13/2}$ at around 1,060 and 1,349 nm wavelengths, respectively. However, its $^4F_{3/2} \rightarrow {}^4I_{9/2}$ laser transition near 900 nm has also attracted much attention for its applications especially in display technologies, as an efficient way for generation of continuous blue laser radiation utilizing an all-solid-state frequency doubling laser.

The glasses with a molar composition of 25.53Li_2O–21.53Ta_2O_5–35.29SiO_2–17.65Al_2O_3 doped with Nd_2O_3 (0.5 wt% in excess) were melted at 1,600°C for 2 h in air followed by quenching and annealing at 600°C for 4 h. The precursor glass samples were then heat-treated at 650°C, for 2 h for nucleation, followed by a second treatment at 680°C (below the glass transition temperature, $T_g = 702$°C) for 0, 3, 5, 10, 20, 50, and 100 h for crystallization.

The FESEM images showed that with progression of heat-treatment, nano-crystals of $LiTaO_3$ in the size range of 14–36 nm precipitated within the nano-size Li–Ta-rich phase regions in the initially phase-separated glass matrix.

The measured NIR photoluminescence spectra have exhibited emission transitions of $^4F_{3/2} \rightarrow {}^4I_j$ ($j = 9/2$, 11/2, and 13/2) from Nd^{3+} ions upon excitation at 809 nm.

It was also found that the dielectric constant of Nd^{3+}-doped glass ceramic exhibited a gradual increase with the heat-treatment time up to 5 h, after which it attained a saturated value of ~30.5 with a small decrease for any further heat-treatment time (sample 11, Table 3.2). The increase in the dielectric constant (ε_r) value upon the heat-treatment time was mainly caused by the increase of volume fraction of $LiTaO_3$ crystal phase, possessing high dielectric constant and spontaneous polarization values.

More recently, Tarafder et al. (2013b) also reported the preparation of Dy^{3+}-doped nano-glass ceramics in the $K_2O-SiO_2-SrO-Bi_2O_3-Ta_2O_5$ system. The glasses with a molar composition of $17.5K_2O-51.3SiO_2-10SrO-10Bi_2O_3-10Ta_2O_5$ doped with Dy_2O_3 (0.5 mol%) and CeO_2 (0.7 mol%) were melted at 1,470°C for 2 h in air with intermittent stirring, followed by quenching and annealing at 530°C for 2 h.

The precursor glass samples were then heat-treated using a two-step schedule, 550°C, 2 h, for nucleation and 580°C for 0, 20, 30, 50, and 100 h for crystallization. It is interesting to note that the chosen crystallization temperature was below the glass transition temperature ($T_g = 622$°C). The formation of $SrBi_2Ta_2O_9$ (SBT) crystal phase was observed after heat treating the glass samples for 50–100 h at the crystallization temperature of 580°C.

The FESEM images showed that with progression of heat treatment, the hexagonal nano-rod-like crystals of SBT gradually precipitated within the initially phase-separated glass (Figure 3.17). The micrograph depicted that the glass sample, which was XRD amorphous, was phase-separated after heat treatment at 580°C for 30 h. The measured Vicker's hardness values were found to be increased with progression of heat treatment due to the interlocking of hexagonal SBT nano-rods in the glassy matrix.

The Dy^{3+}-doped glass ceramics with progression of heat treatment exhibited three emission bands at 482 nm (blue), 576 nm (yellow), and 666 nm (red), upon excitation at 455 nm,

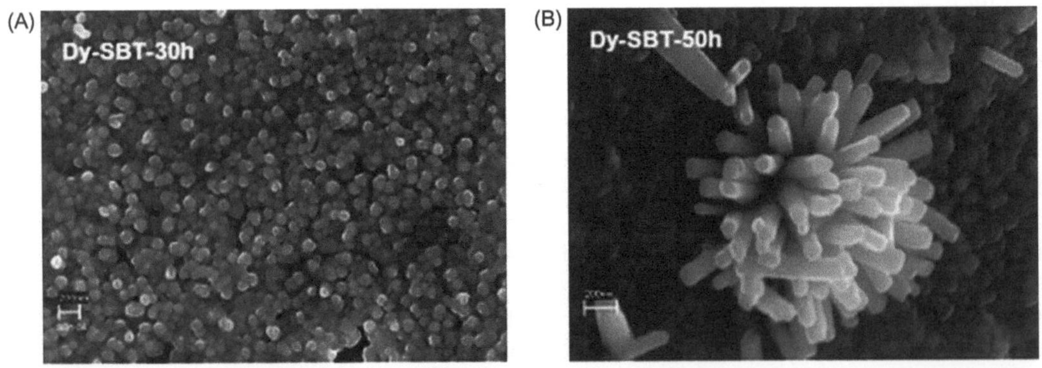

Figure 3.17 FESEM images of glass-ceramic samples (A) Dy-SBT-30h and (B) Dy-SBT-50h. (After Tarafder et al., 2013b. Reproduced with permission of Elsevier Science B.V.)

that were attributed to the $^4F_{9/2} \rightarrow {}^6H_j$ ($j = 11/2, 13/2, 15/2$) transitions of Dy^{3+}. Among $^4F_{9/2} \rightarrow {}^6H_j$ transitions, the $^4F_{9/2} \rightarrow {}^6H_{13/2}$ (at ~482 nm) and $^4F_{9/2} \rightarrow {}^6H_{15/2}$ (at ~576 nm) emission transitions are relatively more intense than the other transition ($^4F_{9/2} \rightarrow {}^6H_{11/2}$) observed at ~666 nm.

The emission spectra of glass ceramics demonstrated that the Dy^{3+} ions entered into the SBT crystalline phase and located in the place of Bi^{3+} sites. The enhanced photoluminescence was observed in the heat-treated glass ceramics due to the low phonon environment of Dy^{3+} ions in the SBT crystals. The enhanced emission and hardness of these glass ceramics would make this material suitable for potential applications such as solid-state yellow lasers and optical display systems.

3.3 Concluding Remarks

The ferroelectric nano-glass ceramics have attracted considerable attention in the recent years, owing to the advantages of the glass ceramic route over the other fabrication methods, e.g., the lower cost, the high speed of fabrication process, and the possibility and ease of fabricating articles of complex shapes and close tolerance in dimensions. These materials also exhibit some interesting peculiarities as their properties can readily be changed by the variation of the volume fraction of the nanoparticles dispersed in the glass matrix as well as by changing the details of their nano-structure.

Among the above-mentioned glass ceramics, those containing lithium/sodium/potassium niobate crystals, because of their outstanding properties such as ferroelectricity, electro-optic, and piezoelectric effects, as well as possible NLO behavior, and the possibility of realization of high Curie temperature and large second-order NLO response in them, are very promising candidates for various applications. These applications include electro-optical, acousto-optical, piezoelectric, and NLO devices.

In recent years, we have witnessed a particular resurgence of interest in the transparent ferroelectric glass ceramics with superior NLO and electro-optical properties. Numerous research projects have been conducted with the aim of selectively creating NLO structures in these glass ceramics, which will be useful in the development of the integrated optical devices. Although the crystals dispersed in a glass ceramic are usually randomly oriented, the glass ceramics subjected to the special synthesis procedures could exhibit aligned microstructures. These materials, possessing large second-order NLO response, are promising candidates for laser hosts, tunable waveguides, tunable fiber gratings, and so on.

The successful candidates for integrated optical devices should combine the ability of generating strong second-order NLO response with the appropriate ferroelectric behavior.

Hence, it is expected that the great potential of the aforementioned nano-glass ceramics for various applications in the field of multifunctional electro-optical devices, which are anticipated to create a major breakthrough in the field, would be further realized in coming years. This would stimulate more interest for comprehensive research and development activities in this field.

4

MAGNETIC PROPERTIES OF NANO-GLASS CERAMICS

CHAPTER OUTLINE

In recent years, research into, and the manufacturing of, **magnetic nanoparticles** has been introduced into many different applications such as magnetic seals in motors, magnetic inks for bank cheques, magnetic recording media, and biomedical applications such as magnetic resonance contrast media and therapeutic agents in cancer treatment (Tartaj et al., 2003).

V.K. Marghussian, Nano-Glass Ceramics, ISBN: 978-0-323-35386-1.

DOI: http://dx.doi.org/10.1016/B978-0-323-35386-1.00004-9

Magnetic nanoparticles show remarkable new phenomena such as superparamagnetism, high-field irreversibility, high saturation field, extra anisotropy contributions, or shifted loops after field cooling. These phenomena arise from finite size and surface effects that dominate the magnetic behavior of individual nanoparticles (Batlle and Labarta, 2002).

Each potential application requires the magnetic nanoparticles to have different properties. For example, in data storage applications, the particles need to have a stable, switchable magnetic state to represent bits of information, a state that is not affected by temperature fluctuations. For biomedical applications, the use of particles that present superparamagnetic behavior at room temperature (no remanence along with a rapidly changing magnetic state) is preferred (Tartaj et al., 2003).

The synthesis of magnetic nanoparticles using different techniques such as micelles, co-precipitation, and dispersion in matrices such as silica, alumina, and polymers, has attracted considerable research interest in recent years (Zysler et al., 2001, 2004; Tadic et al., 2007). These systems proved to be convenient for providing controllable particle size and morphology, as well as for obtaining narrow size distribution of nanoparticles. Out of these, synthesis of nanostructured materials via the glass ceramic route has the advantage over the others as their preparation is much easier and economical, and fine-grained microstructures could be achieved by controlling the crystal nucleation and growth processes.

In this chapter, the fabrication, properties, and application of some of the most important types of magnetic nano-glass ceramics are discussed.

4.1 Theoretical Background and Definitions

Magnetic Dipoles

The magnetic properties of solids originate in the motion of the electrons and in the permanent magnetic moments of the atoms and electrons.

By analogy with electric dipoles (see Section 3.1), the existence of magnetic dipoles may be imagined, composed of north and south poles instead of positive and negative electric charges.

Magnetic dipoles are affected by magnetic fields, in a similar manner as electric dipoles are affected by electric fields, i.e., the magnetic field exerts a torque that tends to orient the dipoles with the field.

Magnetic Field Vectors

The strength of externally applied magnetic field, known as **magnetic field strength**, is usually designated by H. If the magnetic field is generated by means of a solenoid, consisting of N closely spaced turns, then

$$H = \frac{NI}{l} \tag{4.1}$$

where l and I are the length of solenoid and the magnitude of current passing through it, respectively.

The magnetic induction, or **magnetic flux density**, B, represents the magnitude of the internal field strength of a substance that is subjected to an H field. Both B and H are vectors, which are characterized by the magnitude, as well as the direction in space.

The magnetic field strength and flux density are related according to the equations below:

$$B_o = \mu_o H \quad \text{(in vacuum)} \tag{4.2}$$

$$B = \mu H \quad \text{(in a solid material)} \tag{4.3}$$

where μ_o and μ are called the **permeability values** in vacuum and in a solid material, respectively. They are the measures of the degree to which the material can be magnetized, or the ease with which a B field can be induced in the presence of an external H field, hence they are the property of the specific medium through which the H field passes and in which B is measured.

B_o and B are the flux densities in the vacuum and in the solid material, respectively.

A number of other parameters may be used to describe the magnetic properties of solids, e.g., the ratio of the permeability in a material to the permeability in the vacuum is represented as

$$\mu_r = \frac{\mu}{\mu_o} \tag{4.4}$$

where μ_r is called the relative permeability, which is unitless.

Another field quantity, M, called the **magnetization** of the solid is defined by the expression:

$$B = \mu_o(H + M) \tag{4.5}$$

In the presence of an H field, the magnetic moments within a material tend to become aligned with the field and to reinforce it by the extra magnetic induction field $\mu_o M$, in addition to the free-space induction $\mu_o H$.

The magnitude of M is proportional to the applied field as follows:

$$M = \chi_m H \tag{4.6}$$

The factor of proportionality, χ_m, is called the **magnetic susceptibility**, which is unitless. The magnetic susceptibility and the relative permeability are related as follows:

$$\chi_m = \mu_r - 1 \tag{4.7}$$

The magnetic materials depending on how they interact with magnetic fields can be classified as **diamagnetic, paramagnetic, ferromagnetic, antiferromagnetic, and ferrimagnetic.**

Diamagnetism

Diamagnetism is a very weak form of magnetism that is induced by a change in the orbital motion of electrons due to an applied magnetic field. This magnetism is nonpermanent and persists only in the presence of an external field. The magnitude of the induced magnetic moment is very small, and its direction is opposite to that of the applied field. The relative permeability values of diamagnetic materials are <1, and their magnetic susceptibility is negative. The volume susceptibility χ_m for diamagnetic solid materials is in the order of -10^{-5}. Diamagnetism is found in all materials; but because it is so weak, it can be observed only when other types of magnetism are totally absent; hence it is of no practical importance.

Paramagnetism

In some solid materials, each atom possesses a permanent dipole moment arising from spinning and orbiting electrons. In the absence of an external magnetic field, the orientation of the magnetic moments are random and the material possesses no net macroscopic magnetization. An applied field tends to orient the moments, giving rise to a relative permeability that is greater than unity, and to a relatively small but positive magnetic susceptibility, usually in the range 10^{-3}–10^{-6}.

Both diamagnetic and paramagnetic materials are considered to be nonmagnetic because they exhibit magnetization only in the presence of an external field, and their flux density is almost the same as would be in the vacuum.

Ferromagnetism and Antiferromagnetism

Ferromagnetic materials possess a permanent magnetic moment in the absence of an external field and exhibit very large, permanent magnetizations. The spontaneous magnetization is not apparent in materials which have not been exposed to an external field, because of the presence of domains in the material each having its own direction of magnetization. When a field is applied, the domains in which the magnetization is more nearly parallel to the field grow at the expense of the domains with less favorable alignments. Since the spontaneous magnetization may be several orders of magnitude greater than the applied field, ferromagnetic materials have very high permeabilities, e.g., as high as 10^6.

When the applied field is removed, a part of the induced domain alignment may be preserved so that the body acts as a permanent magnet.

Spontaneous magnetization is due to the alignment of uncompensated electron spins by the strong quantum-mechanical "exchange" force.

In **antiferromagnetic** materials, the uncompensated electron spins associated with neighboring cations orient themselves, below a temperature known as the Néel temperature, in such a way that their magnetizations neutralize one another so that the overall magnetization is zero. Metallic manganese, chromium, and manganese oxide (MnO) are some of the materials exhibiting this behavior.

The susceptibilities of these materials are low ($\sim 10^{-3}$) except when the temperature is close to the Néel point when the antiferromagnetic coupling breaks down and the materials become paramagnetic (Moulson and Herbert, 2003, p. 478).

Ferrimagnetism

Some ceramic materials also exhibit a permanent magnetization, termed ferrimagnetism. In these materials, there is antiferromagnetic coupling between cations occupying crystallographically different sites, and the magnetization of one sublattice is antiparallel to that of another sublattice. Because the two magnetizations are of unequal strength, there is a net spontaneous magnetization. As the temperature is increased from 0 K, the magnetization decreases, reaching zero at the Néel point (Moulson and Herbert, 2003, p. 478).

The principles of ferrimagnetism have been illustrated for the **cubic ferrites** which have the **spinel structure** with the general chemical formula AB_2O_4.

The spinel structure can generally be described as a cubic close-packed arrangement of oxygen anions, with A^{2+} and B^{3+} cations located in two different crystallographic sites. These sites have tetrahedral and octahedral oxygen coordination (termed as A- and B-sites, respectively). The spinel structure contains 8 A-sites and 16 B-sites. When the A-sites are occupied by A^{2+} cations and the B-sites are occupied by B^{3+} cations, it is called a normal spinel. If the A-sites are completely occupied by B^{3+} cations and the B-sites are randomly occupied by A^{2+} and B^{3+} cations, the structure is referred to as an inverse spinel. In most spinels, the cation distribution possesses an intermediate degree of inversion where both sites contain a fraction of the A^{2+} and B^{3+} cations.

In the case of ferrites, the general formula may be shown by MFe_2O_4, in which M is a bivalent cation (M^{2+}) and iron is the trivalent cation (Fe^{3+}).

In the inverse ferrites where one half of the Fe^{3+} cations is placed in A-sites and another half in B-sites, their magnetic moments are mutually compensated and the resulting moment of the ferrite is due to the magnetic moments of bivalent cations M^{2+} in the B-positions. The prototype inverse ferrite is the mineral magnetite, Fe_3O_4.

The formula for Fe_3O_4 may be written as $Fe^{2+}O^{2-}-(Fe^{3+})_2(O^{2-})_3$ in which the Fe ions exist in both +2 and +3 valence states in the ratio of 1:2. A net spin magnetic moment exists for each Fe^{2+} and Fe^{3+} ion. There are antiparallel spin–coupling interactions between the Fe ions, similar in character to antiferromagnetism. However, the net ferromagnetic moment arises from the incomplete cancelation of spin moments.

Ceramic materials with different crystal structures may also be ferrimagnetic, e.g., the hexagonal ferrites and garnets. Hexagonal ferrites have a crystal structure similar to the cubic inverse spinel, but with a hexagonal symmetry. The chemical formula of these materials may be represented by $AB_{12}O_{19}$, in which A is a bivalent cation such as Ba^{2+}, Sr^{+2}, Pb^{+2} and B is a trivalent cation such as Al^{3+}, Cr^{3+}, or Fe^{3+}. The two most common examples of the hexagonal ferrites are $BaFe_{12}O_{19}$ and $PbFe_{12}O_{19}$.

The garnets possess very complicated crystal structure, which may be represented by the general formula $M_3Fe_5O_{12}$, in which M denotes a rare earth ion such as Y, Sm, Eu, and Gd. Yttrium iron garnet ($Y_3Fe_5O_{12}$), denoted by YIG, is the most common material of this type.

Magnetocrystalline Anisotrophy

A spinning electron, free from any restraints, can be aligned by an infinitely small field, implying an infinite permeability. A restraint leading to finite permeabilities in magnetic materials is caused by a coupling between the spins and the crystal lattice through the agency of the orbital motion of the electron. This spin–orbit lattice coupling results in orientation of the spins relative to the crystal lattice in a minimum energy direction, the so-called easy direction of magnetization.

Aligning the spins in any other direction leads to an increase in energy, the anisotropy energy EK (Moulson and Herbert, 2003, p. 481).

Magnetostriction

Because of the spin–orbit lattice coupling, changes in the spin directions result in changes in the orientation of the orbits which, because they are restrained by the lattice, have the effect of slightly altering the lattice dimensions. This effect is known as magnetostriction.

The magnetostriction constant λ_m is defined as the strain induced by a saturating field; it is given a positive sign if the field causes an increase in dimensions in the field direction. For single crystals, λ_m varies with the crystallographic direction, and so for a ceramic body is an average of the single-crystal values (Moulson and Herbert, 2003, p. 482).

The Effect of Temperature on Magnetic Behavior

For ferromagnetic, antiferromagnetic, and ferrimagnetic materials, rising the temperature resulted in the increased atomic thermal motions which may counteract the coupling forces between the adjacent atomic dipole moments, causing some randomization of dipole alignments. This results in a decrease of the saturation magnetization value of ferromagnetic and ferrimagnetic materials. With increasing temperature, the saturation magnetization decreases gradually and at Curie temperature T_c abruptly drops to zero. At this temperature, the mutual spin coupling forces are completely destroyed, hence above T_c these materials become paramagnetic. The same effect is observed in antiferromagnetic materials at temperatures above **Néel temperature**.

Domains and Hysteresis

Any ferromagnetic or ferrimagnetic material at temperatures below T_c is composed of small-volume regions in which there is a mutual alignment in the same direction of all magnetic dipole moments. Such regions are called domains. Adjacent domains are separated by domain boundaries or Bloch walls, across which the direction of magnetization gradually changes. Although each domain is magnetized, the material as a whole will have zero magnetization. To give the material a net magnetization, one direction must predominate in all domains. This may be realized either by the rotation of the individual domains or the growth of domains, initially parallel to the applied field, at the expense of their less favorably oriented neighbors. In the M versus H curve, known as the magnetization curve, initially a very slow increase in B is observed, since little domain growth occurs upon the increase of H; however, when the favorably oriented domains begin to grow, the magnetic induction B increases rapidly. Finally, domain growth stops and the saturation stage begins in which the rotation of the remaining unfavorably aligned domains occurs. Since domain rotation requires higher energy than domain growth, the slope of the B versus H curve decreases. When the field is removed, the specimen remains magnetized. Although the domains tend to rotate back, the large aligned domains do not easily revert to the original random arrangement. If a reverse field ($-H$) is applied, the domain structure may be changed to produce a resultant zero magnetic induction. The magnitude of the applied field required is equal to coercive force H_c. Once magnetized, the state $H=0$, $B=0$ is no longer attainable by simply changing the applied field. When the specimen is repeatedly magnetized first in one direction and then the other, the graph of magnetization versus magnetizing field describes the **hysteresis loop** as shown in Figure 4.1.

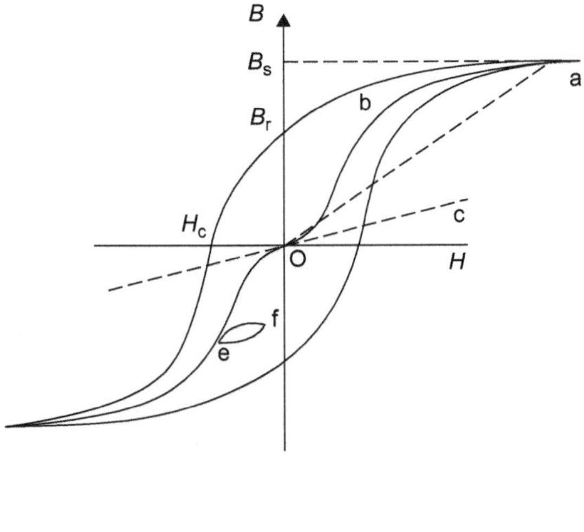

Figure 4.1 Magnetic B–H hysteresis loop. (After Moulson and Herbert, 2003, Chichester, West Sussex, England. Reproduced with permission of John Wiley & Sons.

The line **oba**—the "virgin curve"—represents the relationship determined experimentally when the specimen is magnetized before each measurement of the induction for a given field. The change in B, very near to the origin, represents magnetization by reversible Bloch wall displacements, and the tangent **oc** to this initial magnetization curve is called the initial permeability μ_i. The steep rise in B represents magnetization by irreversible Bloch wall displacements as the walls break away from their pinning points, and the region **ba** represents magnetization by reversible and irreversible domain rotations from one easy direction to another more favorably aligned with the applied field. The latter process requires high field strengths because the magnetization within a domain is rotated against the anisotropy field. If, after the material has been magnetically saturated to the value Bs, the field is reduced to zero, the magnetization vectors rotate out of line with the field toward the nearest preferred direction which is determined in part by magnetocrystalline anisotropy. The magnetization is thus prevented from complete relaxation to the "virgin" curve and hence, for a zero field, there is a remanent induction B_r. In order to reduce the induction to zero, a reverse field H_c has to be applied. The coercive field or "coercivity" H_c depends in part on crystalline anisotropy, as might be expected.

Because of hysteresis, energy is dissipated as heat in a magnetic material, as it is taken round a complete B–H loop, and the hysteresis energy loss W_h per unit volume of material is $W_h = \text{f}B\,dH$ (Moulson and Herbert, 2003, p. 485).

The amount of the hysteresis depends on how freely the domain walls can move, which, in turn, depends on the physical and chemical constitution of the magnetic materials, particularly on the number of impurities present that can inhibit Bloch wall movement. **Hard magnetic materials** usually exhibit fat hysteresis loops, with large coercive forces, while **soft magnetic materials** have narrow hysteresis loops.

Soft and Hard Magnets

Most of the magnetic materials of industrial interests are ferromagnetic materials. The ferromagnetic materials can be categorized into two classes, soft and hard magnetic materials. As shown in the magnetization curve, ferromagnetic materials with the demagnetized state do not show magnetization, although they have spontaneous magnetization. This is because the ferromagnetic materials are divided into many magnetic domains as described above. Within the magnetic domains, the direction of magnetic moment is aligned. However, the direction of magnetic moments varies at magnetic domain walls so that it can reduce the magnetostatic energy in the total volume. In the demagnetized state, total magnetization is canceled because of the random orientation of the magnetizations in magnetic domains. When external magnetic field is applied, domain walls migrate and disappear when all magnetic moments are aligned to the direction of the magnetic field.

When the domain walls can easily migrate, the ferromagnetic material can be easily magnetized at a low magnetic field. This type of ferromagnetic material, which is called soft magnetic material, is suitable for applications of magnetic cores or recording heads. Since soft magnetic materials can be demagnetized at a low magnetic field, coercivity H_c is low. As they can be easily magnetized, permeability is high. For ferromagnetic materials to be soft, their magnetocrystalline anisotropy and magnetostriction constant must be low. In addition, for easy migration of magnetic domains, they must have a small number of defects such as crystal grains. When domain walls are difficult to migrate, magnetization of the ferromagnetic material occurs only when a high magnetic field is applied. In other words, this type of ferromagnetic material is difficult to magnetize, but once magnetized, it is difficult to demagnetize. These materials are called hard magnetic materials and are suitable for applications such as permanent magnets and magnetic recording media. Hard magnetic materials have high magnetocrystalline anisotropy. Since a large magnetic field is required to demagnetize these materials, their coercivity H_c is usually high, but the coercivity is highly sensitive to the microstructure.

Superparamagnetism

Superparamagnetism is a form of magnetism which appears in small ferromagnetic or ferrimagnetic nanoparticles. In sufficiently small nanoparticles, magnetization can randomly flip direction under the influence of temperature. The typical time between two flips is called the **Neel relaxation time**. In the absence of an external magnetic field, when the time used to measure the magnetization of the nanoparticles is much longer than the Neel relaxation time, their average value of magnetization appears to be zero: they are said to be in the superparamagnetic state. In this state, an external magnetic field is able to magnetize the nanoparticles, similar to a paramagnet. However, their magnetic susceptibility is much larger than that of paramagnets.

Normally, any ferromagnetic or ferrimagnetic material undergoes a transition to a paramagnetic state above its Curie temperature.

Superparamagnetism is different from this standard transition since it occurs below the Curie temperature of the material.

Superparamagnetism occurs in nanoparticles which are single domain, i.e., composed of a single magnetic domain. This is possible when their diameter is below 3–50 nm, depending on the materials. In this condition, it is considered that the magnetization of the nanoparticles is a single giant magnetic moment, the sum of all the individual magnetic moments carried by the atoms of the nanoparticle.

4.2 Application of Soft Magnetic Nano-Glass Ceramics

As described above, in recent years, the nano-magnetic materials including the **magnetic nano-glass ceramics** have attracted great attention as promising candidates for application in a variety of fields. The magnetic nano-glass ceramics, according to their properties, can be classified in two main groups, namely **soft** and **hard magnets**, each group having its specific applications. In this section, the applications of soft magnetic nano-glass ceramics are discussed.

The most common soft magnets are cubic spinel ferrites, with the general formula AB_2O_4, the structure of which has previously been explained in Section 4.1.

These types of nano-magnets prepared via the glass-ceramic route, according to their application, can be classified in the following groups.

4.2.1 Biomedical Applications

The application of magnetic nanoparticles in the biomedical field can further be classified into inside (*in vivo*) or outside (*in vitro*) the body. *In vivo* applications could be further separated into therapeutic (hyperthermia and drug targeting) and diagnostic applications (magnetic resonance imaging, MRI), while for *in vitro* applications the main use is indiagnostic, i.e., separation/selection and agnetorelaxometry (Tartaj et al., 2003).

Cancer cells have poorly developed nervous systems and insufficient matrix of blood vessels. The heat dissipation in tumors is poor due to vascular damage and insufficient blood flow.

Hyperthermia destroys cancer cells by raising the tumor temperature to a "high fever" range; they generally perish around 43°C. So this method is an efficient technique to treat the tumors as healthy body cells are unaffected by this heat treatment due to proper heat dissipation caused by regular blood flow (Oh et al., 2001; Arcos et al., 2002). Various methods like hot water, infrared rays, ultraviolet rays, and microwaves have been utilized in attempts to heat the tumors, but these methods are ineffective to treat the deep-seated cancers locally. Magnetic induction hyperthermia is expected to heat the deep-seated tumors effectively (Oh et al., 2001).

For this purpose, ferrimagnetic materials were developed. When placed in an alternating magnetic field of high frequency, these materials can produce sufficient heat due to the hysteresis loop of the magnetic material and induce eddy currents when a variable external magnetic field is applied.

In this method, the use of subdomain magnetic particles (nanometer-sized) is preferred instead of multidomain (micron-sized) particles because nanoparticles absorb much more power at tolerable AC magnetic fields (Jordan et al., 1999; Rosensweig, 2002). The heating potential is also strongly dependent on the particle size and shape, and uniform particles are essential for the precise control of temperature.

The aforementioned application also requires that the magnetic materials selected for the fabrication of nanoparticles possess specific properties such as relatively high saturation magnetization, low-coercive force values, and narrow hysteresis loops (Amirahmadi, 2011).

The major disadvantage of most chemotherapy methods is that they are relatively nonspecific. The therapeutic drugs readministered intravenously leading to general systemic distribution, resulting in deleterious side effects as the drug attacks normal, healthy cells in addition to the target tumor cells. However, if such treatments could be localized, e.g., to the site of a joint, then the continued use of these very potent and effective agents could be made possible (Pankhurst et al., 2003).

Recognition of this point led researchers to propose the use of magnetic carriers to target-specific sites (generally cancerous tumors) within the body. The objectives are twofold: (i) to reduce the amount of systemic distribution of the cytotoxic drug, thus reducing the associated side effects and (ii) to reduce the dosage required by more efficient, localized targeting of the drug. In magnetically targeted therapy, a cytotoxic drug is attached to a biocompatible magnetic nanoparticle carrier.

These drug/carrier complexes, usually in the form of a biocompatible ferrofluid, are injected into the patient via the circulatory system. When the particles have entered the bloodstream, external, high-gradient magnetic fields are used to concentrate the complex at a specific target site within the body.

Magnetic Drug Carriers

A variety of magnetic nanoparticle and microparticle carriers have been developed to deliver drugs to specific target sites *in vivo*. Generally, the magnetic component of the particle is coated by a biocompatible polymer such as polyvinyl alcohol (PVA) or dextran, although recently inorganic coatings such as silica have been developed.

The coating acts to shield the magnetic particles from the surrounding environment and can also be functionalized by attaching carboxyl groups, biotin, avidin, carbodiimide, and other molecules.

The carriers typically have one of two structural configurations: (i) a magnetic particle core **(usually magnetite, Fe_3O_4, or maghemite, γ-Fe_2O_3)** coated with a biocompatible polymer or (ii) a porous biocompatible polymer in which magnetic nanoparticles are precipitated inside the pores. Recent developmental work on carriers has largely focused on new polymeric or inorganic coatings on magnetite/maghemite nanoparticles, although noble metal coatings such as gold were also considered (Pankhurst et al., 2003).

The process of drug localization using magnetic delivery systems is based on the competition between forces exerted on the particles by the blood compartment, and magnetic forces generated from the magnet, i.e., the applied field. When the magnetic forces exceed the linear blood flow rates in arteries or capillaries, the magnetic particles are retained at the target site and may be internalized by the endothelial cells of the target tissue. For this application, the use of nanoparticles favors the transport through the capillary systems of organs and tissues avoiding vessel embolism (Tartaj et al., 2003).

Magnetic nanoparticles offer abundant attractive possibilities in biotechnology. First, they have sizes that place them at dimensions comparable to those of a virus (20–500 nm), a protein (5–50 nm), or a gene (2 nm wide and 10–100 nm long). Second, the nanoparticles are magnetic, which means that they obey Coulomb's law, and can be manipulated by an external magnetic field gradient. Third and finally, nanoparticles have a large surface that can be properly modified to attach biological agents.

Applications in biotechnology impose strict requirements on the particles' physical, chemical, and pharmacological granulometric uniformity, crystal structure, magnetic behavior, surface structure, absorption properties, solubility, and low toxicity.

For example, in order for the magnetic drug targeting to become safe and effective (with the minimum amount of magnetic particles, a maximum amount of drug is easily administered and transported to a specific site), the following parameters of the nano-magnets are critical: (a) particle size (as small as possible to improve tissular diffusion and to have long sedimentation times and high effective surface areas), (b) surface characteristics (easy encapsulation of the magnetic nanoparticles protects them from degradation and endows biocompatibility), and (c) good magnetic response (possibility of decreasing the concentration of nano-magnets in blood and therefore diminishing the associated side effects) (Tartaj et al., 2005).

In biomedicine, it is often advantageous to separate out specific biological entities from their native environment in order that concentrated samples may be prepared for subsequent analysis or other use. Magnetic separation using biocompatible nanoparticles is one way to achieve this. It is a two-step process, involving (i) the tagging or labeling of the desired biological entity with magnetic material and (ii) the separating out of these tagged entities via a fluid-based magnetic separation device (Pankhurst et al., 2003).

Tagging is made possible through chemical modification of the surface of the magnetic nanoparticles, usually by coating with biocompatible molecules such as dextran, PVA, and phospholipids—all of which have been used on **iron oxide nanoparticles** (Sangregorio et al., 1999; Pardoe et al., 2001).

While providing a link between the particle and the target site on a cell or molecule, coating has also the additional advantage of increasing the colloidal stability of the magnetic fluid. Specific binding sites on the surface of cells are targeted by antibodies or other biological macromolecules such as hormones or folic acid. As antibodies specifically bind to their matching antigen, this provides a highly accurate way to label cells. For example, magnetic particles coated with immune-specific agents have been successfully bound to red blood cells, lung cancer cells, bacteria, and urological cancer cells. Magnetic separation has been successfully applied to many aspects of biomedical and biological research. It has proven to be a highly sensitive technique for the selection of rare tumor cells from blood and is especially well suited to the separation of low numbers of target cells. This has, e.g., led to the enhanced detection of malarial parasites in blood samples either by utilizing the magnetic properties of the parasites or through labeling the red blood cells with an immune-specific magnetic fluid (Pankhurst et al., 2003).

Imaging of soft tissue structure of the musculoskeletal system has become the domain of MRI due to its superiority over other imaging techniques. The technique measures changes in the magnetization of hydrogen protons in water molecules sitting in a magnetic field after a pulse of radio frequencies has hit them. Protons from different tissues react differently, giving a picture of anatomical structures. These images can be enhanced by adding "contrast agents" which sharpen the contrast by affecting the behavior of protons in their proximity. In standard clinical MRI scans, contrast agents travel through the bloodstream and tissues, increasing contrast wherever they go. It has been known for many years that the inclusion of magnetic particles within tissue enables a very large signal to be obtained from an MRI scanner. To date, a wide variety of particles have been produced, differing in size and type of coating material used.

Iron oxide nanoparticles play an important role as MRI contrast agents. They are classified into two main groups according to their size. The first group are termed **SPIOs (superparamagnetic iron oxides)** where nanoparticles have a size >50 nm (coating included) and the second type are termed **USPIOs (ultrasmall superparamagnetic iron oxides)** where nanoparticles are <50 nm. The particle size influences both their physicochemical and pharmacokinetic properties. The main applications using the particles are imaging of the gastrointestinal tract, liver and spleen, and lymph nodes. However, USPIO particles are also blood pool agents, which could be used for perfusion imaging (i.e., brain or myocardial ischemic diseases). Both types of particles are now available commercially (Berry and Curtis, 2003).

4.2.1.1 Magnetite Glass Ceramics

From the reports discussed above, it can be inferred that iron oxides, especially **magnetite**, as a ferrimagnetic material, possess the essential magnetic properties for the aforementioned biomedical applications and that it is one of the most recommended materials in the field.

On the other hand, among the fabrication methods of magnetite, the **glass-ceramic route** has attracted great attention in recent years.

The preparation of glass ceramics is much simpler and economical and these materials can be made with controllable particle size and morphology and narrow size distribution of nanoparticles, which are the essential specifications of these applications.

Moreover, there is the possibility of fabricating some magnetic glass ceramics that are also bioactive, i.e., have the ability of bonding to natural tissues via a hydroxyapatite layer. These glass ceramics can be used to fill bone defects which resulted from surgical operations carried out on tumors.

The preparation of **magnetite via the glass-ceramic route** and the study of their magnetic properties have been the subject of numerous studies in recent years as a suitable method to produce nano-crystalline particles: (Lee and Choi, 1996; Ebisawa et al., 1997; Oh et al., 2001; Arcos et al., 2002; Bretcanu et al., 2006; Woltz et al., 2006; Singh et al., 2008; Sharma et al., 2009; Abdel-Hameed et al., 2011; Amirahmadi et al., 2011).

4.2.1.1.1 Processing of Magnetite Nano-Glass Ceramics

Table 4.1 summarizes the chemical composition and heat-treatment conditions of some of these glass ceramics.

Ebisawa et al. (1997) prepared ferrimagnetic glass ceramics by heat treatment of glasses of the system (FeO,Fe_2O_3)–$CaO \cdot SiO_2$ with the addition of Na_2O, B_2O_3, and P_2O_5 (sample 1, Table 4.1) and evaluated their bioactivity. It was found that glass ceramics with Na_2O or B_2O_3 added in combination with P_2O_5 show bioactivity. All the examined glasses started to precipitate magnetite (Fe_3O_4) at ~700°C and β-wollastonite ($CaO \cdot SiO_2$) at ~800°C, similar to the additive-free glass. The content of magnetite increased with increasing heat-treatment temperature, reaching ~36 wt% at ~1,000°C as in the case of the additive-free glass.

Ferrimagnetic glass ceramics in the Fe_2O_3–CaO–SiO_2 system containing P_2O_5/B_2O_3 additives were also prepared by Oh et al. (2001). According to the authors, P_2O_5 or B_2O_3 could improve the crystallization capability of magnetite crystallites and prohibit the generation of other crystalline phases by lowering the crystallization temperature of magnetite. After nucleation at 690°C for 60 min and growth at 940°C for 2 h, the samples were crystallized to magnetite. The maximum amount of crystallized magnetite was 31.1 vol% (sample 2, Table 4.1).

Table 4.1 Composition and Properties of Some Soft Magnetic Nano-Glass Ceramics

Sample No.	Chemical Composition	Major Magnetic Phase/Size	Heat-Treatment Temp. (°C)/Time	Saturation Magnetization (emu/g)/Coercitive Force (Oe)	Remanence Magnetization (emu/g)	Reference
1	$60CaO \cdot SiO_2$, $40Fe_2O_3$, $3B_2O_3$, $3P_2O_5$ (wt ratio)	Fe_3O_4/200 nm	1,050	32/120	–	Ebisawa et al. (1997)
2	$40Fe_2O_3$, 30CaO, $30SiO_2$ ($1B_2O_3/P_2O_5$) wt%	Fe_3O_4/90 nm	690/60 min and 940/2 h	100 emu/cm^3 382.0–388.2 Oe	–	Oh et al. (2001)
3	$24.7SiO_2$, $13.5Na_2O$, 13.5CaO, $3.3P_2O_5$, 14FeO, and $31Fe_2O_3$ (wt%)	Fe_3O_4/50 nm	Melt quenched	34/85	2.4	Bretcanu et al. (2006)
4	41CaO, $(52-x)SiO_2$, $4P_2O_5$, xFe_2O_3, $3Na_2O$ ($2 \geq x \leq 10$) mol%	Fe_3O_4/56 nm ($x=10$)	1,050/3 h	7.95/91	0.71	Singh et al. (2008)
5	$25SiO_2$, 50CaO, $15P_2O_5$, $(10-x)$ Fe_2O_3, xZnO ($x=0, 2, 5$) mol%	Fe_3O_4/40 nm ($x=2$)	800/6 h	3.34/8	0.056	Sharma et al. (2009)
6	$56.60Fe_2O_3$, 13.49CaO, 9.79ZnO, $14.46SiO_2$, $2.83B_2O_3$, $2.83P_2O_5$ (wt%) containing 30 g CuO/100 g batch	Fe_3O_4/10 nm	Melt quenched	19.63/21.88	0.765	Abdel-Hameed et al. (2011)
7	$35Na_2O$, $24Fe_2O_3$, $20B_2O_3$, $20SiO_2$, 1ZnO (mol%)	Fe_3O_4/20 nm	590/1 h	35/60	–	Amirahmadi et al. (2011)
8	33.3CaO, $4.9Na_2O$, $17.1Fe_2O_3$, $44.7B_2O_3$ (mol%)	Fe_3O_4/10 nm	570/7 h	16	–	Woltz et al. (2006)
9	$x(Fe_2O_3)$, $(60-x)B_2O_3$, 40ZnO ($x=12.5$ and 15) mol%	$ZnFe_2O_4$/36nm ($x=15$)	Melt quenched	~0.014 μB/f.u. (at 18 K)/ ~0.05 μB/f.u.	–	Pascuta et al. (2012)
10	$x(ZnO,Fe_2O_3)$,$(65-x)SiO_2$, $20(CaO, P_2O_5)$,$15Na_2O$ ($x=6$, 9, 12, 15, 18, and 21) mol%	$ZnFe_2O_4$/25 nm ($x=21$)	800/1 h	19.60/44	0.618	Singh and Srinivasan (2011)

(Continued)

Table 4.1 (Continued)

Sample No.	Chemical Composition	Major Magnetic Phase/Size	Heat-Treatment Temp. (°C)/Time	Saturation Magnetization (emu/g)/Coercitive Force (Oe)	Remanence Magnetization (emu/g)	Reference
11	xZnO, 25Fe_2O_3, (40 − x)SiO_2, 25CaO,7P_2O_5, 3Na_2O (x = 4, 6, 8, 10)(wt%)	$ZnFe_2O_4$/58 nm (x = 10)	1,100/3 h	21.8/210	4.9	Shah et al. (2010)
12	32Li_2O, 8Fe_2O_3, 60B_2O_3 (mol%)	$LiFe_{2.5}O_4$/50 nm	770/2 h	12/40	–	Rezlescu et al (1998)
13	25Li_2O, 8MnO_2, 20CaO, 2P_2O_5, 45SiO_2 + 4 and 8% Fe_2O_3 (mol%)	$LiMn_2O_4$/ 5 nm–few μm (Fe_2O_3 = 8%)	850/4 h	0.7/63 (at 300 K)	–	Hsi et al. (2007)
14	(glass 1) 39.76Fe_2O_3, 0.24MnO_2, 40CoO, 20B_2O_3 (mol%)	$CoFe_2O_4$/14 nm	800/4 h	23.11/307	4.79	Abdel-Hameed (2012)
	(glass 2) 12.8Fe_2O_3, 4MoO_3, 63.2NiO, 20B_2O_3 (mol%)	$NiMoO_4$/56 nm	800/4 h	0.76/1,012	58×10^{-3}	
	(glass 3) 39.2Fe_2O_3, 1.6V_2O_5, 39.2CoO, 20B_2O_3 (mol%)	$CoFe_2O_4$/6–8 nm	800/4 h	1.83/585	4.4	

Bretcanu et al. (2006) fabricated ferrimagnetic glass ceramics with the composition located in the system SiO_2–Na_2O–CaO–P_2O_5–FeO–Fe_2O_3 by melting the powders obtained by co-precipitation of the reagents (sample 3, Table 4.1).

The magnetite crystallites were precipitated during cooling from the melting temperature. Since the maximal quantity of magnetite crystals was produced during cooling, these glass ceramics did not require any further nucleation and growth heat treatment.

The authors noted that among the synthesis process parameters, the melting temperature played a fundamental role in magnetite crystallization. Increasing the melting temperature first resulted in the increase of the saturation magnetization (M_S), in the range 1,400–1,500°C, but upon further elevation of temperature to 1,550°C, the value of M_S decreased.

This was attributed to the increase of the quantity of magnetite, the sole magnetic phase, from 1,400 to 1,500°C. For melting temperatures higher than 1,500°C, the value of M_S decreased due to the decomposition of magnetite.

On the other hand, when the melting temperature was increased, both the coercivity and remanence values initially decreased at 1,500°C and then increased at 1,550°C.

This was attributed to the possible initiation of internal stresses during rapid cooling of glass, because of the thermal expansion differences existing between the glass matrix and the crystals. Such internal stresses might inhibit the rotation of the magnetic moments and increase the coercive force. This factor prevailed when the melt was cooled from the temperature of 1,550°C.

The authors concluded that by knowing the influence of the melting temperature on the structure and magnetic properties of these glass ceramics, it is possible to design and realize materials having different characteristics that can be modulated according to the given application fields.

Singh et al. (2008) studied the magnetic properties of glass ceramics derived from glasses with composition 41CaO–$(52 - x)$ SiO_2–$4P_2O_5$–xFe_2O_3–$3Na_2O$ ($2 \leq x \leq 10$ mol%). The glass samples were prepared by melting the calcined batch material at 1,550°C in air. The melt was then poured on a copper plate at room temperature. Pieces of as-quenched glass were heat-treated at 1,050°C for 3 h in air to form the glass ceramics.

X-ray diffraction (XRD) investigation revealed the presence of a nano-crystalline magnetite phase in the heat-treated samples containing $x \geq 2$ mol% Fe_2O_3. Magnetic hysteresis cycles of the glass-ceramic samples with $x \geq 2$ mol% of iron oxide exhibited magnetic behavior similar to soft magnetic materials with low coercivity. The evolution of magnetic properties in these samples as a function of iron oxide molar concentration was correlated with the amount and crystallite size of magnetite phase present in them.

Table 4.1 depicts the properties of the specimen containing 10 mol% Fe_2O_3. This specimen exhibited the highest value of saturation magnetization among the samples studied by these authors (sample 4).

The effect of ZnO addition on the fabrication process and properties of magnetite glass ceramics was also studied (Sharma et al., 2009). The chemical composition and properties of the glass-ceramic samples are given in Table 4.1 (sample 5).

The glasses were prepared by the melt-quench technique followed by heat treatment at 800°C for 6 h. This resulted in spherical nano-size crystallites of calcium phosphate, hematite, and magnetite (the only magnetic phase) evenly dispersed in a glassy silica matrix. It was shown that with the increase of ZnO content, the particles size of magnetite was reduced, and the samples showed improved magnetic properties due to the formation of magnetite nanoparticles.

The crystallization behavior and magnetic properties of ferromagnetic glass ceramics in the system Fe_2O_3–CaO–ZnO–SiO_2, with gradual addition of CuO, were also studied (Abdel-Hameed et al., 2011). The aim of their study was the crystallization of a high quantity of magnetite in these glass ceramics and determination of the influences of CuO addition on the sequence of crystallization and magnetic properties of the glass ceramics. The XRD patterns showed the presence of nano-magnetite crystallites in a glassy matrix after direct cooling from the melt, without any additional heat treatments. Increasing the melting temperature resulted in an increase in the magnetite crystallization.

In general, increasing the CuO addition greatly enhanced the crystallization of magnetite. A significant amount of delafossite ($CuFeO_2$) was detected and increased by increasing both the added amount of CuO and the heat-treatment temperature. The composition and properties of these glass ceramics are depicted in Figure 4.1 (sample 6).

Amirahmadi et al. (2011) prepared a glass with the composition of $35Na_2O - 24Fe_2O_3 - 20B_2O_3 - 20SiO_2 - 1ZnO$ (mol%) (sample 7, Table 4.1) by quenching the melt between twin rollers and subsequently heat-treated the resulting glass flakes in the range 485–750°C for 1–2 h under a graphite powder bed. The heat treatment at a lower temperature range (485–590°C) produced some specimens containing magnetite particles of 7–20 nm in diameter. Increasing the heat-treatment temperature resulted in the increased amounts of magnetite and the growth of its particle sizes.

The nano-crystalline magnetite powders were also prepared from glass melts in the Fe_2O_3–FeO–CaO–Na_2O–B_2O_3 system, by firstly reducing them by flushing with nitrogen and subsequently roller quenching (Woltz et al., 2006). The obtained flakes, with a thickness of ~150 μm subsequently, were heat-treated at temperatures in the range 550–620°C (sample 8, Table 4.1).

Recently, Sandu et al. (2012) prepared and determined the structural and magnetic properties of nano-sized magnetite crystallites obtained by the crystallization of Fe-containing borosilicate glasses. Several glass compositions with different ratios of Fe_2O_3/SiO_2 in the range of 0.37–0.67 were investigated, in which the nucleants, Cr_2O_3 and P_2O_5, and modifiers, intermediates (Al_2O_3 and MgO), were also systematically varied. In addition to magnetite, the presence of large amounts of Fe-rich paramagnetic phases was also observed. The effect of residual Fe ions, dispersed in the glassy matrix, on magnetic properties of glass-ceramic samples was also analyzed. The role of nucleants in initiating disorders in both tetrahedral and octahedral sites of the magnetite was also discussed.

Redox Equilibrium of Iron

One of the problems associated with the crystallization of magnetite from glasses is the polyvalency of iron, which occurs in glasses and melts as Fe^{2+} and Fe^{3+}. At high temperatures, these ionic species are in equilibrium with the physically dissolved oxygen of the melt. In principle, the Fe^{2+}/Fe^{3+} redox ratio can be adjusted by appropriate reducing or oxidizing conditions.

While for low iron concentrations (<2 mol%), the thermodynamics of redox equilibria can easily be measured using electrochemical techniques; at large iron concentrations these methods cannot be applied. Hence, the redox ratio can hardly be predicted and has to be adjusted empirically (Woltz et al., 2006).

Woltz et al. (2006) prepared glass melts in the system Fe_2O_3–FeO–CaO–Na_2O–B_2O_3 from the raw materials, by first reducing them by flushing with nitrogen and subsequently roller quenching.

According to Oh et al. (2001), the magnetite crystal is easily oxidized above 900°C under air atmosphere. The mechanism of oxidation involves two consecutive steps, supersaturation and diffusion. Atmospheric oxygen, initially absorbed at the surface of the magnetite particles, is ionized by the electron initiated from the reaction $Fe^{2+} \rightarrow Fe^{3+} + e$.

The supersaturation of Fe^{3+} and O^{-2} at the surface result in the growth of a thin layer of hematite. A continuous film of hematite would prevent oxidation and diffusion would become the controlling step in oxidation.

Since carbon, due to its great chemical affinity for reaction with oxygen, was often recommended as a very useful oxygen barrier, the authors used carbon powder to prevent the oxidation of magnetite. It was confirmed that carbon was effective in protecting the magnetite particles from oxidation in this study.

Amirahmadi et al. (2011) also stated that using a reductive atmosphere, e.g., heat treatment under a graphite bed, proved to be essential to avoid the oxidation of magnetite particles that could lead to a drastic reduction in the quantity of magnetite and the magnetization values.

Ebisawa et al. (1997) in their investigation on (FeO, Fe_2O_3)–$CaO \cdot SiO_2$ glasses (as discussed previously) showed that when their specimens were covered with active carbon powders, the precipitated magnetite was not converted into hematite up to 1,050°C, but when the specimens were not covered with carbon powder, the magnetite crystals present in them were converted into hematite at >1000 °C.

In summary, on the basis of the above reports, it can be inferred that the studied glass ceramics were mainly based on calcium–sodium–iron silicate/borate glasses. It is also evident that due to the marked crystallization tendency of these glass melts, they were usually prepared by quenching methods.

It also seems that, despite the high quenching rates employed, some of these glasses developed nano-crystalline magnetite particles during the cooling process. Hence, these glasses were not subjected to further crystallization heat treatment and due to the high crystal nucleation rates, usually no nucleating agents were used in their crystallization process.

Moreover, one of the problems associated with the crystallization of magnetite from glasses is the Fe^{2+}/Fe^{3+} redox ratio, that should be adjusted by the appropriate reducing or oxidizing conditions.

4.2.1.1.2 Magnetic Properties of Magnetite Nano-Glass Ceramics

The magnetic properties of some of the aforementioned magnetite glass ceramics are given in Table 4.1.

The reported saturation magnetization value of the glass ceramics prepared by Ebisawa et al. (1997) containing B_2O_3 and P_2O_5 additives and heat-treated at 1,050°C (sample 1) was equal to that of the additive-free glass heat-treated at 950°C (32 emu/g), whereas its coercive force was lower than the latter glass (500 Oe). The lower coercive force of these specimens was attributed to the larger size of the magnetite crystallites.

The hysteresis loop of this specimen showed a typical ferrimagnetic effect (Figure 4.2).

The magnetic properties of the as-prepared glass ceramics fabricated by Bretcanu et al. (2006) are given in Table 4.1 (sample 3).

The prepared glass-ceramic samples exhibited narrow hysteresis loops with small coercive fields which are the characteristics of soft magnetic materials.

Figure 4.3 depicts the transmission electron microscope (TEM) micrograph of an as-prepared sample of the above study. In this figure, the nano-magnetite crystallites in the 5–10 nm size range, embedded in the amorphous phase, can be observed.

The studies of Sharma et al. (2009) as given in Table 4.1 (sample 5) revealed the formation of magnetic particles and the increase in saturation magnetization with the addition of 2 mol% ZnO. These glass ceramics showed superparamagnetic and ferromagnetic behavior. Further increase in ZnO content resulted in weaker dipolar interactions and the specimens exhibited paramagnetic behavior.

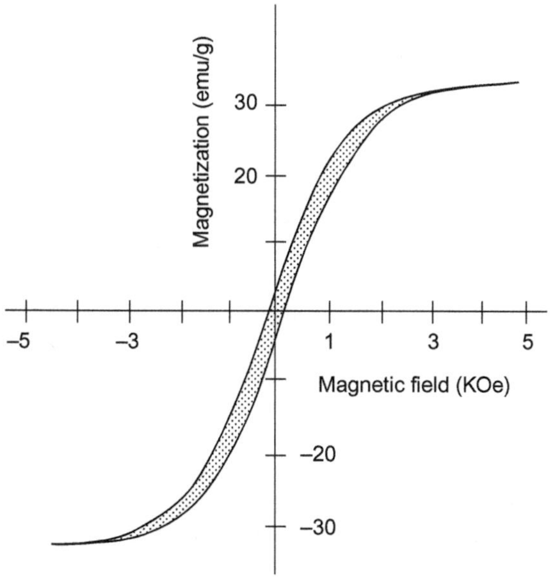

Figure 4.2 Hysteresis loop of the glass ceramic designated as sample 1 in Table 4.1 after the heat treatment at 1,050°C. (After Ebisawa et al., 1997. Reproduced with permission of Elsevier Science B.V.)

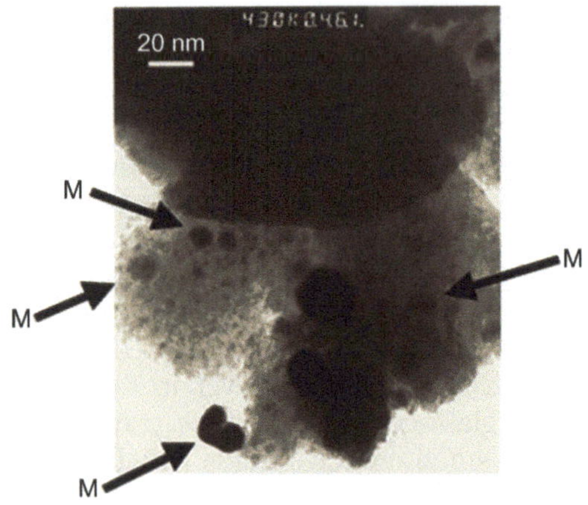

Figure 4.3 TEM micrograph of the glass ceramic designated as sample 2 in Table 4.1. *M* represents magnetite particles. (After Bretcanu et al., 2006. Reproduced with permission of Springer Science + Business Media, Inc.)

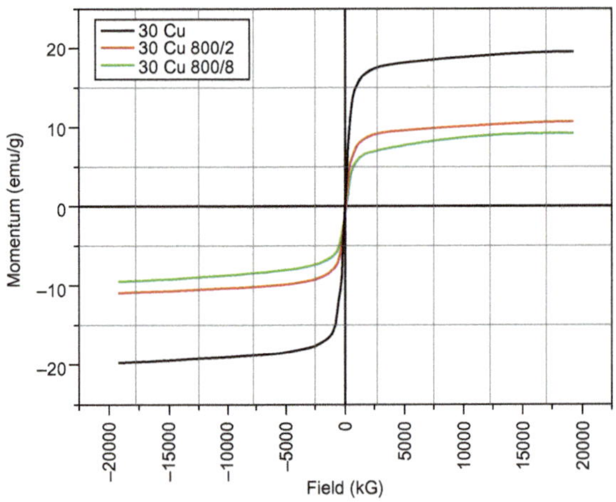

Figure 4.4 The hysteresis loops of specimen containing 30 g Cu (sample 4, Table 4.1) without heat treatment and heat-treated at 800°C for 2 and 8 h. (After Abdel-Hameed et al. (2011). Reproduced with permission of Elsevier Science B.V.).

The study of magnetic properties prepared by Abdel-Hameed et al. (2011) revealed that the samples containing a 30 g CuO/100 g batch, without heat treatment, exhibited the maximum saturation magnetization values and increasing Cu ions to the level of the >10 g/100 g batch and reduced the saturation magnetization values of the heat-treated specimens. It was also shown that the application of a reducing atmosphere increased the crystallite size and consequently decreased the coercivity.

The properties of the specimen containing a 30 g CuO/100 g batch, without heat treatment, is depicted in Table 4.1 (sample 6) and its hysteresis loop is shown in Figure 4.4. The TEM micrograph of these samples revealed the presence of agglomerates of nanoparticles of magnetite, mostly in the size of <10 nm in them.

Amirahmadi et al. (2011) fabricated glass ceramics exhibiting relatively high magnetization values and low-coercive forces of 6–48 emu/g and 0–60 Oe, respectively, by heat treatment of glass specimens for 1–2 h at the relatively low temperature range of 485–590°C. These glass-ceramic specimens contained magnetite particles of 7–20 nm in diameter. Table 4.1 gives the specifications related to one of these specimens heat-treated at 590°C for 1 h (sample 7).

Magnetization measurements on the glass-ceramic specimens prepared by Woltz et al. (2006), at room temperature (sample 8, Table 4.1) showed ferromagnetic behavior and no hysteresis. Temperature-dependent measurements also showed a monotonic decrease of the saturation magnetization with temperature and a Curie temperature of 553°C.

4.2.1.2 Zinc Ferrite–Based Glass Ceramics

Bulk $ZnFe_2O_4$ is an antiferromagnetic material and its nanoparticles show ferromagnetism or ferrimagnetism combined with superparamagnetism. $ZnFe_2O_4$ and Fe_3O_4 form a complete series of solid solutions with the chemical formula of $Zn_xFe_{1-x}[Fe_{1-x}Fe_{1+x}]O_4$ or $Zn_xFe_{3-x}O_4$ (Singh and Srinivasan, 2011). The solid solution $Zn_{0.4}Fe_{2.6}O_4$, which forms for $x = 0.4$, exhibits higher saturation magnetization than magnetite (Anantharaman et al., 1998; Tanaka et al., 1998). Therefore, glass ceramics containing $Zn_{0.4}Fe_{2.6}O_4$ are expected to generate a higher amount of heat for the same applied magnetic field as compared with other systems (Singh et al., 2008). This higher magnetization of $Zn_{0.4}Fe_{2.6}O_4$ is attributed to the occupancy of the tetrahedral sites by Fe^{3+} ions accompanied with Zn^{2+} ions in the octahedral sites in the surface region of the ultrafine particles (Singh and Srinivasan, 2011; Jeyadevan et al., 1994).

Hence, $ZnFe_2O_4$-containing glass ceramics have been suggested as suitable candidates for thermoseeds in **the hyperthermia treatment of bone cancer**.

4.2.1.2.1 Processing of Zinc Ferrite–Based Nano-Glass Ceramics

Pascuta et al. (2012) synthesized two different glass ceramics with the composition of $x(Fe_2O_3)$, $(60 - x)$ B_2O_3, 40(ZnO), where $x = 12.5$ and 15 mol%, using the melt-quench method. The presence of nano-metric zinc ferrite ($ZnFe_2O_4$) crystals in a glassy matrix after cooling from the melting temperature was observed. The estimated amount of crystallized zinc ferrite was varied in the range 16–35% as a function of the chemical composition.

Singh and Srinivasan (2011) prepared glass ceramics with finely dispersed zinc ferrite ($ZnFe_2O_4$) nano-crystallites by the heat treatment of glasses having the composition:

$x(ZnO,Fe_2O_3)$, $(65-x)SiO_2$, $20(CaO,P_2O_5)$, $15Na_2O$ ($x=6$, 9, 12, 15, 18, and 21 mol%). The glass samples were prepared by quenching the melt into preheated graphite moulds. The as-quenched glasses were then heat-treated at 1,073 K for 1 h to form glass ceramics and then slowly cooled to room temperature.

All glass-ceramic samples contained calcium sodium phosphate ($NaCaPO_4$) and zinc ferrite ($ZnFe_2O_4$) as the two major crystalline phases.

According to the authors, calcium sodium phosphate is a bone mineral and its presence indicated the biocompatible nature of the glass-ceramic samples. The average crystallite size of $ZnFe_2O_4$ is increased from 10 nm for the $x=6$ sample to 25 nm for the $x=21$ sample.

Pascuta et al. (2011) fabricated glass ceramics in the $x(Fe_2O_3)$, $(60-x)B_2O_3$, $40(ZnO)$ ($x=17.5$ and 20 mol%) system by the melt-quench method.

The samples contained a unique magnetic crystalline phase, the zinc ferrite ($ZnFe_2O_4$), embedded in an amorphous matrix. The $ZnFe_2O_4$ crystals were precipitated during cooling from the melting temperature.

The average crystallite size of zinc ferrite was increased from 41 nm in the samples containing 17.5 mol% Fe_2O_3 to 98 nm in the sample with 20 mol% Fe_2O_3.

The degree of crystallinity was also increased from 68 to 74 wt% upon the increment of Fe_2O_3 content from 17.5 to 20 mol%.

Singh and Srinivasan (2010) studied the crystallization behavior and properties of glasses with the composition $x(ZnO,Fe_2O_3)$, $(65-x)SiO_2$, $20(CaO,P_2O_5)$, $15Na_2O$ ($6 \leq x \leq 21$) mol%.

Fine particles of zinc ferrite ($ZnFe_2O_4$) and calcium sodium phosphate ($NaCaPO_4$) were crystallized in a glassy matrix by heat treatment. Initial magnetization curves revealed that samples with $x=6$ and 9 mol% zinc-iron oxide exhibited both ferrimagnetic and paramagnetic contributions, whereas samples with $x>9$ mol% zinc-iron oxide exhibited only ferrimagnetic contribution. The apatite-forming ability of the glass-ceramic samples was also investigated by examining apatite formation on the surface of the samples treated in simulated body fluid (SBF). Increase in apatite-forming ability was observed with an increase in zinc-iron oxide content. Therefore, the authors suggested that these materials are expected to find application as thermoseeds in hyperthermia treatment of bone cancer.

Shah et al. (2010) prepared glass ceramics of the composition xZnO–25Fe_2O_3–(40 – x)SiO_2–25CaO–7P_2O_5–3Na_2O (x = 4, 6, 8, 10) by the melt-quench method using oxy-acetylene flame. Glass powder compacts were sintered at 1,100°C for 3 h and then rapidly cooled.

XRD revealed three main crystalline phases: $ZnFe_2O_4$, $CaSiO_3$, and $Ca_{10}(PO_4)_6(OH)_2$.

The crystallite size of $ZnFe_2O_4$ were in the range of 30–58 nm and increased with the increase in ZnO concentration.

4.2.1.2.2 Magnetic Properties of Zinc Ferrite–Based Nano-Glass Ceramics

The magnetic properties of some of the aforementioned zinc ferrite glass ceramics are also given in Table 4.1.

Magnetic properties of the glass ceramics prepared by Pascuta et al. (2012) are depicted in Table 4.1 (sample 9). The measurements were carried out in the temperature range 4.2–300 K. DC magnetization was also recorded under zero field cooled (ZFC) and field cooled (FC) sequences.

From the magnetization curves for glass ceramic containing 15 mol% Fe_2O_3, it was found that the nanoparticles exhibited ferromagnetic interactions combined with superparamagnetism with a blocking temperature, TB = 18 K. The blocking temperature is a measure of the thermal energy required to overcome the anisotropy energy barrier and is defined as the average temperature at which the nanoparticles do not relax during the time of measurement; they are blocked. With the increase of temperature, the ZFC curve for the aforementioned glass ceramic came to a maximum (TB), which corresponds to the case that the majority of particles show superparamagnetic characteristics. Above TB, the magnetization decreases as the temperature increases.

The hysteresis curves for both samples containing 12.5 and 15 mol% Fe_2O_3 above and below TB were obtained. It was observed that the studied glass ceramics exhibited magnetic behavior characteristic for soft magnetic particles, with small coercivity.

For example, the coercivity was ~0.05 μ_B/f.u. for measurements performed in maximum 0.4 T.

Singh and Srinivasan (2011) reported the magnetic properties of the glass-ceramic specimens containing nanoparticles of zinc ferrite (sample 10, Table 4.1).

Figure 4.5 depicts the room-temperature magnetic hysteresis (M–H) loops of the heat-treated glasses with different zinc-iron oxide concentrations for magnetic field strength of ±20 kOe. It can be seen that the magnetic field necessary to saturate the samples increases with an increase in zinc-iron oxide concentration. The initial magnetization curves of the samples with $x = 6$ and 9 mol% zinc-iron oxide, shown as inset (B), indicate that these samples do not show a tendency to saturate even at ±20 kOe. It can also be inferred from the nature of these two curves that the samples with $x = 6$ and 9 mol% zinc-iron oxide exhibit a combination of linear (paramagnetic) and nonlinear (spontaneous magnetization) behavior.

It was also shown by the authors that saturation magnetization (M_S) increases with increases in zinc-iron oxide concentration from $x = 6\%$ to 21% and shows a trend to saturate for $x = 21$ with the maximum value of about 19.6 emu/g. The increase in saturation magnetization with increases in zinc-iron oxide concentration was attributed to the development of $ZnFe_2O_4$ phase in the samples.

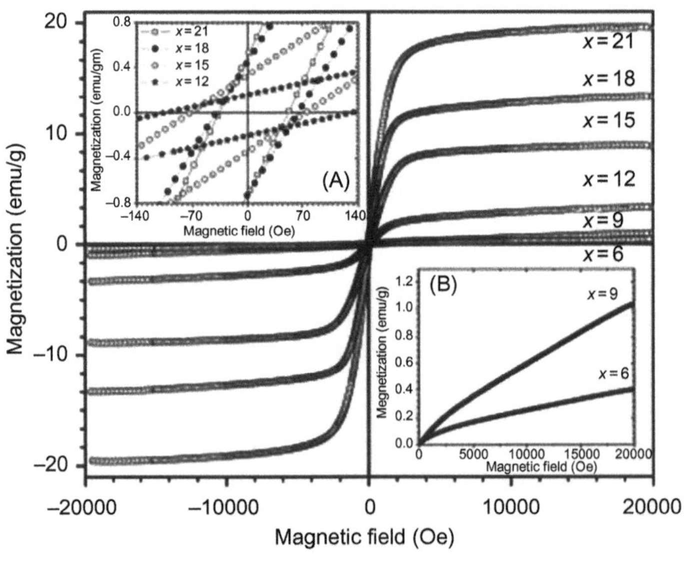

Figure 4.5 Room-temperature magnetic hysteresis loops of glass ceramics with different zinc-iron oxide concentration under ±20 kOe field sweep. Inset (A) shows an expanded view of hysteresis loops close to the origin. Inset (B) displays the initial magnetization curves of the sample with $x = 6$ and 9 mole% zinc-iron oxide. (After Singh and Srinivasan, 2011. Reproduced with permission of Elsevier Science B.V.)

Magnetic anisotropy, shape and dimension of the crystals, residual stress, and crystal imperfections influence the coercivity (H_c) and remanence magnetization.

The coercivity of the samples is decreased with increases in zinc-iron oxide concentration and slowly decreased with the increase in average crystallite size of $ZnFe_2O_4$. The area of the hysteresis loop is also increased with the increase in zinc-iron oxide content. The largest area was obtained for the samples with $x = 21$ mol%, which also exhibits the highest saturation magnetization and lowest coercivity among the samples studied. Since the area under the hysteresis loop is proportional to the energy loss, the samples with higher zinc-iron oxide concentrations are capable of generating more heat.

The magnetic properties of the glass-ceramic samples (sample 11) prepared by Shah et al. (2010) are provided in Table 4.1.

The authors showed that saturation magnetization, coercivity, and hence hysteresis area are increased with the increase in ZnO content (Figure 4.6).

Nano-sized $ZnFe_2O_4$ crystallites were of pseudo-single-domain structure and thus coercivity was increased with the increase in crystallite size.

Improvements in saturation magnetization as well as in crystallite size of $ZnFe_2O_4$ (which affects the coercivity and hysteresis area) were observed at 8% of ZnO.

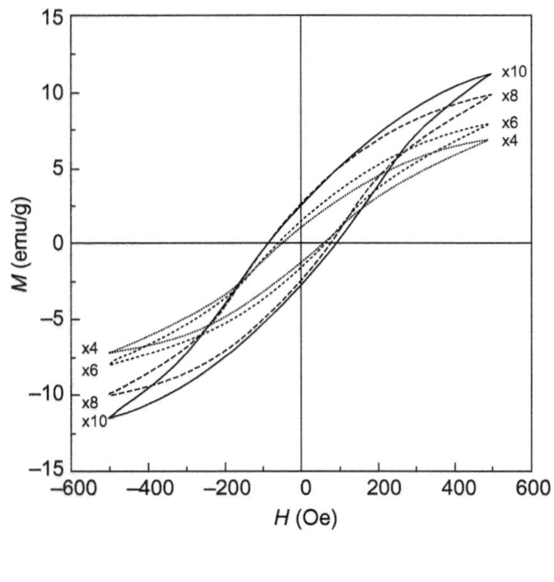

Figure 4.6 The Vibrating Sample Magnometer (VSM) hysteresis graphs of glass-ceramic samples X4, X6, X8, and X10 at 500 Oe. (After Shah et al., 2010. Reproduced with permission of Elsevier Science B.V.)

According to the authors, $ZnFe_2O_4$ exhibited ferrimagnetism due to the random distribution of Zn^{2+} and Fe^{3+} cations at tetrahedral A-sites and octahedral B-sites. This inversion/random distribution of cations was probably due to the surface effects of nano-$ZnFe_2O_4$ and rapid cooling of the material from 1,100°C (thus preserving the high-temperature state of the random distribution of cations). Calorimetric measurements showed that maximum specific power loss and temperature increase after 2 min were 26 W/g and 37°C, respectively, for the sample containing 10% ZnO.

It was also claimed by the authors that the prepared glass ceramics were bioactive and could bond to the living tissues in a physiological environment.

Pal et al. (1996) fabricated Ni–Zn ferrite glass ceramics in which $NiO \cdot 0.5Zn0.5Fe_2O_4$ crystals with dimensions in the range 15–29 nm were grown within a borate glass by a two-stage heat treatment in the temperature range 873–953 K.

The chosen glass composition was 10NiO, 10ZnO, $20Fe_2O_3$, and $60B_2O_3$ (mol%).

The glass samples were prepared by quenching the melt between two brass plates.

4.2.2 Other Applications of Soft Magnetic Nano-Glass Ceramics

In addition to the aforementioned biomedical applications, spinel ferrite nanoparticles, prepared by the glass-ceramic method, have intensively been investigated in recent years because of their remarkable electrical and magnetic properties and wide practical applications, such as in the information storage systems, ferrofluid technology, and magnetocaloric refrigeration (Abdel-Hameed et al., 2012).

The following glass ceramics are typical examples of the most studied types of materials.

4.2.2.1 Lithium Ferrite ($LiFe_{2.5}O_4$)–Based Glass Ceramics

Lithium-based ferrites have attracted considerable attention for several decades owing to their high Curie temperature, high saturation magnetization, excellent hysteresis loop squareness, and economical price. These offer performance advantages over other spinel structures and at the present time they are important components of microwave devices such as isolators, circular gyrators, phase shifters, and memory cores (Qi et al., 2003; Fu and Hsu, 2005; Widatallah et al., 2001).

$LiFe_{2.5}O_4$ is an inverse spinel with the Li^+ and three-fifths of the Fe^{3+} ions occupying the octahedral B-sites of the cubic spinel structure of the general formula AB_2O_4 (Verma and Joy, 2005).

Since the number of ferric ions on A and B sites is unequal in lithium ferrite, the calculated magnetic moment is not just that of lithium ions but is given by the difference in the magnetic moment of ions on A and B sites. Consequently, lithium ferrite possesses a higher Curie temperature than other spinel ferrites (Fu, 2006).

The preparation of lithium ferrites via the glass-ceramic route has also been attempted by some researchers.

Rezlescu et al. (1998) investigated the magnetic and microstructural properties of quenched $Fe_2O_3–Li_2O–B_2O_3$ glass strips, which were subjected to various heat treatments in the 200–770°C range. Crystals of ferrite ($LiFe_{2.5}O_4$), with sizes ranging from 3.5 to 50 nm, depending on the annealing temperature, were obtained.

It was shown that the crystallization of Li ferrite began at 430°C and annealing at lower temperature range, 430–530°C, resulted in the nucleation and a small increase in the number of isolated grains.

The variation of specific saturation magnetization as a function of the average grain size exhibited a sharp decrease for grain sizes <6 nm. At larger grain sizes (or high annealing temperatures), specific saturation magnetization increased slowly and began leveling off (to a value of ~12 emu/g) at grain sizes of 40–50 nm. The coercivity value is also increased with increasing grain size, reached a maximum value of ~60 Oe at ~28 nm and then decreased sharply. This was attributed to the transition from the single- to multidomain magnetization mechanism for larger sizes (~28 nm).

The magnetic properties of a sample heat-treated at 770°C for 2 h is depicted in Table 4.1 (sample 12).

The authors finally concluded that the glass crystallization route is a convenient method to produce materials of predetermined properties, since their structural and magnetic properties can be adjusted by the manipulation of the heat-treatment conditions. Also, by this method, powders with very fine magnetic particles, with grain sizes of 10–50 nm, can be prepared for many practical applications.

Rezlescu, N. (1999) investigated the formation of fine ferrite particles in glass ribbons by isothermal treatments and analyzed their influence on the magnetic properties of glass-ceramic systems. Two glass systems were studied: $Li_2O-Fe_2O_3-B_2O_3$ (glass 1) and $BaO-Fe_2O_3-B_2O_3-Sb_2O_3$ (glass 2). The glass ribbons were treated successively at various temperatures in the range 300–800°C to crystallize the ferrite particles within the matrix.

$LiFe_5O_8$ (lithium ferrite) and $BaFe_{12}O_{19}$ (barium ferrite) magnetic phases began to appear at 430 and 600°C in glasses 1 and 2, respectively.

Hsi et al. (2007) studied the crystallization of $25Li_2O-8MnO_2-20CaO-2P_2O_5-45SiO_2$ (mol%) glass samples. Two additional glass samples were also prepared by addition of 4 and 8 mol% Fe_2O_3 to the above base glass composition.

On heat treatment of the glass samples at 850°C for 4 h, the (Li,Mn) ferrite phase was obtained in the iron oxide containing glass ceramics and the $Li_2FeMn_3O_8$ phase was found in samples containing 8 mol% Fe_2O_3. TEM investigations showed the presence of (Li,Mn) ferrite particles, with an average size of 40 nm, dispersed in the β-wollastonite matrix in glass ceramics containing 4 mol% Fe_2O_3. The (Li,Mn) ferrite particle sizes in the glass ceramics containing 8 mol% Fe_2O_3 varied from a few μm to 5 nm. The Superconducting Quantum Interface Device (SQUID) result showed that only the glass ceramic containing 4 mol% Fe_2O_3 exhibited superparamagnetic behavior at temperature 300 K and ferromagnetic behavior at 4 K. The glass ceramics containing 8 mol% Fe_2O_3 exhibited ferromagnetic behavior at both temperatures. Sample 13, Table 4.1, provides the magnetic properties of the latter specimen.

4.2.2.2 Cobalt Ferrite ($CoFe_2O_4$)–Based Glass Ceramics

Among spinel ferrites, $CoFe_2O_4$ is of particular interest because of its reasonable saturation magnetization, excellent chemical stability, mechanical hardness, and high cubic magnetocrystalline anisotropy and has intensively been investigated for various applications such as gas sensors, magnetic fluids, catalysts, and thin films for applications in magneto-optical devices or high-density recording media (Abdel-Hameed et al., 2012).

Abdel-Hameed et al. (2012) studied the crystallization behavior and magnetic properties of ferrimagnetic glass ceramics in the systems $Fe_2O_3-CoO-MnO_2$, $Fe_2O_3-NiO-MoO_3$, and $Fe_2O_3-CoO-V_2O_5$.

The aim of their work was the preparation of glass ceramics containing single-domain magnetic particles of $CoFe_2O_4$ and

$NiMoO_4$ spinels. The studied glasses had the following compositions:

Glass 1, $39.76Fe_2O_3-0.24MnO_2-40CoO-20B_2O_3$ (mol%); **Glass 2**, $12.8Fe_2O_3-4MoO_3-63.2NiO-20B_2O_3$; **Glass 3**, $39.2Fe_2O_3-1.6V_2O_5-39.2CoO-20B_2O_3$.

The glasses that were prepared by quenching the melt between two steel plates consequently were heat-treated at 800 and 1,000°C for 4 h.

The main crystalline magnetic phases developed were $CoFe_2O_4$, with crystallite sizes of $\approx$14–20 nm, $NiMoO_4$, with crystallite sizes of $\approx$56–79 nm and $CoFe_2O_4$, with crystallite sizes of $\approx$6–8 nm in glasses 1, 2, and 3, respectively.

Magnetic Properties

In glass samples 1 and 3, M_S value decreased by increasing the heat-treatment temperature from 800°C to 1,000°C. This was attributed to the decrease of the $CoFe_2O_4$ phase by increasing the temperature of heat treatment.

In glass 2 which generally showed lower values of M_S, according to the authors, the small amount of $NiMoO_4$, the only phase exhibiting ferromagnetic properties, was responsible for the lower M_S values in these glass ceramics.

Table 4.1 depicts the magnetic properties of the prepared glass ceramics (sample 14).

4.3 Application of Hard Magnetic Nano-Glass Ceramics

Materials with high coercivity, called **hard magnetic materials**, are used in the fabrication of permanent magnets. Permanent magnets have application in electric motors, magnetic recording media (e.g., hard drives, floppy disks, or magnetic tape), and magnetic separation. Materials with nano-size particles have more extended and far reaching functional characteristics than ordinary coarse-grained materials. For example, these materials are used for manufacturing high-coercive permanent magnets with a wide range of applications from motor rotors to magnetic holders, and in ultrahigh-density magnetic recording media.

The creation of devices with ultrahigh recording densities approaching several TB/Cm^2 necessitated the development of processes for preparing chemically stable uniform nanoparticles with sizes in the order of 10 nm and coercive forces of several KOe (Kazin et al., 2009). The lower limit of the particle size is determined by the appearance of supermagnetism in sufficiently small nanoparticles, (with diameters <3–50 nm, depending on the material) in which the magnetization can randomly flip direction under the influence of temperature. For magnetic hardness to be conserved in particles with the smallest possible sizes, it is necessary to obtain single-domain particles with high magnetic magnetocrystalline anisotropy constant and shape anisotropy (Kazin et al., 2009).

In this regard, magnetically hard **M-type barium and strontium hexaferrites** have attracted considerable interest. These materials usually have great uniaxial magnetic anisotropy, low electrical conductivity and magnetostriction, good thermal stability, and are relatively inexpensive. In addition, they are chemically inert, a characteristic which is very important for nanoparticles, making a protective coating unnecessary. A protective coating, which is usually applied to enhance the chemical resistance, can induce particle coarsening and deterioration of functional characteristics (Kazin et al., 2009).

4.3.1 Barium Hexaferrite ($BaFe_{12}O_{19}$)–Based Glass Ceramics

Barium hexaferrite ($BaFe_{12}O_{19}$) is the model for a family of "M-type ferrites," so called because they are based upon the hexagonal **magnetoplumbite** structure.

Its crystal structure, though related to the spinel structures, is very much more complex. The large unit cell contains two formula units, i.e., a total of 64 ions. The Ba^{2+} and O^{2-} ions together form a close-packed structure with some of the layers cubic close-packed and others hexagonal close-packed. The origin of the magnetic properties are basically the same as those already discussed and can be summarized as follows: of the $12Fe^{3+}$ ions in a formula unit, nine are on octahedral sites, two on tetrahedral sites, and one on a five-coordinated site. Seven of the ions on octahedral sites and the one on a five-coordinated site have their spins in one sense and the remainder are oppositely directed. Thus, there are four more ions with spins in the one sense than there are with spins in the other sense. Since there are five electrons with parallel spins in each Fe^{3+} ion, there are 20 unpaired spins per formula unit, leading to a saturation magnetization of 20 μ_B per cell volume. Thus, $BaFe_{12}O_{19}$ has a high magnetic anisotropy with its "preferred direction of magnetization" ("easy" direction) along the *c*-axis (Moulson and Herbert, 2003, p. 489). Various substitutions are also made in the $BaFe_{12}O_{19}$ structure to tailor intrinsic magnetic properties, e.g., Sr for Ba, and partial substitution of Al for Fe, to increase coercivity.

Many researchers have prepared glass ceramics containing barium hexaferrite nanoparticles and studied their magnetic properties (Görnert et al., 1990, 1994; Lee and Speyer, 1994; Por et al., 1996; Pal et al., 1997; Taubert et al., 1997; Klupsch et al., 1999; Müller et al., 1999a,b; 2002; Rezlescu, L. 1999).

Görnert et al. (1990) prepared barium hexaferrite powders in which iron was partially replaced by cobalt and titanium via the crystallization of $BaO-B_2O_3-Fe_2O_3-CoO-TiO_2$ glasses. The glass melts were quenched between two rotating rollers to form amorphous flakes or tapes, 30–100 μm thick, and subsequently were annealed in the 300–900°C range for 10 s to several days. Under suitable heat-treatment conditions, in the 740–860°C range, both barium hexaferrite and a borate phase were obtained.

The ferrite particles were extracted by dissolving the borate phase in dilute (20%) acetic acid. The remaining slurry was carefully rinsed and dried. The resulting powder consisted only of the $BaFe_{12-2x}Co_xTi_xO_{19}$ particles with a narrow size distribution and an average size of ~50 nm.

The coercive force of the doped hexaferrite particles was far lower than the undoped particles; the respective values were 120 and 378 kA/m. Table 4.2 provides the characteristics of a typical sample of the above glass ceramics (sample 1).

Lee and Speyer (1994) fabricated glass ribbons of approximate composition $0.13Na_2O-0.30BaO-0.30Fe_2O_3-0.27SiO_2$, using roller quenching, after melting at 1,400°C. By annealing the glass samples at 710°C for 2 h, the crystallization of $BaFe_{12}O_{19}$ particles was observed. It was revealed that the barium ferrite crystallites, with a mean diameter of 55.88 nm, were uniformly distributed in the glass matrix. $BaFe_2O_4$ and $Na_2Ba_2Si_2O_7$ were also formed on heat treatment. Figure 4.7 depicts the TEM microstructure of a typical glass-ceramic sample. The magnetic properties of the aforementioned glass ceramic are given in Table 4.2 (sample 2).

Müller et al. (1999a, 2002) prepared barium hexaferrite, $BaFe_{12-2x}Ti_xCo_xO_{19}$ ($x = 0.8$ and 1.2) ferrofluids, using oleic acid as a surfactant and dodecane as a carrier liquid. The ferrite particles were prepared by crystallization of a $40BaO-33B_2O_3-27(Fe_2O_3 + CoO + TiO_2)$ (mol%) glass.

The glass melt first was quenched with the rate of ~04 K/s to obtain amorphous flakes, which subsequently were heat-treated at 580°C for 10 days. The resulting ferrite particles were

Table 4.2 Composition and Properties of Some Hard Magnetic Nano-Glass Ceramics

Sample No.	Chemical Composition	Major Magnetic Phase/Size	Heat-Treatment Temp. (°C)/Time	Saturation Magnetization (emu/g)/Coercitive Force (Oe)	Remanence Magnetization (emu/g)	Reference
1	$BaO—B_2O_3—Fe_2O_3—CoO—TiO_2$ glass system	$BaFe_{12-2x}—Ti_xCo_xO_{19}$/50 nm	750°C/8 h (for $x = 0.775$)	~50/1,500	—	Görnert et al. (1990)
2	$13Na_2O$, 30BaO, $30Fe_2O_3$, $27SiO_2$ (mol%)	$BaFe_{12}O_{19}$/55 nm	710/2 h	24.4/3,060	—	Lee and Speyer (1994)
3	35BaO, $25Fe_2O_3$, $40B_2O_3$ (mol%)	$BaFe_{12}O_{19}$/17.6 nm	530/1.5 h +580/1 h	0.75/40	—	Pal et al. (1997)
4	$Fe_2O_3—BaO—B_2O_3—Sb_2O_3$ Glass system	$BaFe_{12}O_{19}$/~100 nm	700/2 h	~18/2,200	—	Rezlescu, L. (1999b)
5	$27Fe_2O_3—40BaO—(33-x)B_2O_3-xSiO_2$ ($5<x<20$) mol%	$BaFe_{12}O_{19}$/~50–500 nm ($x = 20$)	Melt quenched + sintered at 1,000/0.5 h	~61/5,000	—	Müller et al. (1999b)
6	$SrFe_{12}O_{19} + 12SrSiO_3$	$SrFe_{12}O_{19}$/260 × 50 nm	1,000/2 h	~17/4,800	—	Zaitsev et al. (2005)
7	$SrFe_{12}O_{19} + 12SrB_2O_4$	$SrFe_{12}O_{19}$/20 × 110 nm	700/2 h	51/3,900	—	Zaitsev et al. (2006a)
8	$SrFe_{12}O_{19} + (12-n)SrB_2O_4 + nSrSiO_3$, $n = 3, 6, 9$	$SrFe_{12}O_{19}$/420 × 100 nm	850/2 h	~17/4,400	—	Zaitsev et al. (2006b)
9	$14SrO—6Fe_2O_3—12B_2O_3$	$SrFe_{12}O_{19}$/<100 nm	680/no holding	~21/3,200	—	Kazin et al. (2010)

Figure 4.7 TEM microstructure of a quenched ribbon of a barium hexaferrite glass ceramic after crystallization at 710°C for 2 h. (After Lee and Speyer, 1994. Reproduced with permission of Springer Science + Business Media, Inc.)

isolated by dissolving the matrix in acetic acid. The obtained powder with an average particle size of 8 nm showed almost pure superparamagnetic behavior.

Pal et al. (1997) prepared Ba ferrite ($BaFe_{12}O_{19}$) containing glass ceramics from the glass of composition $35BaO \cdot 25Fe_2O_3 \cdot 40B_2O_3$ (mol%). The glass melt was quenched between two brass plates and the obtained glass specimens were heat-treated by a two-step crystallization schedule at 530°C and 580°C temperatures, respectively. It was revealed that $BaFe_{12}O_{19}$ was the sole crystalline phase present in the resulting glass ceramics. The crystallite sizes of the latter phase depended on the time of heat treatment and varied in the 8.2–17.6 nm range.

The magnetic properties of the glass ceramics were also determined in the temperature range 20–300 K. It was found that the magnetization value did not attain saturation at the highest magnetic field applied during the measurements. The rather low value of magnetization was attributed to the small-volume fraction of the hexaferrite phase in the glass ceramics. The obtained coercive forces were also much smaller than that of bulk barium hexaferrite for all the specimens and its value increased as the particle size of the barium hexaferrite phase was increased. Table 4.2 depicts the properties of one of the aforementioned glass ceramics (sample 3).

Glass strips of the system Fe_2O_3–BaO–B_2O_3 with the addition of Sb_2O_3 were also prepared by rapid quenching of the melt between two copper rollers (Rezlescu, L. 1999), the obtained strips were treated successively at various temperatures, in the range 200–780°C, for 2 h.

The crystallization of $BaFe_{12}O_{19}$ ferrite with particle sizes 50–330 nm was observed at temperatures >550°C.

The effect of annealing temperature and grain size on the specific magnetization and coercivity was also investigated. The specific magnetization showed a sharp increase for small grain size after annealing at >600°C. This was attributed to the onset of the crystallization of $BaFe_{12}O_{19}$ ferrite and the increase of its amount in the glass host and the possibility of decreasing the number of superparamagnetic particles. The value of coercivity was also increased with increasing the grain size and then, for a further increase of the size, it slightly decreased. The magnetic properties of some of the aforementioned samples are depicted in Table 4.2 (sample 4).

Müller et al. (1999b) also described a new method of magnetic powder preparation from which compact glass ceramics containing Ba ferrite particles can be fabricated. In this method, Ba ferrite particles can maintain the single-domain behavior of the starting powder.

First, single-domain Ba ferrite particles were obtained by the heat treatment of a quenched melt of $27Fe_2O_3$–40BaO–$(33-x)$ B_2O_3–$xSiO_2$ ($5<x<20$) mol%, at temperatures from 780°C to 820°C for 4–24 h.

By milling the crystallized glass and leaching out the borate phase, a homogeneous mixture of Ba ferrite plate-like crystals with a preferential diameter range of 50–500 nm and much finer SiO_2 particles (diameter ~5 nm) was achieved. The powder mixture was then isostatically pressed at 150 MPa and sintered in 980–1,000°C temperature range. By this method, glass ceramics with >90 mass% ferrite content could be prepared, in which the ferrite particles mostly maintained their single-domain behavior.

Magnetic properties of the prepared glass ceramics were also determined, and it was shown that the coercivity values of the samples were much higher than the values of typical conventional ferrite samples. Figure 4.8 depicts the hysteresis loop of a typical SiO_2-barium hexaferrite glass-ceramic sample and Table 4.2 gives its magnetic properties (sample 5).

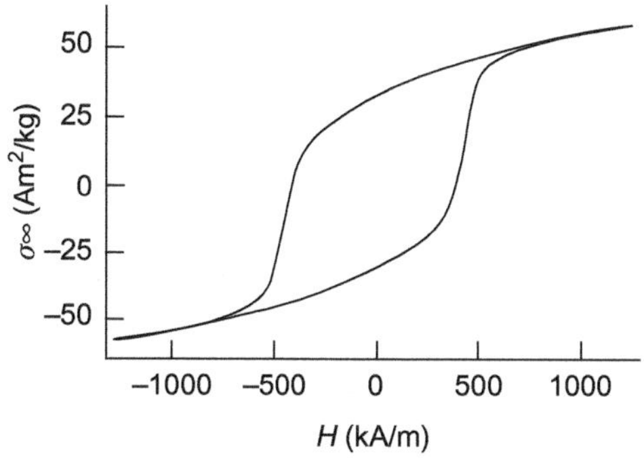

Figure 4.8 Hysteresis loop of a barium hexaferrite-SiO_2 glass ceramic. (After Müller et al., 1999b. Reproduced with permission of Elsevier Science B.V.)

In summary, on the basis of the above reports it can be inferred that the barium ferrite glass ceramics were mainly crystallized from borate glasses, in the $BaO-Fe_2O_3-B_2O_3$ system, and less frequently from silicate or mixed silicate–borate glasses.

It should be stated that the borate-based glasses are vulnerable to moisture attack. This characteristic would be an advantage in the cases in which the borate phase is to be leached out, e.g., in the preparation of particulate magnetic materials.

However, when the glass ceramics are to be used in the bulk form in which the magnetic phase is dispersed in a glassy matrix, the matrix phase should have high resistivity to moisture attack. This is one of the main factors that stimulated the studies on the silicate or mixed borate–silicate glasses as more moisture-resistant hosts for ferrite crystals.

It is also evident that due to the marked crystallization tendency of these glass melts, they were usually prepared by quenching methods. Unlike some glass, ceramic materials in which the crystalline particles are developed during the cooling process, the barium ferrite glass ceramics are invariably subjected to further crystallization heat treatment.

Moreover, due to the high crystal nucleation rates of these glasses, usually no nucleating agents are used in their crystallization process.

4.3.2 Strontium Hexaferrite ($SrFe_{12}O_{19}$)–Based Glass Ceramics

Zaitsev et al. (2005) prepared glass samples with nominal composition: ($SrFe_{12}O_{19} + 12SrSiO_3$). Precursor samples, in the form of pellets, were first prepared from chemicals by heat treatment at 800°C for 24 h. The pellets were then melted at 1,400–1,500°C and the melt was quenched between two copper plates.

The obtained glass samples were subsequently heat-treated at temperatures of 700–1,250°C. The magnetic phase started to develop at 700°C and the fully crystalline $SrFe_{12}O_{19}$–$SrSiO_3$ material formed at 1,000°C. The particles of the $SrFe_{12}O_{19}$ phase demonstrated plate-like shape, with the size and aspect ratio dependent on the heat-treatment conditions. The average particle size for the sample annealed at 1,000°C was 50–260 nm. With the elevation of the annealing temperature, the size of the particles constantly increased, while the aspect ratio first slightly increased (1,100°C), then dropped almost by a factor of 4.

High coercivity of the samples suggested that the particles were mostly monodomain. Table 4.2 gives the characteristics of a sample of the aforementioned glass ceramics (sample 6).

Glass samples were also prepared in the system SrO–Fe_2O_3–B_2O_3 with the nominal composition: $SrFe_{12}O_{19}$ + $12Sr_nB_2O_{n+3}$ ($n = 1$, 1.5, 2) (Zaitsev et al., 2006a). The precursors were prepared from heat treatment of the appropriate chemicals at 700°C for 24 h. The powders obtained were then melted in a platinum crucible at 1,250°C for 2 h. The melt was quenched between two steel rollers forming a glass plate with a thickness of ~0.1 mm.

The prepared glass samples were then heat-treated in 600–950°C range for 2 h. It was revealed that during the crystallization process, in addition to $SrFe_{12}O_{19}$, several nonmagnetic (mostly borate) phases were also formed.

The latter phases were subsequently dissolved out in a hot (10%) acetic acid solution. In some glass-ceramic samples (with $n = 1$ and 1.5), $SrFe_{12}O_{19}$ remained as the sole crystalline phase.

The magnetic properties of some samples which annealed at different temperatures were also reported. It was observed that by increasing the temperature, the magnetization and coercive force of all the samples were generally increased. Samples of the nominal composition $SrFe_{12}O_{19} + 12SrB_2O_4$ ($n = 1$) were characterized by the highest coercive force, reaching 6,300 Oe after annealing at 900°C. The value of the magnetization at 9,100 Oe for this sample was ~17 emu/g. The magnetic properties of a powder consisting of $SrFe_{12}O_{19}$ nanoparticles is given in Table 4.2 (sample 7). This powder was prepared from the latter sample ($n = 1$) after annealing at 700°C and leaching in acetic acid.

Zaitsev et al. (2006b) also prepared glass samples in the same system SrO–Fe_2O_3–B_2O_3 with the nominal compositions: $SrFe_{12}O_{19} + (12 - n)SrB_2O_4 + nSrSiO_3$, $n = 3$, 6, 9 by rapid quenching of the melt. The resulting samples were subsequently annealed at temperatures of 600–900°C. It was shown that $SrFe_{12}O_{19}$ crystallized above 700°C and formed nano- and submicron platelet particles. The aspect ratio of these particles depended on the heat-treatment conditions.

Magnetic properties of the samples were also presented. It was revealed that the magnetization of the samples increased with the annealing temperature up to 850°C. Maximum magnetization values observed for the annealing temperature of 850°C corresponded to almost complete crystallization of $SrFe_{12}O_{19}$ in all the samples. The coercive force of all the samples also increased with the heat-treatment temperature and the specimens with the highest n value, (9) generally exhibited higher coercivities.

Table 4.2 provides the characteristics of a glass-ceramic sample of the latter group (sample 8).

Kazin et al. (2010) synthesized strontium hexaferrite single-domain particles embedded in the borate matrix by the glass-ceramic method. Two glass samples with the compositions $14SrO{-}6Fe_2O_3{-}12B_2O_3$ (1) and $13SrO{-}5.5Fe_2O_3{-}4.5Al_2O_3{-}4B_2O_3$ (2) were prepared by melting a mixture of oxides at 1,250°C for 2 h and quenching the melt between two steel rollers. The samples of glass 1 were inserted into a cold furnace and taken out, as soon as the desired annealing temperature, T_{an}, was reached, while glass 2 sample was annealed in two stages: first at 750°C for 2 h and then at 970°C for 24 h with subsequent air quenching.

The glass ceramics exhibited various morphologies and magnetic characteristics depending on the initial glass composition and the thermal treatment conditions.

After annealing glass 1 at temperatures $<550°C$, the samples showed paramagnetic behavior suggesting the absence of any ferromagnetic phase. At $T_{an} = 640–680°C$, both the magnetization and coercivity values abruptly increased indicating the formation of a major quantity of the hexaferrite phase at this temperature range. The presence of strontium hexaferrite was also confirmed by XRD analysis. The coercivity value continuously increased with the annealing temperature and reached 6 kOe at $T_{an} = 950°C$.

The mean size of plate-like hexaferrite particles, that were in the order of tens of nm at low T_{an} temperatures (680°C), increased to hundreds of nm in 680–950°C temperature range, and their thickness/diameter ratio was also increased.

Most interestingly, the two-stage annealing of glass 2 resulted in the submicron-grained glass ceramics with a coercivity value of 12.5 kOe, the highest value reported for hexaferrites. This was attributed to the partial substitution of Al ions for Fe ions in the hexaferrite structure.

The magnetic properties of a sample of glass 1 heat-treated at 680°C and containing nanoparticles of $SrFe_{12}O_{19}$ is depicted in Table 4.2 (sample 9).

More recently, Liu et al. (2012) prepared nano-ferrimagnetic glass ceramics with the main crystal phase of $SrFe_{12}O_{19}$. The parent glass samples, located in the $SrO-Al_2O_3-Fe_2O_3-B_2O_3$ system, were produced from carefully prepared batch materials, using the molecular-scale homogenously mixed compounds, obtained from aqueous solutions, that after solvent evaporation were melted and naturally cooled. The glass samples then were heat-treated at the temperatures as determined by Differential Thermal Analysis (DTA). It was found that Fe_3O_4 crystal precipitated during naturally cooling the melt from 1,500°C to annealing temperature; and $SrFe_{12}O_{19}$ nanoparticles, with particle sizes 20–50 nm, appeared as the main crystal phase, after heat-treating at 865°C for 2 h. The measurement of magnetic properties of the aforementioned ferromagnetic glass ceramics indicated that the saturation magnetization and the intrinsic coercivity values were 32 A m^2/kg and 236 KA/m, respectively.

4.4 Concluding Remarks

In recent years, magnetic nanoparticles have attracted much attention as very promising candidates for a broad spectrum of industrial applications such as magnetic seals in motors, magnetic inks for bank cheques, magnetic recording media, and biomedical applications such as magnetic resonance contrast media and therapeutic agents in cancer treatment. On the other hand, the synthesis of magnetic nanoparticles using the glass-ceramic route has also been the subject of numerous studies. In addition to its well-known advantages, such as the ease of preparation and the economic considerations, the glass-ceramic route is also very convenient for providing controllable particle size and morphology, as well as for obtaining narrow size distribution of nanoparticles.

The glass-ceramic route has already been applied to the preparation of some ferrite nanoparticles such as magnetite and $ZnFe_2O_4$. The most promising applications of these soft magnetite nanoparticles, that have recently attracted much interest, are medical applications, such as the magnetic drug targeting, magnetic separation of leukemia cells from the blood, magnetic hyperthermia, and contrast agents in the MRI process.

Recently, many research activities were also conducted to utilize these magnetic nanoparticles in the less common areas of biomedical applications. These nanoparticles were tested for tissue engineering applications, e.g., in the mechanical conditioning of the cells growing in culture. In such systems, the magnetic particles attach to the cell membrane and, upon the application of a magnetic field, activate the membrane and initiate some biochemical reactions within the cell, thereby promoting the growth of functional bone and cartilage.

Another example is the application of magnetic liposomes (lipid vesicles, containing magnetic nanoparticles in their structure either in the lipid bilayer or in the aqueous compartment) as "vehicles" for targeted drug delivery.

On the other hand, in recent years, there has been a resurgence of interest in the field of hard magnetic particles, such as barium hexaferrite, prepared by the glass crystallization route. This method enables the precipitation of nanoparticles, with the mean size of 5–50 nm, with a fairly narrow crystallite size distribution. The crystallite size can be adjusted by the temperature and time of heat treatment.

Nano-crystalline ferrites enable the preparation of stable ferrofluids for a large number of carrier liquids due to their lack of sedimentation in the gravitational field.

They have also great potential for manufacturing ultrahigh-density magnetic recording media and high-coercive permanent magnets with a wide spectrum of applications.

The fabrication of devices with ultrahigh record densities approaching several TB/cm^2 requires the preparation of chemically stable uniform nanoparticles with sizes in the order of 10 nm and coercive forces of several kOe. In order to conserve the magnetic hardness in such small nanoparticles (against the supermagnetism effect), it is necessary to obtain single-domain particles with high magnetic magnetocrystalline anisotropy constant and shape anisotropy. Barium and strontium hexaferrites are very promising materials for the above applications because of their characteristics such as the great uniaxial magnetic anisotropy and low electrical conductivity.

5

BIOMEDICAL APPLICATIONS OF NANO-GLASS CERAMICS

CHAPTER OUTLINE

During the past four to five decades, advances in **biomedical materials** or **biomaterials** have made significant contributions to the development of the modern health-care industry. These materials are defined as "natural or synthetic materials suitable for introduction into living tissue especially as part of a medical device."

As in other fields of application, the ceramic materials have also developed and gradually gained more importance in various fields of biomedical applications in recent years.

These materials, now widely known as **bioceramics**, are usually manufactured in the form of bulk materials (porous or dense), granules, or coatings, and are primarily used owing to their

V.K. Marghussian, Nano-Glass Ceramics, ISBN: 978-0-323-35386-1.

DOI: http://dx.doi.org/10.1016/B978-0-323-35386-1.00005-0

biocompatibility, chemical stability, high wear resistance, and low density.

On the other hand, especially after the pioneering work of Hench et al. (1971) and introduction of the first bioactive glasses, the glasses and glass ceramics possessing bioactive characteristics have attracted much attention for clinical applications. Hence, a new class of glass ceramics, **bioglass ceramics**, has been defined, encompassing the glass ceramics with biomedical applications.

When bioactive materials are implanted in the human body, a biologically active carbonated hydroxyl apatite layer is formed on their surfaces. Subsequently, the implant bonds to and integrates with living bone. *In vivo* implanting study revealed that bioactive glasses show no local or systemic toxicity, no inflammation, no foreign-body response, and bond to both soft and hard tissues without an intervening fibrous layer (Hench, 1998).

Glass ceramics have the added advantage of better mechanical properties than glass. Furthermore, in addition to the general economic and technical priorities of glass ceramics, they can provide greater possibility for the manipulation of the properties such as strength, degradation rate, and Coefficient of Thermal Expansion (CTO) by heat treatment.

In recent decades, **bioglass ceramics** were intensively used as bone grafts or osseous fillers for local defects in orthopedic and dental surgery.

The recent trend in bioceramic research is mainly concentrated on bioactive and bioresorbable calcium phosphates-based ceramics, i.e., hydroxyapatite (HA), bioactive glasses, tricalcium phosphates, and biphasic calcium phosphates, as they exhibit superior biological properties over other materials. However, these bioceramics usually show poor mechanical performance, which restricts their uses in load-bearing applications.

In recent years, the nano-structured materials have been extensively studied by various researchers to overcome the existing limitations of calcium phosphates.

Nano-structured materials, including **nano-glass ceramics**, are expected to improve the fracture toughness and other mechanical properties. Nanoscale ceramics can exhibit significant ductility before failure contributed by the grain-boundary phase. It has been reported that a brittle ceramic with nano-sized grains could sustain a large plastic strain up to 100% (Kalita et al., 2007).

Nano-structured biomaterials also promote osteoblast adhesion and proliferation, osseointegration, and the deposition of calcium-containing minerals on the surface of these materials (Xu et al., 2004).

Nano-structured ceramics can also be sintered at lower temperatures; thereby problems associated with high-temperature sintering processes are also eliminated.

In recent years, many researchers have investigated the fabrication and properties of nano-structured glass ceramics, mainly calcium phosphates, as potential substitutes for hard tissue, in orthopedic and dental surgery.

In this chapter, the fabrication, properties, and application of some of the most important types of **nano-structured bioglass ceramics** are discussed.

5.1 Definitions

Biocompatibility

The ability of a material to perform with an appropriate host response, in a specific application, and not having toxic or injurious effects on biological systems.

Bone Grafting

Bone grafting is a surgical procedure that replaces missing bone in order to repair bone fractures or other problems, which pose a significant health risk to the patient, or which fail to heal properly.

Bone grafts may be autologous (bone harvested from the patient's own body), allograft (bone usually obtained from a bone bank), or synthetic.

Cellular Differentiation

The process by which less specialized cells become more specialized cell types is called cellular differentiation. Differentiation occurs numerous times during the development of a multicellular organism, e.g., adult stem cells divide and create fully differentiated daughter cells during tissue repair. Differentiation dramatically changes a cell's size, shape, membrane potential, metabolic activity, and responsiveness to signals.

Cellular Proliferation

Unlike blood cells, muscle cells, or nerve cells, which do not normally replicate themselves, stem cells may replicate many times, or proliferate via dividing and renewing themselves for long periods.

Osteoconduction

Osteoconduction occurs when the bone graft material serves as a scaffold for new bone growth that is perpetuated by the native bone. **Osteoblasts** (mononucleate cells that are responsible for bone formation), from the margin of the defect that is being grafted, utilize the bone graft material as a framework, upon which to spread and generate new bone.

Osteoinduction

Osteoinduction involves the stimulation of osteoprogenitor cells to differentiate into osteoblasts that then begin new bone formation. A bone graft material that is osteoconductive and osteoinductive will not only serve as a scaffold for currently existing osteoblasts, but will also trigger the formation of new osteoblasts, theoretically promoting faster integration of the graft.

Osteogenesis

Osteogenesis occurs when vital osteoblasts, originating from the bone graft material, contribute to new bone growth, along with bone growth generated via the other two mechanisms.

Bioinert Materials

Although no material implanted in living tissues is completely inert, and all materials elicit a response from living tissue, the bioinert materials evoke a weak response.

Hence, when a bioinert material is used as an implant, the interface between the implant and living tissues is not chemically or biologically bonded and there is the possibility of relative movement. In this case, usually a progressive development of a nonadherent fibrous capsule occurs in the interface, in the presence of both soft and hard tissues. Movement at the biomaterial–tissue interface eventually leads to deterioration in the function of the implant or of the tissue at the interface or both of them.

Bioactive Materials

Bioactive material is one that elicits a specific biological response at the interface of the material, which results in the formation of a bond between the tissues and the material.

Certain bioactive glasses, ceramics, glass ceramics, and composites have been shown to form a mechanically strong bond to bone, and some even more specialized compositions of bioactive glasses bond to soft tissues as well as bone.

A common characteristic of bioactive glasses and bioactive ceramics is a time-dependent, kinetic modification of the surface that occurs upon implantation. The surface forms a biologically active hydroxycarbonate apatite (HCA) layer, which is equivalent chemically and structurally to the mineral phase in bone and provides the bonding interface with the tissues

Resorbable Biomaterials

Resorbable biomaterials are degraded gradually over time and are replaced by the natural host tissue. This leads to a very thin or nonexistent interfacial layer. This is the optimal solution to biomaterial problems, if the requirements of strength and short-term performance can be met. The requirements for a successful resorbable bioceramic are:

1. Maintenance of strength and stability of the interface during the degradation period and replacement by the natural host tissue.
2. Matching resorption rates to the repair rates of body tissues, which themselves vary enormously, depending upon the type of tissue and its age and health.

As described above, in recent years, the nano-glass ceramics have attracted great attention as promising candidates for various applications in the field of biomedical materials.

5.2 Applications

The main applications of nano-glass ceramics in this field, except the application of magnetic nano-glass ceramics, which has been discussed in Section 4.2.1, can be classified as follows:

1. **Nano-structured bioglass-ceramic coatings**
2. **Nano-glass ceramics in implantology and dentistry**

5.2.1 Nano-Structured Bioglass-Ceramic Coatings

The main weakness of most of the metallic alloy implants, despite their favorable mechanical properties, is the formation of a dense fibrous tissue at the implant–bone interface due to their bioinertness, which may result in the loosening of the device. Coating these implants with a bioactive coating is an effective way to solve this problem, as the coating can accelerate new bone formation at the bone–implant interface and thereby interlock the implant with the surrounding skeletal tissue. This may lead to the formation of a more stable interface strong enough to support lifelong functional loading.

In order to overcome the existing limitations of some coatings, the nano-structured coatings in biomaterials have also been extensively studied by various researchers, in recent years.

The nano-structured materials are expected to improve the fracture toughness and other mechanical properties of biomaterials. These materials can also promote osteoblast adhesion and proliferation, osseointegration, and the deposition of calcium containing minerals on their surface (Xu et al., 2004).

Various techniques have been used for the application of the biomedical coatings, such as conventional enameling, sputtering, sol-gel processing, ion beam deposition, plasma spraying, electrophoretic deposition, and pulsed laser deposition.

All the coatings that are produced as amorphous or glassy layers in the first stage of their fabrication process, and subsequently are subjected to a controlled heat treatment, in order to transform them into partially crystalline products, can structurally be classified as glass-ceramic coatings.

Bioglass-ceramic coatings, according to the above definition, have attracted considerable attention, as they provide great possibilities to manipulate their properties by post-treatments, including strength, degradation rate, and Coefficient of Thermal Expansion (CTE).

Considering the superior properties of nano-structured materials, as discussed above, many researchers have also investigated the fabrication and properties of **nano-structured glass-ceramic coatings on biomaterials**, utilizing various techniques.

Some typical examples of the research carried out in recent years in the field of fabrication of nano-structured glass-ceramic coatings for biomaterials are discussed here.

The studied coatings have been classified according to their fabrication techniques.

5.2.1.1 Enamels

Lopez-Esteban et al. (2009) fabricated graded bilayered glass-ceramic composite coatings on titanium alloy (Ti–6Al–4V) substrates using an **enameling technique**. These alloys are widely used as orthopedic and dental implant materials due to their excellent biocompatibility, low corrosion rates, and favorable mechanical properties, but suffer from the aforementioned drawback of the possibility of formation of dense fibrous tissue at the implant–bone interface due to its bioinertness.

Bilayered coatings were fabricated by the sequential deposition of two layers, the first layer, in contact with the alloy, consisted of a higher-silica glass in the SiO_2–CaO–MgO–Na_2O–K_2O–P_2O_5 system, and the second layer a mixture of a lower-silica glass, in the same system, and 0, 5, 10, or 15 wt% CP (CaO–P_2O_5) particles. The particles were HA (calcium hydroxyapatite), tricalcium phosphate (β-TCP), and dicalcium phosphate anhydrate. The final coatings that were fired in 800–820°C temperature range were of good quality and did not crack or delaminate. Glass–metal adhesion was achieved through the formation of a nano-structured Ti_5Si_3 layer. A surface layer, containing a mixture of a low-silica glass (<53 wt% SiO_2) and synthetic HA particles, promoted the precipitation of new apatite during *in vitro* tests, whereas the coatings containing other CPs failed to promote *in vitro* apatite precipitation. In all cases, the high-silica layer in contact with the metal remained intact and maintained good adhesion even after 12 months in SBF (simulated body fluid), providing long-term stability.

5.2.1.2 Plasma-Sprayed Coatings

Wu, et al. (2009b) prepared a Ca–Si–Ti-based sphene ($CaTiSiO_5$) coating on titanium alloy (Ti–6Al–4V) for orthopedic applications using the **conventional plasma spray** method. Sphene powders were synthesized by the sol-gel process. The calcined sphene powders with the particle size of 45–200 μm were sprayed onto Ti–6Al–4V disks using an atmosphere plasma spray system.

Plasma-sprayed sphene coating on Ti–6Al–4V exhibited a significantly improved bonding strength (33.2 ± 2.4 MPa) and chemical stability compared with plasma-sprayed HA coating. This coating also showed excellent chemical stability and cellular bioactivity. It supported human osteoblast (HOB)-like cell attachment and significantly enhanced HOB proliferation and differentiation compared with plasma-sprayed HA coating and uncoated Ti–6Al–4V that were the indications of its potential application for orthopedic implants.

Wang et al. (2011) prepared nano-structured hardystonite (HT), ($Ca_2ZnSi_2O_7$), and sphene (SP; $CaTiSiO_5$) glass-ceramic coatings for titanium alloys (Ti–6Al–4V).

Glass-ceramic coatings were deposited on Ti–6Al–4V alloy using the **conventional plasma spray technique**.

The HT ($Ca_2ZnSi_2O_7$) and SP ($CaTiSiO_5$) powders were synthesized by high-temperature solid reaction and the as-prepared powders, after grinding and sieving, were mixed with 6 wt% polyvinyl alcohol (PVA) solution in order to make it more flowable for plasma spraying.

An atmospheric plasma spray system was utilized to deposit HT and SP coatings on commercial Ti–6Al–4V disks.

The bonding strength of both coatings was higher than that of plasma-sprayed HA coatings and their hardness values were also higher compared with Ti–6Al–4V alloys and HA coatings.

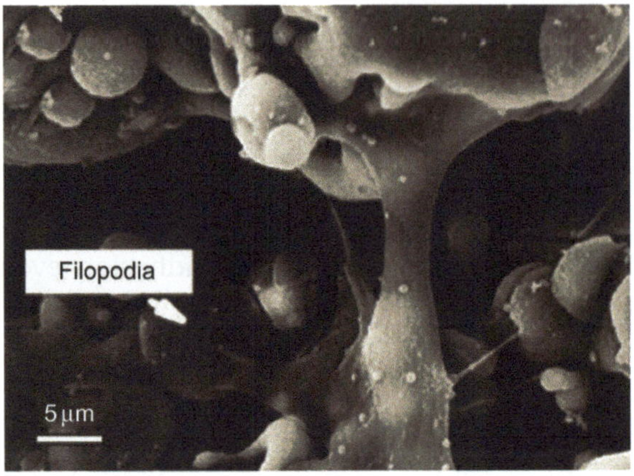

Figure 5.1 SEM micrographs depicting the attachment, spreading, and proliferation of HOBs on the surface of HT coating cultured for 24 h. (After Wang et al., 2011. Reproduced with permission of the Royal Society Publishing.)

Scanning electron microscopy (SEM) micrographs of the HT coating depicted in Figure 5.1 showed the attachment, spreading, and proliferation of primary HOBs on the surface of the coating cultured for 24 h. Similar behavior was reported for SP coating.

Both types of coatings were capable of enhancing the expression of bone-related genes, which was ascribed to the Ca and Si ions released from the coatings.

It was also shown that, HOBs on HT coatings exhibited a higher proliferation rate than SP coatings and Ti–6Al–4V alloys, which was attributed to the release of Zn ions from the HT coatings.

Xiao et al. (2011) used **the liquid precursor plasma spraying technique** to prepare P_2O_5–Na_2O–CaO–SiO_2 bioactive glass-ceramic coatings (BGCCs). Two different precursors, in the form of sol and suspension, were used as **liquid feed stocks** for plasma spraying. The nominal composition of the starting glass was SiO_2 45%, Na_2O 24.5%, CaO 24.5%, and P_2O_5 6% (wt%).

Similar materials were used to prepare the sol and suspension precursors except the catalysts used for the hydrolysis reaction.

It was observed that though, after heat treatment at 800°C, $Na_2Ca_2Si_3O_9$ was the dominant phase in both samples, prepared using different precursors, but substantial differences in the phase content and microstructure existed between the samples. Whereas highly crystallized glass-ceramic coatings with large splat size were obtained using the sol precursor, the coatings with a major amorphous phase and much finer splat size were obtained using the suspension precursor. The observed difference was attributed to the gelation process of the sol precursor in the plasma plume, which hindered the droplet breakup and led to the insufficient melting of the precursor feedstock and lack of formation of glassy phases upon supercooling.

The nano-structured bioactive glass coatings synthesized using the suspension precursor contained crystalline granules with the size of 20–100 nm.

The *in vitro* bioactivity studies of the aforementioned coatings indicated a fast HA formation capability of the synthesized BGCCs. This was attributed to the existence of the nano-granules in the coating which accelerated the dissolution of the coatings and resulted in the faster formation of the apatite layer.

5.2.1.3 Coatings Produced by Sol-Gel Technique

Kim et al. (2004) studied the preparation and application of HA coatings on titanium (Ti) substrates utilizing the sol-gel technique. In order to improve the bonding strength between the HA layer and Ti substrate, a titania (TiO_2) buffer layer was also inserted.

The bonding strength of the HA/TiO_2 double layer coating on Ti was markedly improved in comparison with the HA single coating. The highest strength value was 55 MPa after heat treatment at 500°C for 1 h. The improvement in bonding strength with the insertion of TiO_2 was attributed to the coating structure, exhibiting a dense and uniform microstructure comprising of tiny nano-sized crystallites (Figure 5.2).

The authors suggested that the tight bonding of the TiO_2 layer, to both the HA layer and the Ti substrate, is another important factor responsible for the high bonding strength values. This was because of the favorable chemical affinity of TiO_2 with respect to HA as well as to Ti.

More recently, Sebdani and Fathi (2012) used the sol-gel technique to prepare HA–forsterite–bioactive glass nano-coating for 316L stainless steel substrate.

The nano-composite coatings were prepared by adding 10 wt% bioactive glass nano-powder and forsterite nano-powder (10, 20, and 30 wt%) as reinforcements into the HA sol.

The 316L stainless steel samples were coated by dipping into the sol and after aging at ambient temperature and drying, heat-treated at different temperatures. The optimum heat-treatment temperature, at which a homogeneous and crack-free coating could be obtained, was determined as 600°C. The prepared coatings were comprised of crystallites, with the mean size of <100 nm.

During the immersion of coated samples in the SBF, at predetermined time intervals, an apatite layer was formed at their

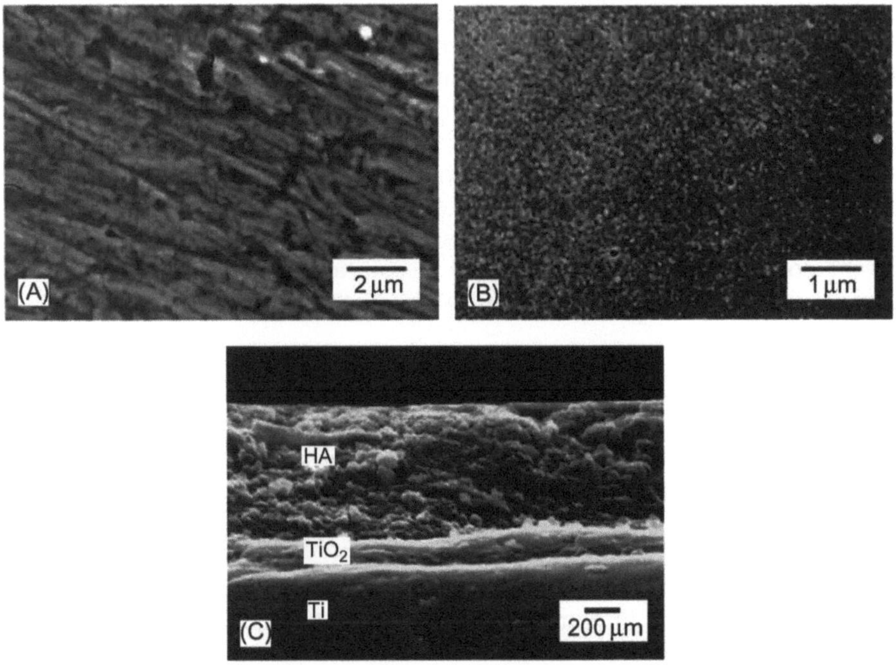

Figure 5.2 SEM images of the various coating systems deposited onto Ti: (A) TiO_2 coating surface, (B) HA/TiO_2 double layer coating surface, and (C) HA/TiO_2 double layer coating cross-sectional views. The heat treatment for each coating was performed at 500°C for 1 h in air. (After Kim et al., 2004. Reproduced with permission of Elsevier Science B.V.)

surface, and the stimulation for apatite formation was increased, by increasing the forsterite amounts. The authors suggested that the HA–forsterite–bioactive glass nano-composite coatings might be good candidates for biomedical applications.

5.2.1.4 Coatings Produced by the Magnetron Sputtering Technique

Berbecaru et al. (2010) prepared coatings by the magnetron sputtering technique to be applied on 0.3 mm thick medical-grade Ti–6Al–4V foils, with smooth surfaces as deposition substrates. The bioglass sputtering target was manufactured by cold pressing from high-purity 45S5 bioactive glass (Bioglass®) powder with a 7.5 μm average grain size. The deposition was carried out by the magnetron sputtering technique in radio frequency mode. A constant argon pressure of 0.45 Pa was used for all the depositions. The as-deposited thin layers were annealed for 2 h at 700°C in air, followed by a slow cooling in order to induce crystallization. The main crystalline phases identified by X-ray diffraction (XRD) after annealing were complex silicates: $Na_4Ca_4Si_6O_{18}$ (combeite); Ca_2SiO_4; $CaSiO_3$ (wollastonite), and phosphates: $NaCaPO_4$, and $Ca(PO_3)_2$ as well as titanium dioxide, TiO_2 (rutile and anatase).

The behavior of the coatings was investigated by soaking the samples in SBF. The polymerization reaction in the surface layer and the formation of a silica-rich layer appeared for the sample immersed for 72 h in SBF, as a major condition of subsequent precipitation of Ca- and P-rich phases.

According to the authors, the formation of the silica-rich layer, along with the presence of the Si—O bonds with nonbridging oxygens (which can accelerate the ionic exchange processes between the bioglass surface and SBF solution), as well as the presence of water and carbonates at the film surface, are the indications of the possibility of the formation of a carbonated HA layer on the top of the film for prolonged immersion times. As discussed before, this can be taken as an indication of the bioactivity of the coated samples.

5.2.2 Nano-Glass Ceramics in Implantology and Dentistry

In recent years, many researchers have investigated the fabrication and properties of **nano-structured glass ceramics**, mainly calcium phosphates, as potential substitutes for hard tissue, in orthopedic and dental surgery.

Magallanes-Perdomo et al. (2011) prepared an apatite–wollastonite (A/W) glass ceramic by the crystallization of a glass having the eutectic composition of the $Ca_3(PO_4)_2$–$CaSiO_3$ binary system.

The mixed powders of the chemicals were melted for 30 min at 1,600°C and after casting in the form of glass blocks, annealed at 795°C. The glass samples were then heat-treated according to a two-step schedule, at 820°C for 2 h and at 1,100°C for 1 h.

The resulting glass ceramics contained 44.8 wt% apatite, 28.0 wt% wollastonite-2 M, and 27.2 wt% of glass phase.

The SEM micrograph of glass-ceramic samples revealed the existence of a dense and homogeneous microstructure consisting of $\sim 2\,\mu m$ domains. The domains were composed of the residual glass phase that enclosed a mixture of nano-sized apatite and wollastonite-2 M crystals. Transmission electron microscope (TEM)/Energy Dispersive Spectroscopy (EDS) analyses showed that this nano-structure consisted of ellipsoidal apatite and wollastonite-2 M nanoparticles in the range of ~20–50 nm in length and 20 nm in diameter (Figure 5.3). The wollastonite particles displayed twin contrast, demonstrated by the banding of different spacing (inside the black circle).

The prepared glass ceramics exhibited improved mechanical properties, e.g., the measured micro-hardness of the glass ceramics was higher than the parent glass.

The bioactivity of this material was also studied by soaking the samples in SBF for several days at 36.5°C. After 7 days, a porous HCA layer was formed at the SBF–glass-ceramic interface, indicating their good bioactivity.

Denry et al. (2012) prepared niobium-doped fluorapatite (FAp) glass ceramics for biomedical applications. The nominal composition of the glass was SiO_2 30.5, MgO 13.8, Na_2O 2.3, K_2O 5.8, CaO 14.4, Al_2O_3 15.9, F 4.9, P_2O_5 11.4, and 1 Nb_2O_5, as dopant (wt%).

The glass-ceramic specimens were prepared by casting and heat treatment of the glasses in the range 700–1,200°C for 1 h.

The microstructural examination of the as-cast glass specimens by atomic force microscopy (AFM), revealed the occurrence of amorphous phase separation (APS) in them.

XRD confirmed the presence of FAp in all specimens, together with forsterite and enstatite at higher temperatures. Both heating rate and heat-treatment temperature strongly influenced microstructure and crystallinity.

The phase separation was the basis for the development of nanoscale FAp crystals. Crystallization of FAp occurs within the phase-separated droplets and is limited by their size, as crystal growth is hindered by the depletion in key elements of the surrounding glassy matrix. It was also proposed that phase-separated droplets facilitate the crystallization of nano-sized FAp by lowering the activation energy for crystal growth and by

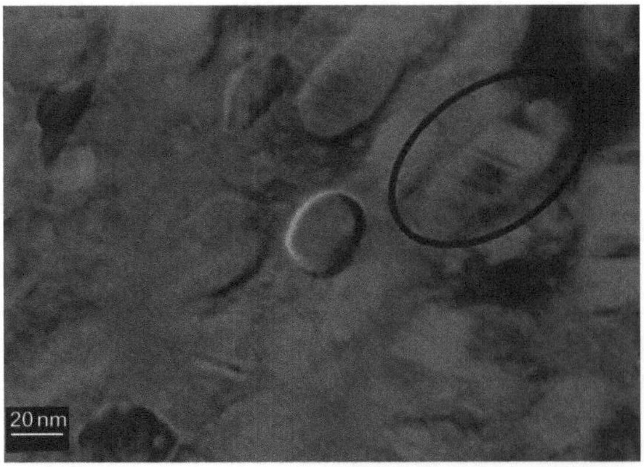

Figure 5.3 TEM micrograph showing the ellipsoidal nanocrystals of apatite and wollastonite-2 M, the latter phase displaying twin contrast. (After Magallanes-Perdomo et al., 2011. Reproduced with permission of Elsevier Science B.V.)

acting as heterogeneous nucleation sites. The authors postulated that doping with niobium oxide promoted glass-in-glass phase separation and affected the crystallization of phase-separated droplets as discussed above.

According to the authors, the crystallization of forsterite at temperatures >950°C favorably affected the biomedical characteristics of the resulting glass ceramics.

A dual microstructure with submicrometer FAp crystals and polygonal forsterite crystals was obtained when slow heating rates and crystallization temperatures of 950–1,100°C were used. Needle-shaped FAp crystals appeared after heat treatment at >1,100°C.

Considering the favorable osteoconductive and osteoinductive characteristics of these glass ceramics, they were anticipated to be promising candidates for the fabrication of macroporous scaffolds.

Aguiar et al. (2008) investigated the influence of the glass structure on the *in vitro* bioactive behavior of two glass samples with the compositions of 25.3SiO_2, 10.9P_2O_5, 32.6CaO, 31.2MgO (mol%) and 33.6SiO_2, 6.40P_2O_5, 19.0CaO, 41.0MgO (mol%).

The glasses frits, prepared according to a double-melting procedure, were poured into a mould and the obtained glass blocks were finally heat-treated at 730°C for 30 min. They were proved to be amorphous on XRD examination.

In order to clarify the contribution of phosphate groups to the glass structure, the glass samples were subjected to high-resolution TEM (HRTEM) analyses (Figure 5.4).

Figure 5.4A shows the typical electron contrast of amorphous materials. By selecting an area of interest (B) and applying Fourier transform (FT), a diffuse diffraction pattern was obtained (C), indicating the existence of an amorphous material. But after filtering the image, a characteristic diffraction pattern of a monocrystalline structure was found (D). Inverting the FT, a reconstructed image of a nano-crystalline area (E), with interplanar distance of 0.22 ± 0.03 nm, was obtained. Repeating this procedure for other interest areas of the two glass samples, interplanar distances in the 0.21–0.26 (±0.03) nm range, characteristic of phosphate networks, were found. These experimental results indicated that phosphate groups formed isolated **orthophosphate nano-crystalline islands** in the amorphous matrix of the glass samples.

In vitro tests revealed the formation of two differentiated calcium phosphate layers (CaP and CaP + Si_xO_y) on surfaces of the two glass samples after immersion for 72 h at 37°C in SBF. The presence of CaP-rich film and the absence of a silica-rich

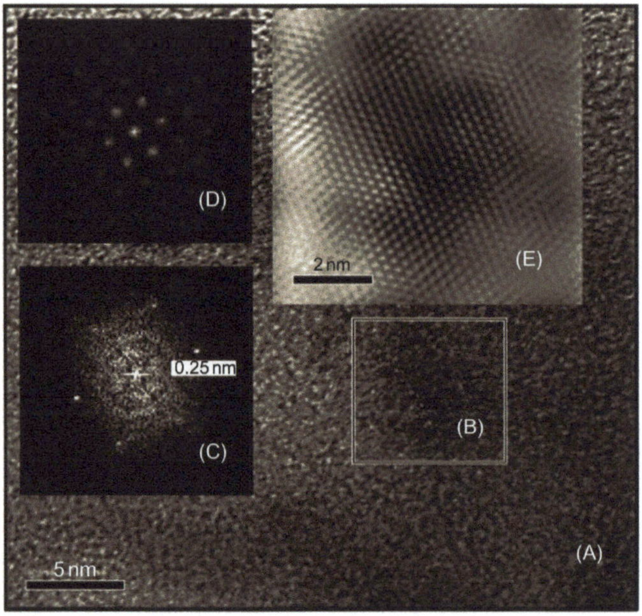

Figure 5.4 HRTEM images of the glass containing 33.6 mol% SiO_2: (A) amorphous matrix, (B) area of interest, (C) FT diffuse diffraction pattern with bright dots, (D) filtered FT, and (E) reconstructed nano-crystalline areas. (After Aguiar et al., 2008. Reproduced with permission of Elsevier Science B.V.)

layer after immersion in SBF suggested that the aforementioned phosphate nano-regions played a key role in the initial stages of the bioactive process, acting as nucleation sites for a calcium phosphate-rich film.

Hill et al. (2010) studied the crystallization of a FAp $Ca_5(PO_4)_3F$, glass ceramic.

The details of the composition, preparation, and characterization of this glass ceramic have been explained in Section 1.1.2.2.1.

Figure 5.5 shows the SANS (small-angle neutron scattering) plot of the glass and the sample heat-treated at 850°C for 8 h. The glass shows no significant small-angle scattering indicating a homogenous single-phase material, whereas the heat-treated sample exhibits a well-defined peak in $I(q)$ with a maximum at $q = 0.036\ \text{Å}^{-1}$ corresponding to a spacing of ~17 nm which can be attributed to the FAp crystallites.

The crystal nucleation in the glass ceramic occurred via prior APS. The APS probably resulted in one of the phases having a composition close to FAp. This resulted in lower activation energy for crystal nucleation and promoted this process.

The APS also inhibited the subsequent crystal growth process, because the FAp crystals nucleated within the droplet phase grew rapidly to the size of the droplet phase, but then

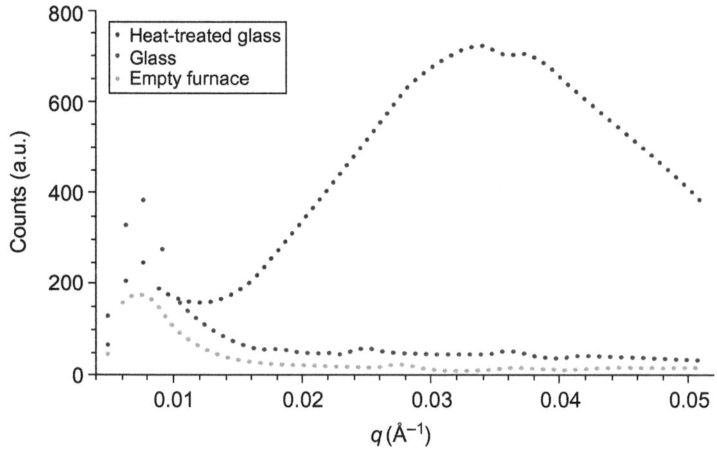

Figure 5.5 SANS patterns for the glass and the sample heat-treated for 8 h at 850°C. (After Hill et al., 2010. Reproduced with permission of Elsevier Science B.V.)

were retarded by the calcium–phosphate depleted matrix glass. Another important factor was the reduction of glass viscosity by Ca^{2+}, PO_4^{3-}, and especially fluorine in this glass. Therefore, the crystallization of FAp by removing Ca^{2+}, PO_4^{3-}, and F^- from the glass raised the viscosity of the residual glass and retarded further crystal growth.

One of the advantages of prepared glass ceramics in the aforementioned research, besides their nano-scale structure, was their relatively high, ~20–30 vol%, content of the FAp crystal phase, making them suitable candidates for possible applications in the fields such as biomedicine and dentistry.

Höland et al. (2003) represented a review of the controlled nucleation and crystallization for the most widely used glass ceramics in the field of dental restoration, namely leucite, lithium disilicate, and leucite–apatite glass ceramics.

The authors described the process of crystallization of leucite ($KAlSi_2O_6$) from a glass powder sample having the composition 63.0SiO_2, 17.7Al_2O_3, 11.2K_2O, 4.6Na_2O, 0.6B_2O_3, 0.4CeO_2, 1.6CaO, 0.7BaO, and 0.2TiO_2 (wt%) as a surface crystallization process. Leucite glass ceramics are primarily used for fabricating ceramic veneering layers to mask metal framework.

The authors also described the crystallization behavior of high-strength, machinable glass ceramics belonging to the system SiO_2–Li_2O–Al_2O_3–K_2O–P_2O_5.

It was suggested that in the above glass ceramics, the nano-scaled Li_3PO_4 phase, which crystallized first, acted as a heterogeneous nucleating agent for both Li_2SiO_3 and $Li_2Si_2O_5$ nano-crystals.

Moreover, the lithium metasilicate (Li_2SiO_3) phase which appeared as the faster growing crystals in the lower temperature range of 530–750°C decomposed at 780–820°C, according to the following solid-state reaction that resulted in a drastic increase of the lithium disilicate phase:

$$Li_2SiO_3(\text{crystal}) + SiO_2(\text{glass}) \rightarrow Li_2Si_2O_5(\text{crystal})$$

A highly crystalline glass ceramic with an interlocking microstructure of $Li_2Si_2O_5$ crystals was formed by heat treatment at 850°C.

This glass ceramic exhibited the remarkably high values of flexural strength and K_{1c} of 740.8 ± 79.7 MPa and 3.3 MPa $m^{1/2}$, respectively.

The authors also described the fabrication and properties of leucite ($KAlSi_2O_6$)–fluorapatite ($Ca_5(PO_4)_3F$) glass ceramics. It was shown that leucite was formed by the mechanism of surface crystallization, while the crystallization of FAp nanocrystals took place in the volume of the glass particles. This crystallization process occurred via the phosphate phase separation and the formation of an intermediate precursor phase (rhenanite, $NaCaPO_4$) at 800°C.

5.3 Concluding Remarks

Coating the metallic alloy implants with a bioactive coating is an effective way to solve the problems arising from the formation of a dense fibrous tissue at the implant–bone interface, which may result in the loosening of the device and other complications. The coating can accelerate new bone formation at the bone–implant interface and thereby interlock the implant with the surrounding skeletal tissue. This may lead to the formation of a more stable interface, strong enough to support lifelong functional loading.

In order to overcome the existing limitations of some coatings, the nano-structured coatings have also been extensively studied by various researchers, in recent years.

The nano-structured coatings have shown the ability to improve the adhesion of the coating to the metallic implants, resulting in higher bonding strength and better mechanical properties. It has also been found that the existence of the nano-granules in the coating, by accelerating the dissolution rate of the coatings, may result in the enhanced bioactivity, i.e., higher formation rate for the apatite layer on the implant surface. They could also promote osteoblast adhesion and significantly enhance HOB proliferation and differentiation compared with the uncoated materials and those implants coated with the more conventional coarse-grained coatings.

In recent years, many researchers have also investigated the fabrication and properties of **nano-structured glass ceramics as coatings**, and as potential substitutes for hard tissue in **orthopedic and dental surgery**.

The nano-structured bioglass ceramics are expected to combine the general technical and economical advantages of glass ceramics (as discussed previously), with the characteristics of nano-structured biomaterials, such as improved mechanical and biological properties. For example, nanoscale ceramics can exhibit significant ductility before failure. It has been reported that a brittle ceramic with nano-sized grains could sustain a large plastic strain up to 100%.

The nano-structured bioceramics have also shown the ability of promoting the osteoblast adhesion and proliferation, and the deposition of calcium-containing minerals on their surface.

Despite the numerous attempts made in the field of fabrication of nano-structured bioglass ceramics in recent years, it seems that the great potential of these materials, for biomedical applications, has not appreciated enough. Considering the superior mechanical properties of these materials and their favorable bioactive, osteoconductive, and osteoinductive characteristics, they are especially promising candidates for implantology (medical and dental prostheses) and the fabrication of scaffolds for tissue engineering.

Therefore, it seems that more extensive and comprehensive research activities are needed in order to realize the great potential of these materials in the areas of medical and dental applications.

6

OTHER APPLICATIONS OF NANO-GLASS CERAMICS

CHAPTER OUTLINE

6.1 Nanoporous Glass Ceramics

Microporous ceramics have attracted considerable attention in recent years and are used in various engineering applications such as membranes, supporting materials for immobilization of enzymes, catalyst supports, and sensors (Hosono and Abe, 1995).

On the other hand, microporous and nanoporous glass-ceramic materials have many advantages over conventional porous ceramics. More convenient production technology, better control over the size and distribution of pores, and the possibility of producing crystalline skeletons of differing compositions, structures, and functionalities are some of these advantages.

These materials have also some advantages in comparison with the earlier Vycor® porous glasses, regarding the chemical and mechanical properties, easier production routes, and more versatility in the skeletal material properties.

V.K. Marghussian, Nano-Glass Ceramics, ISBN: 978-0-323-35386-1.

DOI: http://dx.doi.org/10.1016/B978-0-323-35386-1.00006-2

These porous glass ceramics are usually prepared utilizing phosphate glasses that are mainly located in the systems $CaO-TiO_2-P_2O_5$ or other related systems. The glasses, after undergoing a spinodal-type phase separation (like Vycor glasses), are heat-treated to develop crystalline phases such as β-$Ca_3(PO_4)_2$ and $RTi_2(PO_4)_3$ (R = Li, Na,1/2Ca). The former phase is usually leached out, leaving behind a porous skeleton of $RTi_2(PO_4)_3$.

6.1.1 Fabrication, Properties, and Application of Porous Glass Ceramics

Hosono and Abe (1995) prepared bulk porous glass ceramics with a skeleton of two types of functional titanium phosphate crystals by controlled crystallization and leaching of $RTi_2(PO_4)_3$ (R = Li, Na, 1/2Ca)$-Ca_3(PO_4)_2$ glasses.

One crystal has a nasicon-type structure, such as $RTi_2(PO_4)_3$, with a three-dimensional network structure comprising TiO_6 and PO_4 units, while the other is a zirconium phosphate-type crystal $Ti(HPO_4)_2 \cdot 2H_2O$, with a two-dimensional layered structure. The mean pore diameter and porosity of 30–60 nm and 40–60 vol%, respectively, were achieved.

Porous $CaTi_4(PO_4)_6$ glass ceramics have excellent characteristics for applications such as supports for immobilization of enzymes, humidity sensors, and porous $LiTi_2(PO_4)_3$ glass ceramics as ion exchange media.

Conspicuous bacteriostatic activities were also found for the partially Ag-exchanged porous $LiTi_2(PO_4)_3$ and $Li_{1.4}Ti_{1.6}Al_{0.4}(PO_4)_3$. Integrated microporous glass ceramics with skeletons of $LiTi_2(PO_4)_3$ (LTP) and $Ti(HPO_4)_2 \cdot 2H_2O$ (TP) crystals were synthesized utilizing a novel discovery that LTP is transformed into TP in acid solutions.

Yamamoto et al. (1997) fabricated a porous glass ceramic with a skeleton of $CuTi_2(PO_4)_3$ crystals by controlled crystallization of a base glass having the nominal composition $50CuO-20TiO_2-30P_2O_5$.

The glass was first heated at 520°C for 20 h for nucleation and subsequently at 550°C for 12 h and then to 590°C for 12 h for crystal growth. These heat-treatment temperatures were determined by differential thermal analysis (DTA) of the base glass.

A dense glass ceramic composed of $CuTi_2(PO_4)_3$ and $Cu_3(PO_4)_2$ crystals, the surface of which was covered by a CuO thin layer, was obtained after the heat treatment. When the resulting glass ceramic was leached in 0.5 N H_2SO_4 aqueous solution at room temperature for 3 days, $Cu_3(PO_4)_2$ and CuO phases were dissolved out selectively, leaving a crystalline $CuTi_2(PO_4)_3$ skeleton.

The specific surface area, porosity, and average pore radius of the porous glass ceramic obtained were ~45 m^2/g, 40 vol%, and 9 nm, respectively, and they showed catalytic activity in the conversion reaction of propene into acrolein.

According to the authors, these porous glass ceramics have great potential in the preparation of novel catalysts. Figure 6.1 depicts the fractured surface of a typical porous glass ceramic.

Yamamoto and Abe (1998) also reported on the possibility of fabrication of a porous cupric nasicon-type titanium phosphate ($CuTi_4(PO_4)_6$) glass ceramic via the controlled crystallization of a glass and the subsequent chemical leaching of the resulting dense glass ceramic. They utilized the same glass composition and heat-treatment procedure as described above for the fabrication of the cuprous nasicon ($CuTi_2(PO_4)_3$) glass ceramics.

The only difference was that in the latter experiment, the chemical leaching had been carried out in 0.5 N HCl aqueous solution instead of the 0.5 N H_2SO_4 solution used in the previous experiment.

According to the authors, when chemically leached in HCl, the Cu^+ ions of the titanium phosphate crystal were oxidized to Cu^{2+} ions, whereas in the case of leaching in dilute H_2SO_4, the oxidation did not occur.

The specific surface area and the average pore radius of the $CuTi_4(PO_4)_6$ porous glass ceramic were ~70 m^2/g and ~7 nm, respectively.

This porous glass ceramic showed high catalytic activities in the dehydration of 2-propanol.

Yamamoto and Abe (2000) in a later work subjected the glass ceramics, as prepared according to the procedure described in

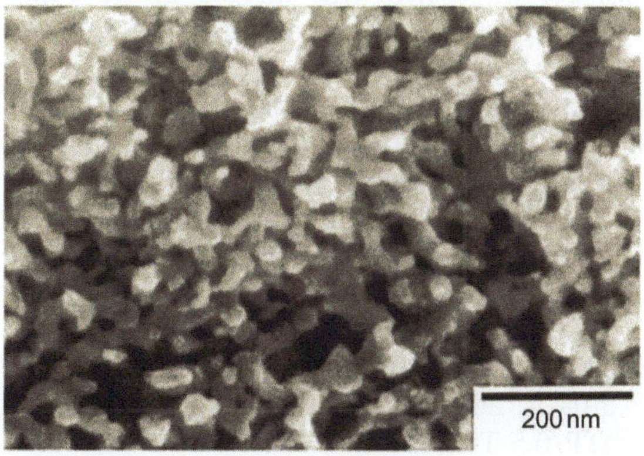

Figure 6.1 Fractured surface of a porous glass ceramic, comprised of a $CuTi_2(PO_4)_3$ skeleton. (After Yamamoto et al., 1997. Reproduced with permission of the American Ceramic Society.)

the aforementioned two works, to a post-crystallization heating in air between 600°C and 900°C for 30 min.

Subsequently, the specimens were leached in 0.5 N H_2SO_4 solution at room temperature for 3 days to obtain porous glass ceramics comprising mainly of a $CuTi_2(PO_4)_3$ phase. The latter phase has the nasicon structure as described previously.

The pore size distribution of the porous glass ceramics was close to a logarithmic Gaussian distribution curve, with a rather narrow width. The average pore diameters varied from 15 to 210 nm for the specimens subjected to the post-crystallization heating temperatures of 600°C and 900°C, respectively.

The specific surface areas were also measured by the BET method. The Brunauer, Emmett, Teller (BET) values which were quite high for the porous glass ceramics decreased monotonically with increasing the post-crystallization temperature.

The large specific surface area and the narrow pore size distribution suggested that a phase separation by the spinodal decomposition mechanism occurred in the glass specimens prior to the crystallization.

The catalytic activity of the prepared porous $CuTi_2(PO_4)_3$ glass ceramics for the oxidation of propene to acrolein was also determined.

It has been known that some silver-bearing materials have bacteriostatic activities and the bacteriostatic materials such as zeolite, calcium phosphates, silica gels, or clay minerals, into which Ag^+ ions are introduced, have been commercially used. These materials have a tendency to release Ag^+ ions. It has also been reported that porous glass ceramics with a skeleton of $LiTi_2(PO_4)_3$ show excellent cation exchange properties and that Li^+ ions in glass ceramics are selectively exchanged with Ag^+ ions even in solutions containing higher concentrations of Na^+ ions.

Considering the above points, Kasuga et al. (1999) prepared porous glass ceramics with the surface phase consisting predominantly of $AgTi_2(PO_4)_3$ crystals and the interior phase of $LiTi_2(PO_4)_3$ crystals.

$Li_{1.4}Al_{0.4}Ti_{1.6}(PO_4)_3$ (LATP) and $Ca_3(PO_4)_2$ (TCP) were selected as the skeleton of the porous glass-ceramic body and the leachable phases, respectively.

The LATP phase exhibits excellent cation exchange properties for substituting Ag^+ ions into the surface and TCP can be readily dissolved out in acids in order to produce porous materials.

The nominal composition of the base glass was 25.0CaO, $11.7Li_2O$, $3.3Al_2O_3$, $26.7TiO_2$, and $33.3P_2O_5$ (mol%), corresponding to 1.0 LATP/0.5 TCP mole ratio.

After melting the batch mixture, it was poured into water to form the base glass frits. The resultant frits were heated at 580°C for 20 h for nucleation, followed by 8 h at 690°C for crystal growth.

The resulting dense glass-ceramic grains were soaked in 1 N HNO_3 at room temperature for 3 days. The acid treatment leached out the calcium phosphate phase, producing porous glass ceramics with an LATP skeleton.

The specific surface area of the resulting porous glass ceramics was estimated to be ~10 m^2/g by the BET method.

A sample of the porous glass-ceramic grains was soaked in an aqueous solution containing 0.1 M $AgNO_3$, at room temperature for 3 days. An appreciable amount of Ag^+ ions were introduced into the porous glass ceramics during this treatment. As a result, by the exchange of Ag^+ ions for Li^+ ions, porous glass ceramics with the surface and interior phases consisting respectively of $AgTi_2(PO_4)_3$ and $LiTi_2(PO_4)_3$ crystals were prepared.

Although the exchange rate of glass ceramics heat-treated at higher temperatures (e.g., at 900°C) was found to decrease drastically, they showed excellent bacteriostatic properties and are expected to have applications as bacteriostatic materials with high thermal resistance and medical safety.

Daiko et al. (2008) studied the fabrication and properties of porous glass ceramics with a skeleton of γ-$Ti(HPO_4)_2 \cdot 2H_2O$ crystals, derived from base glasses of the nominal composition $11.7Li_2O$, $3.3Al_2O_3$, $26.7TiO_2$, $33.3P_2O_5$, and 25CaO (mol%).

The batch mixture was melted in a platinum crucible at 1,350°C for 2 h in air. The melt was poured onto a stainless steel plate to obtain the mother glass.

The plate-shaped glass specimens were first heated at 580°C for 20 h for nucleation and subsequently at 690°C for 8 h for crystal growth. The resulting glass ceramics were leached in 1 M HCl aqueous solution at 100°C for 3 days to obtain porous $Ti(HPO_4)_2 \cdot 2H_2O$ glass ceramics.

It was proposed that the glass specimens underwent a spinodal-type phase separation prior to the final crystallization during the heat-treatment process. This resulted in the formation of crystalline phases such as β-$Ca_3(PO_4)_2$ and $LiTi_2(PO_4)_3$. After the HCl treatment, almost all of the above phases disappeared, and new crystal phases of $Ti(OH)PO_4$ and $Ti(HPO_4)_2 \cdot 2H_2O$ (THP) were formed.

The resulting glass ceramics had both microsized (1–2 μm) and nanosized (5–30 nm) pores.

Subsequently, a porous TiO_2 layer of 200–500 nm thickness was successfully deposited on the surfaces of the glass ceramics in order to prepare a self-cleaning adsorbent, since it had previously been shown that the adsorbed NH_3 molecules are photo-decomposed by photocatalytic reaction of the anatase layers.

The resulting anatase-modified porous glass ceramics showed high activities for NH_3 adsorption as well as for the absorbed NH_3 molecules, under UV radiation.

Kord et al. (2009a) prepared nanoporous glass ceramics belonging to the $CaO-TiO_2-P_2O_5$ system considering a 2:1 mole ratio for β-$Ca_3(PO_4)_2$:$CaTi_4(PO_4)_6$, which was the desired ratio for the crystalline phases in the end product.

The glasses formulated according to the above composition ($31.25P_2O_5$, $43.75CaO$, $25.0TiO_2$ mol%) were melted at 1,350°C for 2 h and were cast onto preheated steel moulds. The glass specimens then were annealed at 600°C for 2 h and furnace-cooled.

The glass specimens after crystallization heat treatments at the appropriate phase-separation-nucleation and crystallization temperatures were leached in 1 M HCl solution in order to dissolve out the more soluble, β-$Ca_3(PO_4)_2$ phase and to produce a porous skeleton of $CaTi_4(PO_4)_6$ phase.

It was concluded that glass-in-glass phase separation with a spinodal decomposition mechanism plays a very decisive role in controlling the microstructure, and the pore size and pore volume of these glass ceramics.

According to the authors, the optimum phase-separation-nucleation temperature and time, determined as 710°C and 24 h in this work, is an almost invariable condition for the fabrication of these types of porous materials (if composition was kept constant) with acceptable properties and nano-size porosity.

The chosen crystallization (growth) condition of 765°C and 24 h was also proved to be quite effective in producing bodies of acceptable values of porosity ($41 \pm 4\%$) and quite small (perhaps the minimum possible) mean pore diameter of 14.3 ± 2 nm.

Figure 6.2 depicts the scanning electron microscopy (SEM) micrograph for a specimen nucleated at 710°C for 24 h and subsequently heated at 765°C for 60 min. The micrograph shows the nucleation and growth of crystals within the phase-separated areas of the glass specimen.

Kord et al. (2009b) also added 2–6 mol% ZrO_2 to the base glass composition as described above ($31.25P_2O_5$, $43.75CaO$, $25.0TiO_2$ mol%) at the expense of TiO_2. The prepared glasses were crystallized to bulk glass ceramics containing the major phases of β-$Ca_3(PO_4)_2$ and $CaTi_4(PO_4)_6$. DTA was utilized to determine the appropriate phase-separation-nucleation and crystallization temperatures. The β-$Ca_3(PO_4)_2$ phase was dissolved out by

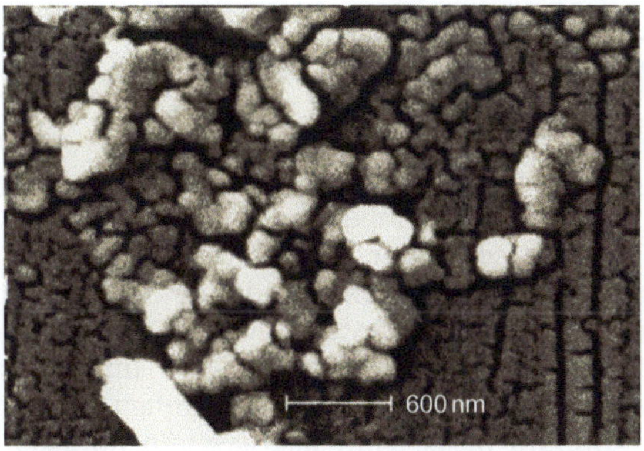

Figure 6.2 SEM micrograph of a specimen nucleated at 710°C for 24 h and subsequently heated at 765°C for 60 min, showing the nucleation and growth of crystals within the phase-separated areas. (After Kord et al., 2009a. Reproduced with permission of Elsevier Science B.V.)

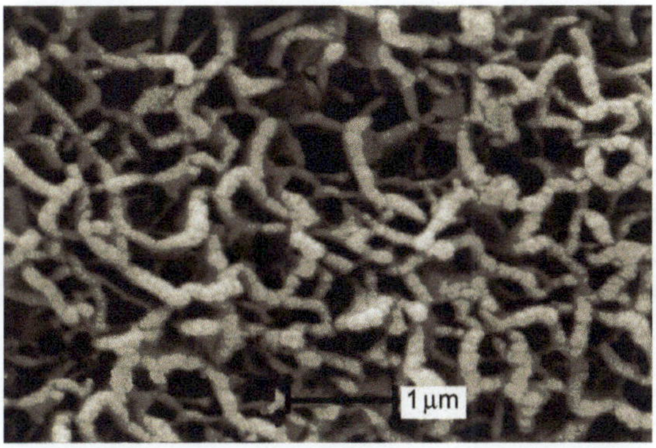

Figure 6.3 SEM micrograph showing the porous skeleton of $CaTi_4(PO_4)_6$ in specimen Zr_6, containing 6 mol% ZrO_2 after leaching. (After Kord et al., 2009b. Reproduced with permission of Elsevier Science B.V.)

leaching the resulting glass ceramics in HCl, leaving a porous skeleton of $CaTi_4(PO_4)_6$. It was shown that the ZrO_2 addition resulted in the reduction of volume porosity and the mean pore diameter while the specific surface area was increased.

The smallest median pore diameter and largest surface area were 8.6 nm and 32 m^2/g, respectively, obtained for the specimen containing 6 mol% ZrO_2. The ZrO_2 addition also improved the chemical durability and bend strength of porous glass ceramics.

The glass ceramics exhibited relatively nonuniform microstructures comprising mainly of flaky $Ca_3(PO_4)_2$ agglomerates and acicular and rod-like agglomerates of both phases. The specimens also demonstrated two distinctly different ranges of pore sizes after leaching, large 100–500 nm pores, and nanopores located on the walls of the large pores. The small size of the median pore diameter (~10 nm) indicates the predominant role of the fraction of small pores. Figure 6.3 depicts a typical microstructure of the porous glass ceramic.

6.2 Tough Nano-Glass Ceramics for Magnetic Memory Disk Substrates

Nano-crystalline glass ceramics in the SiO_2–Al_2O_3–MgO–ZnO–TiO_2 system containing the major phases of spinel and enstatite have been developed for potential use as magnetic disk substrates in computer hard drives. These glass ceramics with high elastic modulus and moderate strength and toughness values offer many advantages for the disk substrate application, including improved flatness and smoothness, better wear resistance, and the ability to manufacture thinner than glass or aluminum disks. They also contain no alkali ions and have excellent chemical durability (Pinckney and Beall, 1997; Beall and Pinckney, 1999). Alkali ions may migrate through the metallic thin films used on magnetic disks. This can lead to catastrophic disk failure and has been a source of concern for other glass ceramic and ion-exchanged glass substrates, all of which contain alkali ions.

A wide range of compositions yielding a nano-crystalline microstructure capable of delivering surface roughness values near 5 Å was developed within the composition area, $SiO_2 = 35–60$, $Al_2O_3 = 10–35$, $MgO = 0–30$, $ZnO = 0–25$ $(MgO + ZnO) > 10$, and $TiO_2 = 5–20$ (wt%) (Pinckney and Beall, 1997).

6.2.1 Nucleation and Crystallization

Compositions within the cordierite liquidus field of the ternary MgO–Al_2O_3–SiO_2 system are easily nucleated with TiO_2 to give a wide range of possible phase assemblages based on various mixtures of cordierite, β-quartz solid solution, α-quartz, magnesium-petalite, spinel, and enstatite. For example, for a typical glass sample with the composition:

47.1SiO_2, 22.1Al_2O_3, 16.9MgO, 1.7ZnO, and 12.3TiO_2 (wt%), the nucleation is preceded by a highly uniform, fine-scale phase separation into SiO_2-rich and TiO_2/Al_2O_3-rich areas. The initial crystalline phase was a solid solution of magnesium aluminum titanate. After a heat treatment at 800°C for 2 h and 1,025°C for 4 h, the resulting glass ceramics mainly consisted of spinel, enstatite, and magnesium dititanate crystalline phases. All three phases occurred as crystals <100 nm in size, with spinel crystals <25 nm (Beall and Pinckney, 1999).

6.2.2 Mechanical Properties

Spinel provides the glass ceramics with high elastic modulus and surface hardness, and the enstatite improves fracture toughness, because of the presence of well-known lamellar twinning in it. The lamellar twinning developed in enstatite during the martensitic transition from protoenstatite to clinoenstatite on cooling. Such twinning is known to deflect cracks and increase fracture toughness.

This material has a Young's modulus value of 144 GPa, a modulus rupture of 107 MPa, and a fracture toughness of $1.3\ MPa \cdot m^{0.5}$.

On the other hand, the strength and fracture toughness of these glass ceramics, even for those with very little enstatite, are higher than the expected values. This can possibly be attributed to the occurrence of the phenomenon known as dispersion strengthening. Spinel crystals have both a higher density and a higher thermal expansion than the surrounding residual glass.

The residual glass upon cooling is, therefore, brought into tangential compression around the crystals and radial tension perpendicular to the interface. These stresses cause any cracks to deflect around the crystals, producing a tortuous fracture pattern, which increases the fracture energy and fracture toughness.

The material can be polished to smoother than those of alkali-containing glasses, which are vulnerable to fracture propagation through stress corrosion and ease of fracture propagation due to the lower toughness. With a strain point of 1,000°C, these glass ceramics also have upper use temperatures, much higher than those of glass or aluminum, enabling the use of the high-temperature sputtering processes required to apply the ultrahigh coercivity barium ferrite magnetic films on their surface.

6.3 Nano-Glass-Ceramic Coatings and Sealants

There has, in recent years, been a major interest in glass ceramic-to-metal seals and coatings for new applications, including, for example, solid oxide fuel cells (SOFCs), thermal barrier coatings (TBC), and seals for battery applications among others.

Glass ceramics combine the generally superior properties of crystalline ceramics with the ease of processing of glass. Major advantages of glass ceramics for the aforementioned applications include more refractory behavior and superior mechanical properties, good bonding characteristics to metals, and the possibility of using a wide range of chemical compositions to control some physicochemical properties such as viscosity, Coefficient of Thermal Expansion (CTE), and glass transition temperature.

6.3.1 SOFC Sealants

Unlike earlier fuel cells, SOFCs are constructed entirely of solid materials. These fuel cells convert chemical energy into electrical energy via electrochemical oxidation of the fuel instead of its combustive oxidation. This results in higher efficiency and cleaner operation compared to conventional combustion processes.

Hydrogen or hydrocarbon-based fuels are usually used in these fuel cells and the operating temperatures are usually high, typically in the range 650–1,000°C.

During operation, a fuel and an oxidizing gas are combined via an ion-conducting electrolyte. If hydrogen is used as the fuel and air as the oxidizer, the only waste product is water. Currently, there are two basic SOFC designs: tubular and planar. The planar design is generally preferred because of its higher operating power and simpler construction. A disadvantage of the planar design compared to the tubular design is that it requires high-temperature gas seals to join the cells together to form the stack (Donald et al., 2011).

An individual cell consists of the anode that may be cubic zirconia or a Ni/cubic zirconia cermet, an electrolyte of yttria-stabilized zirconia (YSZ), Sr- and Mg-substituted lanthanum gallate, or Gd-substituted CeO_2, and a cathode that may be lanthanum manganate, $LaMnO_3$, or a mixed LaSrCoFe-based oxide. Cells are joined together via an interconnect which may be a ferritic stainless steel or, for higher operating temperatures, a doped lanthanum chromite ceramic. The interconnect carries the current between the individual cells and acts as a separator between the oxidant and fuel. Individual cells are generally bonded to the next cell via a glass or glass-ceramic seal, although other media have also been attempted. Brazing has also been employed for sealing SOFC components (Donald et al., 2011).

6.3.1.1 The Glass-Ceramic Materials Employed in SOFC Sealants

Various glass-ceramic materials have been investigated in recent years, as potential sealing media for SOFC systems, including alkali and alkaline earth silicates, aluminosilicates, borosilicates, and aluminophosphates (Reis and Brow, 2006; Ghosh et al., 2008; Meinhardt et al., 2008; Stephens et al., 2009; Milhans et al., 2011; Puig et al., 2011; Khedim et al., 2012). The most promising results have been reported for compositions based on alkaline earth aluminosilicate glasses, especially SiO_2–Al_2O_3–BaO–CaO glasses.

For example, Ghosh et al. (2008) investigated several glass compositions based on the barium aluminosilicate system for their application as a sealant for SOFCs, especially suitable for anode-supported planar SOFCs operating between 750°C and 800°C. To estimate the applicability of these glasses as sealants, their thermal properties, crystallization behavior, microstructure upon different heat-treatment schedules, and the overall bonding characteristics of the glass sandwiched between ceramic (YSZ) electrolyte and metallic (Crofer22APU) interconnect were investigated.

Microstructural investigation revealed a well-adhered bonding between the electrolyte and the metallic interconnect. The optimized glass composition, 46.50BaO, 8.80Al_2O_3, 18.20SiO_2, 12.50B_2O_3, and 14.0La_2O_3 (wt%), also showed a high resistivity at the SOFC operation temperature (<800°C).

The SEM microphotographs of this optimized glass, heat-treated at 800°C for different periods of time, revealed the presence of a large number of nanoscale crystals that grew as spherulites. After prolonged heat treatment at the same temperature, the spherulites grew in size and some of the lath-like crystals formed in the early stage transformed into elongated rod-shaped morphology (Figure 6.4).

According to Meinhardt et al. (2008), the high-temperature glass and glass-ceramic sealants, because of the following characteristics, have been the most attractive materials for SOFC applications:

- They tend to display acceptable stability in the reducing and oxidizing atmospheres of the stack.
- They can be readily applied to the sealing surfaces as a powder dispersed in a paste or a tape cast sheet.

Figure 6.4 Microstructure of the developed barium aluminosilicate glass-ceramic specimens heat-treated at 800°C for (A) 10 h and (B) 50 h. (After Ghosh et al., 2008. Reproduced with permission of Elsevier Science B.V.)

- They typically exhibit good wetting behavior on both sealing surfaces.
- They are electrically insulating.
- They can be engineered to exhibit a coefficient of thermal expansion matching those of the adjacent SOFC components in the final joint.

The aforementioned authors prepared and evaluated a series of barium aluminosilicate glasses modified with CaO and B_2O_3 with respect to their suitability in sealing planar SOFCs. Most interestingly, the authors summarized the effect of various oxide components and additives upon the properties of the resulting glasses such as viscosity, surface tension, thermal expansion coefficient, and their crystallization behavior.

The long-term Coefficient of Thermal Expansion (CTE) of the glass composition (35BaO, 15CaO, $10B_2O_3$, $5Al_2O_3$, and $35SiO_2$ mol%) at a proposed operating temperature of 750°C was found to be particularly stable, due to its partial crystallization and conversion to a glass ceramic. This most promising glass-ceramic sealant exhibited minimal chemical interaction with the YSZ electrolyte, yet formed a strong bond with this material. Interactions with metal components were found to be more extensive and depended on the composition of the metal oxide scale that formed during sealing.

Mechanical measurements conducted on the bulk glass ceramic and on seals indicated that the sealant was anticipated to display adequate long-term strength for most conventional stationary SOFC applications.

More recently, Khedim et al. (2012) prepared glass-ceramic sealants in the $CaO-Al_2O_3-SiO_2-B_2O_3$ system for operating around 800°C. The thermal and physicochemical properties of the glass ceramics and their crystallization behavior were investigated. Bonding characteristics as well as the chemical interactions of parent glass with YSZ electrolyte and ferritic steel-based interconnect (Crofer®) were also investigated.

The comparative study based on physicochemical and thermal characterizations showed that the glass composition 35.80CaO, $18.24Al_2O_3$, $39.30SiO_2$, and $6.03B_2O_3$ (wt%) met most of the requirements for efficient sealing. Further optimization tests on this glass by minor additions of V_2O_5, K_2O, and TiO_2 showed that V_2O_5 improved glass behavior at high temperature by reducing the viscosity, while enhancing crystallization. Addition of K_2O delayed the crystallization but increased the glass viscosity and adding TiO_2 had no significant effect on glass behavior. It was concluded that the aforementioned glass composition (without the minor additives) was the most suitable for such applications. Figure 6.5 demonstrates the micrograph obtained for a cross section of the above glass sample after long-term interaction tests ($T = 850°C$, $t = 100$ h). The morphology of the crystalline phases identified from X-ray diffraction patterns is presented in Figure 6.5. At the metal/glass ceramic and ceramic/glass-ceramic interfaces, the glass shows satisfactory interface quality (bonding, wetting, and adherence behavior) that can satisfy the sealing requirement.

It has also been shown that the optimization of the heating rate was a compromise between delaying the crystallization process and the drop in viscosity.

6.3.2 Glass-Ceramic Coatings as Thermal Barriers

TBC are required to protect Ni-based superalloy components in advanced gas turbine and diesel engines which operate at higher temperatures and with reduced air cooling relative to conventional engine components. More recently, reported coatings

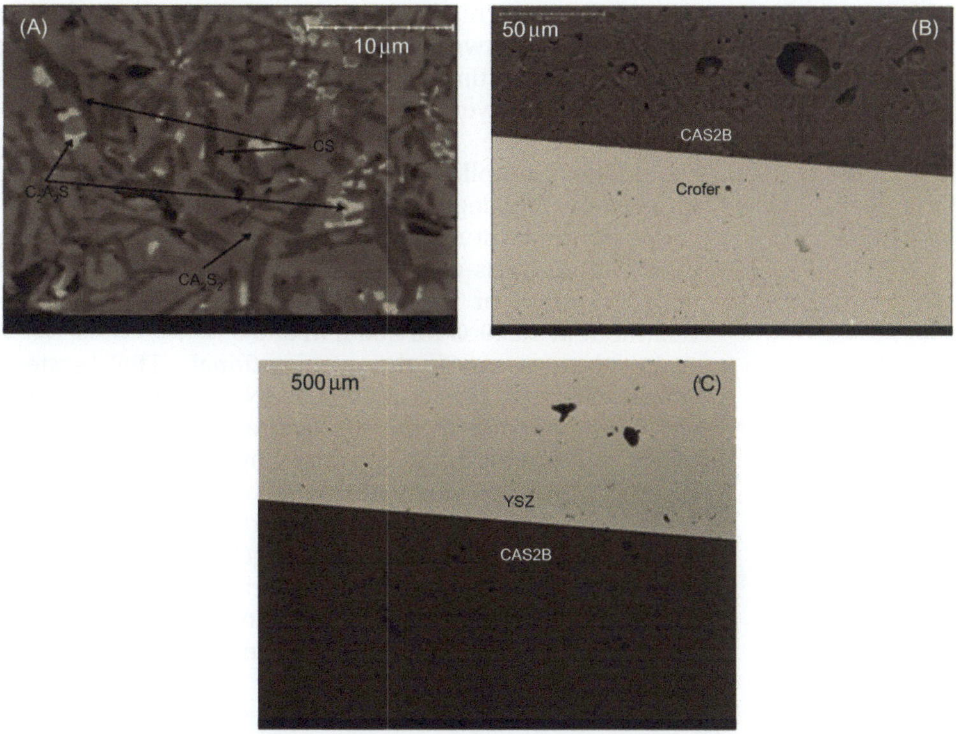

Figure 6.5 Cross section of heat-treated CAS_2B sample (A) bulk glass ($C_2A_2S = Ca_2Al_2SiO_7$, SiO_2; $CA_2S_2 = CaAl_2Si_2O8$; $CS = CaSiO_3$); (B) metal/CAS_2B interface; and (C) ceramic/CAS_2B interface. CAS_2B indicates the glass sample. (After Khedim et al., 2012. Reproduced with permission of Elsevier Science B.V.)

for these applications include those based on magnesium and zinc aluminosilicate glass ceramics, hybrid composite coatings comprised of alkaline earth zinc molybdenum silicate glass ceramic/YSZ, and basalt-based glass-ceramic coatings prepared from waste materials, including fly ash (Donald et al., 2011).

A TBC system for the protection of nickel-base superalloys is generally a three-layered structure of one ceramic top coat, an intermediate NiCoCrAlY/PtAl-based metallic bond coat and a metallic substrate. The metallic bond coat is deposited between the metallic substrate and the ceramic top coat to protect the underlying metal from oxidation and high-temperature corrosion and to enhance the adherence between the dissimilar substrate and top coat (Das et al., 2009).

Das et al. (2009) applied a glass-ceramic coat, as a bond coat, between the YSZ top coat and the nickel-base superalloy substrate.

During service, oxidation of the bond coat at the high temperatures produces a thermally grown oxide (TGO) scale at the bond coat–top coat interface which increases with increasing operation time. This leads to the generation of high stresses in the interface, between the bond coat and the TGO that may result in crack initiation and propagation leading to the spalling of the ceramic layer and its catastrophic failure. Therefore, the formation of the TGO layer and its progressive thickening should be controlled to minimize the bond coat oxidation-induced coating degradation.

Experimental studies were carried out to correlate the performance of the glass-ceramic TBC with the conventional TBC system, in terms of oxidation resistance under identical heat treatment at 1,200°C for 500 h in air.

In the case of the conventional TBC system, YSZ ($8Y_2O_3–92ZrO_2$) and NiCoCrAlY (Ni–24Co–15Cr–8Al–1Y, in wt%) were applied as the top coat and the bond coat, respectively, on the nickel-base superalloy (nimonic–AE 435) substrate by the air plasma spraying technique.

The slurry of bond coat glass ceramic was prepared by ball milling of a $SiO_2–BaO–CaO–MgO–ZnO–MoO_3$ glass frit, along with various mill additions and was applied on the nimonic alloy substrate by the conventional enameling technique. Thereafter, 8 wt% YSZ was air plasma sprayed onto the glass-ceramic coated substrate.

Both the TBC systems were characterized by SEM as well as Energy Dispersive Spectroscopy (EDS) analysis. No TGO layer was found between the bond coat and the top coat in the case of a glass-ceramic-bonded TBC system while the conventional TBC system exhibited a TGO layer of ~16 μm thickness at the bond coat–top coat interface region.

Datta and Das (2005) described the method of preparation and processing of two novel glass-ceramic coating materials, and their process of application, characterization, and evaluation of coatings.

The coating material compositions, based on SiO_2–Al_2O_3–MgO–ZnO–TiO_2–B_2O_3–R_2O system, were prepared by the conventional frit manufacturing process. The frit was further processed to make a thick paint-like slurry to enable its uniform and thin application on the clean metal part, by conventional spraying or dipping processes.

Thermal shock resistance, adherence at 90-bend test, static oxidation resistance at the required working temperature (1,000°C) for continuous service, and abrasion resistance were evaluated using the standard methods.

The evaluation of properties clearly showed strong chemical bonding at the metal–ceramic interface and a smooth glossy impervious defect-free surface finish.

Therefore, the suitability of these coatings for protection of different hot-zone components in different types of engines was proved.

Moskalewicz et al. (2009) studied the production and properties of protective SiO_2–Al_2O_3–CaO–Na_2O glass-ceramic-based coatings on TIMETAL 834 by a slurry technique.

The authors indicated that TIMETAL 834, a near-α titanium-base alloy, offers increased tensile strength and creep resistance up to 600°C, together with improved fatigue strength, when applied as the material for fabrication of the last-stages compressor parts in modern gas turbines. Temperatures in this part of the engine may reach 800°C. One of the major limitations of titanium alloys (including TIMETAL 834), restricting their use at high temperatures, is a poor oxidation resistance at temperatures above 600°C. To improve it, a surface treatment is frequently applied.

According to the authors, glass-ceramic-based coatings are generally very promising materials for protection of titanium alloys against oxidation at high temperature.

It was found that the studied coating, which contained $CaSiO_3$, $NaAlSiO_4$, Na_2CaSiO_4 crystalline phases, and an amorphous phase, essentially improved the selected properties of the TIMETAL 834 alloy such as its hardness and high-temperature oxidation resistance.

6.3.3 Glass-Ceramic Sealants for Solid-State Batteries

The Na/S battery is an energy storage technology that uses sodium and sulfur as the high power and energy density, temperature stability, low cost, and suitability for high-volume mass production (Wen et al., 2008).

However, the Na/S battery operates at a temperature of ~300°C to keep the electrode materials in a molten state. For safety and performance reasons, a satisfactory sealant between the β-alumina and α-alumina is essential. The requirements of a suitable sealant include characteristics such as excellent thermal expansion coefficient match, chemical stability, thermal shock resistance, gas tightness, and bonding behavior (Song et al., 2010).

Song et al. (2010) developed new glass-ceramic sealants for the Na/S battery. The borosilicate glass was chosen as the basic glass and TiO_2 as the nucleating agent. The sintering behavior was also evaluated and the sintering temperature of 750°C and holding time of 15 min were chosen as the suitable formation conditions for the sealants.

The thermal expansion coefficient of the glass-ceramic sealants matched well with that of the α-alumina and β-alumina. The thermal shock resistance and chemical stability of the sealants were also satisfactory.

No microcracking was observed in the sealants after 100 times thermal cycling.

The thickness of the diffusion layer was only ~10 μm after heat-treating at 350°C for 850 h, indicating a favorable chemical match of the sealants with alumina.

However, the densification of the sealants remained to be further improved. The factors influencing the densification of the sealants were clarified, and the possible means of solving the issue were proposed.

6.4 Concluding Remarks

The impressive range of properties and potential applications of some nano-glass ceramics, explained above, can be summarized as follows:

1. Some nanoporous glass ceramics such as $RTi_2(PO_4)_3$ have been developed with excellent characteristics for applications such as supports for immobilization of enzymes, humidity sensors, and as ion exchange media. Furthermore,

quite attractive bacteriostatic activities were found for some nanoporous glass ceramics such as partially Ag-exchanged $LiTi_2(PO_4)_3$ and $Li_{1.4}Ti_{1.6}Al_{0.4}(PO_4)_3$ glass ceramics and some of them also showed high catalytic activities.

2. Nano-crystalline glass ceramics in the SiO_2–Al_2O_3–MgO–ZnO–TiO_2 system containing the major phases of spinel and enstatite have also been developed for potential use as magnetic disk substrates in computer hard drives. These glass ceramics with high elastic modulus and moderate strength and toughness values offer many advantages for the disk substrate application, including improved flatness and smoothness, better wear resistance, and the ability to manufacture thinner than glass or aluminum disks. These glass ceramics also have upper use temperatures, much higher than those of glass or aluminum (which were previously in use), hence, they can withstand the high temperature of sputtering processes utilized in the initiating of magnetic films on their surface.
3. Glass ceramic-to-metal seals and coatings for new applications, e.g., SOFCs, TBC, seals for battery applications, have attracted considerable interest and have recently been the subject of numerous research activities. The advantages of glass ceramics for the aforementioned applications include more refractory behavior and superior mechanical properties, good bonding characteristics to metals, and the possibility of using a wide range of chemical compositions to control some physicochemical properties such as viscosity, coefficient of thermal expansion, and glass transition temperature.

However, the microstructural considerations have been given relatively little attention in the majority of the aforementioned reports. Though some of the prepared glass ceramics seems to be nano-crystalline, in the majority of reported cases, the microstructural details either have been ignored or the prepared glass ceramics were comprised of micron- or submicron-size crystalline phases. Considering the superior properties of the nano-structured glass ceramics, as discussed in various chapters of the present book, it can be proposed that the preparation of nano-glass ceramics for the above-mentioned applications may improve their performance, especially in the mechanical and chemical fields. Therefore, the research activities on nano-glass-ceramic coatings and sealants, with the aim of controlling their crystallization behavior to obtain nano-structured glass ceramics and studying the effect of structure on the properties of the resulting materials, is recommended as a very interesting and novel research field.

It should also be noted that the nano-glass ceramics discussed above are a few prominent examples selected from the numerous varieties of nano-glass ceramics that have been studied and developed during the past decades. Many other application fields related to nano-glass ceramics are presently being explored and a great number of other exciting, novel, and unforeseen properties and applications could possibly be discovered for these materials in the future. Hence it can be speculated that the field of nano-glass ceramics would be a more active and expanding field of research and development activities in the coming years.

REFERENCES

Abdel-Hameed, S.A.M., Margha, F.H., 2012. Preparation, crystallization behavior and magnetic properties of nanoparticles magnetic glass–ceramics in the systems Fe_2O_3–CoO–MnO_2; Fe_2O_3–NiO–MoO_3 and Fe_2O_3–CoO–V_2O_5. J. Non-Cryst. Solids 358 (4), 832–838.

Abdel-Hameed, S.A.M., Marzouk, M.A., Abdel-Ghany, A. E., 2011. Magnetic properties of nanoparticles glass–ceramic rich with copper ions. J. Non-Cryst. Solids 357 (24), 3888–3896.

Abouellell, M.M., Leonberger, F.J., 1989. Waveguides in lithium niobate. J. Am. Ceram. Soc. 72 (8), 1311–1321.

Aguiar, H., Solla, E.L., Serra, J., González, P., Leóna, B., Almeidab, N., et al., 2008. Orthophosphate nanostructures in SiO_2–P_2O_5–CaO–Na_2O–MgO bioactive glasses. J. Non-Cryst. Solids 354 (34), 4075–4080.

Al-Assiri, M.S., El-Desoky, M.M., 2011. Correlation between nanostructural and electrical properties of barium titanate-based glass-ceramic nano-composites. J. Alloys Compd. 509 (36), 8937–8943.

Al-Assiri, M.S., El-Desoky, M.M., 2012. Synthesis, structural and ferroelectric properties of barium titanate based glass-ceramic nano-composites. J. Non-Cryst. Solids. 358 (12–13), 1605–1610.

Al-Assiri, M.S., El-Desoky, M.M., 2013. Grain-size effects on the structural, electrical properties and ferroelectric behaviour of barium titanate based glass-ceramic nano-composite. J. Mater. Sci. Mater. Electron. 24 (2), 784–792.

Al-Assiri, M.S., El-Desoky, M.M., Al-Hajry, A., Al-Shahrani, A., Al-Mogeeth, A.M., Bahgat, A.A., 2009. Study of nanostructural behavior and transport properties of $BaTiO_3$ doped vanadate glasses and glass–ceramics dispersed with ferroelectric nanocrystals. Phys. B. 404 (8–11), 1437–1445.

Alekseeva, I., Dymshits, O., Ermakov, V., Zhilin, A., Petrov, V., Tsenter, M., 2008. Raman spectroscopy quantifying the composition of stuffed β-quartz derivative phases in lithium aluminosilicate glass ceramics. J. Non-Cryst. Solids 354 (45–46), 4932–4939.

Alekseeva, I.P., Dymshits, O.S., Golubkov, V.V., Loiko, P. A., Tsenter, M.Y., Yumashev, K.V., et al., 2014. Influence of NiO on phase transformations and optical properties of ZnO–Al_2O_3–SiO_2 glass-ceramics nucleated by TiO_2 and ZrO_2. Part I. Influence of NiO on phase transformations of ZnO–Al_2O_3–SiO_2 glass-ceramics nucleated by TiO_2 and ZrO_2. J. Non-Cryst. Solids 384, 73–82.

Alizadeh, P., Marghussian, V.K., 2000a. The effect of compositional changes on the crystallization behavior and mechanical properties of diopside–wollastonite glass ceramics in the SiO_2–CaO–MgO (Na_2O) system. J. Eur. Ceram. Soc. 20 (6), 765–773.

Alizadeh, P., Marghussian, V.K., 2000b. Effect of nucleating agents on the crystallization behavior and microstructure of SiO_2–CaO–MgO (Na_2O) glass ceramics. J. Eur. Ceram. Soc. 20 (6), 775–782.

Amirahmadi, Z., Marghussian, V.K., Beitollahi, A., Mirkazemi, S.M., Sarpoolaky, H., 2011. Magnetite nanoparticles prepared by the crystallization of Na_2O–Fe_2O_3–B_2O_3–SiO_2 glasses. J. Non-Cryst. Solids 357 (16–17), 3195–3199.

Anantharaman, M.R., Jagatheesan, S., Malini, K.A., Sindhu, S., Narayanasamy, A., Chinnasamy, C.N., et al., 1998. On the magnetic properties of ultra-fine zinc ferrites. J. Magn. Magn. Mater. 189 (1), 83–88.

Andrews, L.J., Beall, G.H., Lempicki, A., 1986. Luminescence of Cr^{+3} in mullite transparent glass ceramics. J. Lumin. 36 (2), 65–74.

Arcos, D., del Real, R.P., Vallet-Regí, M., 2002. A novel bioactive and magnetic biphasic material. Biomaterials. 23 (10), 2151–2158.

Aronne, A., Sigaev, V.N., Pernice, P., Fanelli, E., Usmanova, L.Z., 2004. Non-isothermal crystallization and nanostructuring in potassium niobium silicate glasses. J. Non-Cryst. Solids. 337 (2), 121–129.

Augis, J.A., Bennett, J.E., 1978. Calculation of the Avrami parameters for heterogeneous solid-state reactions using a modification of the Kissinger method. J. Therm. Anal. 13 (2), 283–292.

Auzel, F., 1990. Up-conversion processes in coupled ion systems. J. Lumin. 45 (1–6), 341–345.

Avrami, M., 1939. Kinetics of phase change: I. General theory. J. Chem. Phys. 7 (12), 1103–1112.

Avrami, M., 1940. Kinetics of phase change: II. Transformation–time relations for random distribution of nuclei. J. Chem. Phys. 8 (2), 212–224.

Avrami, M., 1941. Kinetics of phase change: III. Granulation, phase change and microstructure. J. Chem. Phys. 9 (2), 177–184.

Bansal, N.P., Doremus, R.H., Bruce, A.J., Moynihan, C.T., 1983. Kinetics of crystallization of ZrF.-BaF.-LaF. glass by differential scanning calorimetry. J. Am. Ceram. Soc. 66 (4), 233–238.

Barták, J., Málek, J., 2012. Crystal growth kinetics of Sb_2S_3 in Ge–Sb–S amorphous thin films. J. Therm. Anal. Calorim. 110 (1), 275–280.

Batlle, X., Labarta, A., 2002. Finite-size effects in fine particles: magnetic and transport properties. J. Phys. D Appl. Phys. 35 (6), R15–R42.

Bauchspieβ, K.R., Schneider, H., Kulikov, A., 1996. EXAFS studies of Cr doped mullite. J. Eur. Ceram. Soc. 16 (2), 203–209.

Beall, G.H., 1993. Glass ceramics: recent development and application. In: Weinberg, M.C. (Ed.), Nucleation and Crystallization in Glasses and Liquids. Ceramic Transactions Series, vol. 30. American Ceramic Society, Westerville, OH, pp. 241–266.

Beall, G.H., Duke, D.A., 1969. Transparent glass ceramics. J. Mater. Sci. 4 (4), 340–352.

Beall, G.H., Pinckney, L.R., 1999. Nanophase glass ceramics. J. Am. Ceram. Soc. 82 (1), 5–16.

Beall, G.H., Karstetter, B.R., Rittler, H.L., 1967. Crystallization and chemical strengthening of stuffed β-quartz glass ceramics. J. Am. Ceram. Soc. 50 (4), 181–190.

Berbecaru, C., Alexandru, H.V., Stan, G.E., Marcov, D.A., Pasuk, I., Ianculescu, A., 2010. First stages of bioactivity of glass ceramics thin films prepared by magnetron sputtering technique. Mater. Sci. Eng. B. 169 (1–3), 101–105.

Bergeron, C.G., Russell, C.K., 1965. Nucleation and growth of lead titanate from a glass. J. Am. Ceram. Soc. 48 (3), 115–118.

Berry, C.C., Curtis, A.S.G., 2003. Functionalisation of magnetic nanoparticles for applications in biomedicin. J. Phys. D Appl. Phys. 36 (13), R198–R206.

Best, S., Bonfiel, W., 1994. J. Mater. Sci. Mater. Med. 5, 516.

Bhargava, A., Shelby, J.E., Snyder, R.L., 1988. Crystallization of glasses in the system $BaO–TiO_2–B_2O_3$. J. Non-Cryst. Solids. 102 (1–3), 136–142.

Bhattacharyya, S., Bocker, C., Heil, T., Jinschek, R.J., Höche, T., Rüssel, C., et al., 2009. Experimental evidence of self-limited growth of nano-crystals in glass. Nano Lett. 9 (6), 2493–2496.

Bocker, C., Rüssel, C., 2009. Self-organized nanocrystallization of BaF_2 from $Na_2O/K_2O/BaF_2/Al_2O_3/SiO_2$ glasses. J. Eur. Ceram. Soc. 29 (7), 1221–1225.

Borrelli, N.F., Herczog, A., Maurer, R.D., 1965. Electro-optic effect of ferroelectric microcrystals in a glass matrix. Appl. Phys. Lett. 7 (5), 117–118.

Bretcanu, O., Spriano, S., Brovarone Vitale, C., Verné, E., 2006. Synthesis and characterization of coprecipitation-derived ferrimagnetic glass-ceramic. J. Mater. Sci. 41 (4), 1029–1037.

Buerger, M.J., 1954. The stuffed derivatives of the silica structures. Am. Miner. 39 (7–8), 600–614.

Burgner, L.L., Weinberg, M.C., 2001a. Crystal growth mechanisms in inorganic glasses. Phys. Chem. Glasses 42 (3), 184–190.

Burgner, L.L., Weinberg, M.C., 2001b. An assessment of crystal growth behavior in lithium disilicate glass. J. Non-Cryst. Solids 279 (1), 28–43.

Cahn, J.W., 1961. On spinodal decomposition. Acta Met. 9, 795–801.

Cahn, J.W., Charles, R.J., 1965. The initial stages of phase separation in glasses. Phys. Chem. Glasses 6 (5), 181–191.

Chaliha, R.S., Tiwari, V.S., Gupta, P.K., Annapurna, K., Tarafder, A., Karmakar, B., 2011. Structure and dielectric properties of potassium niobate nano glass–ceramics. J. Mater. Sci. Mater. Electron. 22 (7), 728–734.

Chen, D., Wang, Y., Yu, Y., Ma, E., Bao, F., Hu, Z., et al., 2006. Influence of Er^{3+} content on structure and up-conversion emission of oxyfluoride glass ceramics containing CaF_2 nano-crystals. Mater. Chem. Phys. 95 (2–3), 264–269.

Chen, D.Q., Wang, Y.S., Yu, Y.L., Huang, P., Weng, F.Y., 2008. Near-infrared quantum cutting in transparent nanostructured glass ceramics. Opt. Lett. 33 (16), 1884–1886.

Chen, D.Q., Yu, Y.L., Wang, Y.S., Huang, P., Weng, F.Y., 2009. Cooperative energy transfer up-conversion and quantum cutting down-conversion in Yb^{3+}:TbF_3 nano-crystals embedded glass ceramics. J. Phys. Chem. C. 113 (16), 6406–6410.

Chen, L., Yu, C., Hu, L., Chen, W., 2012. Preparation and spectroscopic properties of nanostructured glass ceramics containing Yb^{3+}, Er^{3+} ions and Co^{2+}-doped spinel nano-crystals. Solid State Sci. 14 (2), 287–290.

Chen, L., Yu, C., Hu, L., Chen, W., 2013. Effect of La_2O_3 on the physical and crystallization properties of Co^{+2}-doped $MgO–Al_2O_3–SiO_2$ glass. J. Non-Cryst. Solids. 360 (1), 4–8.

Christian, J.W., 1975. The Theory of Transformation in Metals and Alloys. second ed. Pergamon Press, London.

Clupper, D.C., Hench, L.L., 2003. Crystallization kinetics of tape cast bioactive glass 45S5. J. Non-Cryst. Solids 318 (1–2), 43–48.

Daiko, Y., Yajima, H., Kasuga, T., 2008. Preparation of porous titanium phosphate glass ceramics for NH_3 gas adsorption with self-cleaning ability. J. Eur. Ceram. Soc. 28 (1), 267–270.

Das, S., Datta, S., Basu, D., Das, G.C., 2009. Glass–ceramics as oxidation resistant bond coat in thermal barrier coating system. Ceram. Int. 35 (4), 1403–1406.

Datta, S., Das, S., 2005. A new high temperature resistant glass–ceramic coating for gas turbine engine components. Bull. Mater. Sci. 28 (7), 689–696.

Dejneka, M.J., 1998. The luminescence and structure of novel transparent oxyfluoride glass ceramics. J. Non-Cryst. Solids 239 (1–3), 149–155.

Deng, D., Ma, H., Xu, S., Wang, Q., Huang, L., Zhao, S., et al., 2011. Broadband infrared luminescence of Ni^{2+}-doped silicate glass–ceramics containing lithium aluminate spinel nano-crystals. J. Non-Cryst. Solids 357 (5), 1426–1429.

Denry, I., Holloway, J.A., Gupta, P.K., 2012. Effect of crystallization heat treatment on the microstructure of niobium-doped fluorapatite glass ceramics. J. Biomed. Mater. Res. Part B Appl. Biomater. 100B (5), 1198–1205.

Deubener, J., 2000. Compositional onset of homogeneous nucleation in (Li, Na) disilicate glasses. J. Non-Cryst. Solids 274 (1–3), 195–201.

Devekey, R.C., Majumdar, A.J., 1975. The role of TiO_2 in the formation of cordierite glass ceramics. Phys. Chem. Glasses 16 (2), 36–43.

Doherty, P.E., Lee, D.W., Davis, R.S., 1967. Direct observation of the crystallization of $Li_2O–Al_2O_3–SiO_2$ glasses containing TiO_2. J. Am. Ceram. Soc. 50 (2), 77–81.

Donald, I.W., 2004. Crystallization kinetics of a lithium zinc silicate glass studied by DTA and DSC. J. Non-Cryst. Solids 345–346, 120–126.

Donald, I.W., Mallinson, P.M., Metcalfe, B.L., Gerrard, L. A., Fernie, J.A., 2011. Recent developments in the preparation, characterization and applications of glass- and glass–ceramic-to-metal seals and coatings. J. Mater. Sci. 46 (7), 1975–2000.

Du, J., Jones, B., Lanagan, M., 2005. Preparation and characterization of dielectric glass-ceramics in $Na_2O–PbO–Nb_2O_5–SiO_2$ system. Mater. Lett. 59 (22), 2821–2826.

Duan, X.L., Yuan, D.R., Cheng, X.F., Sun, Z.H., Sun, H.Q., Xu, D., et al., 2003. Spectroscopic properties of Co^{+2}: $ZnAl_2O_4$ nano-crystals in sol–gel derived glass–ceramics. J. Phys. Chem. Solids 64 (6), 1021–1025.

Duan, X.L., Yuan, D.R., Cheng, X.F., Sun, H.Q., Sun, Z.H., Wang, X.Q., et al., 2005. Microstructure and properties of Co^{2+}: $ZnAl_2O_4/SiO_2$ nanocomposite glasses prepared by sol–gel method. J. Am. Ceram. Soc. 88 (2), 399–403.

Duke, D.A., Chase, G.A., 1968. Glass ceramics for high precision reflective-optic applications. Appl. Opt. 7 (5), 813–817.

Dutkiewicz, J., Stoch, L., Morgiel, J., Kostorz, G., Stoch, P., 2003. Analytical and HREM study of the early stages of $SiO_2–Al_2O_3$–(Mg, Zn)O glass crystallization. Mater. Chem. Phys. 81 (2–3), 411–413.

Duval, D.J., Risbud, S.H., Shackelford, J.F., 2008. Mullite. In: Shackelford, J.F., Doremus, R.H. (Eds.), Ceramic and Glass Materials: Structure, Properties and Processing. Springer, New York, NY, p. 27. (Chapter 2).

Ebisawa, Y., Miyaii, F., Kokubo, T., Ohura, K., Nakamura, T., 1997. Bioactivity of ferrimagnetic glass ceramics in the system $FeO–Fe_2O_3–CaO–SiO_2$. Biomaterials. 18 (19), 1277–1284.

El-Desoky, M.M., Zayed, H.S.S., Ibrahim, F.A., Ragab, H. S., 2009. Electrical conductivity improvement of strontium titanate doped vanadate glasses by nano-crystallization. Phys. B. 404 (21), 4125–4131.

Ferreira, E.A., Cassanjes, F.C., Poirier, G., 2013. Ctystallization behavior of barium titanate tellurite glass doped with Eu^{3+} and Er^{3+}. Opt. Mater. 35 (6), 1141–1145.

Fischer, R.X., Schneider, H., 2000. Crystal structure of Cr-mullite. Am. Miner. 85 (9), 1175–1179.

Fokin, V.M., Zanotto, E.D., Schmelzer, J.W.P., 2003. Homogeneous nucleation versus glass transition temperature of silicate glasses. J. Non-Cryst. Solids 321 (1–2), 52–65.

Fokin, V.M., Zanotto, E.D., Yuritsyn, N.S., Schmelzer, W. P.J., 2006. Homogeneous crystal nucleation in silicate glasses: a 40 years perspective. J. Non-Cryst. Solids 352 (26–27), 2681–2714.

Fokin, V.M., Cabral, A.A., Reis, R.M.C.V., Nascimento, M.L.F., Zanotto, E.D., 2010. Critical assessment of DTA-DSC methods for the study of nucleation kinetics in glasses. J. Non-Cryst. Solids 356 (6–8), 358–367.

Fu, J., Parker, J.M., Flower, P.S., Brown, R.M., 2002. Eu^{2+} ions and CaF_2-containing transparent glass ceramics. Mater. Res. Bull. 37 (11), 1843–1849.

Fu, Y.P., 2006. Microwave-induced combustion synthesis of $Li_{0.5}Fe_{2.5-x}Cr_xO_4$ powder and their characterization. Mater. Res. Bull. 41 (4), 809–816.

Fu, Y.P., Hsu, C.S., 2005. $Li_{0.5}Fe_{2.5-x}Mn_xO_4$ ferrite sintered from microwave-induced combustion. Solid State Commun. 134 (3), 201–206.

Gautam, C.R., Kumar, D., Parkash, O., 2013. Controlled crystallization of (Pb, Sr) TiO_3 borosilicate glass ceramics doped with Nb_2O_5. Glass Phys. Chem. 39 (2), 162–173.

Ghosh, S., Kundu, P., Sharma, A.D., Basu, R.N., Maiti, H. S., 2008. Microstructure and property evaluation of barium aluminosilicate glass–ceramic sealant for anode-supported solid oxide fuel cell. J. Eur. Ceram. Soc. 28 (1), 69–76.

Golezardi, S., Marghussian, V.K., Beitollahi, A., Mirkazemi, S.M., 2010. Crystallization behavior, microstructure and dielectric properties of lead titanate glass ceramics in the presence of Bi_2O_3 as a nucleating agent. J. Eur. Ceram. Soc. 30 (6), 1453–1460.

Görnert, P., Sinn, E., Schüppel, W., Pfeiffer, H., Rösler, M., Rchubert, T., et al., 1990. Structural and magnetic properties of $BaFe_{12-2x}Co_xTi_xO_{19}$ powders prepared by the glass crystallization method. IEEE Trans. Magn. 26 (1), 12–14.

Görnert, P., Pfeiffer, H., Sinn, E., Muller, R., Schüppel, W., Rösler, M., et al., 1994. Nanocrystalline M-type hexaferrite powders: preparation, geometric and magnetic properties. IEEE Trans. Magn. 30 (2), 714–716 (Part 1–2).

Gouveia-Neto, A.S., da Costa, E.B., Bueno, L.A., Ribeiro, S.J.L., 2004. Intense red up-conversion emission in infrared transparent glass ceramic. J. Lumin. 110 (1–2), 79–84.

Graça, M.P.F., da Silva, M.G.F., Valente, M.A., 2008. Structural and electrical characteristics of $LiNbO_3$ embedded in a 34% SiO_2 glass matrix. J. Eur. Ceram. Soc. 28 (6), 1197–1203.

Gránásy, L., 1995. "Anomalous" nucleation prefactors revisited: a diffuse interface analysis of crystal nucleation in oxide glasses. Scr. Metall. Mater. 32 (10), pp. 1611–1617.

Gránásy, L., 1996. Diffuse interface model of volume nucleation in glasses. Thermochim. Acta 280/281 (July), 83–100.

Gránásy, L., James, P.F., 1999. Non-classical theory of crystal nucleation: application to oxide glasses: review. J. Non-Cryst. Solids 253 (1–3), 210–230.

Gránásy, L., Pusztai, T., Hartmann, E., 1996. Diffuse interface model of nucleation. J. Cryst. Growth 167 (3–4), 756–765.

Grossman, D.G., Isard, J.O., 1969a. Crystal clamping in $PbTiO_3$ glass ceramics. J. Mater. Sci. 4 (12), 1059–1063.

Grossman, D.G., Isard, J.O., 1969b. Lead titanate glass ceramics. J. Am. Ceram. Soc. 52 (4), 231–236.

Habelitz, S., Carl, G., Rüssel, C., Marchetti, K., Roeder, E., Eifler, D., et al., 1997. Oriented mica glass-ceramic by extrusion and subsequent heat treatment. Glastechnische Berichte—Glass Sci. Technol. 70, 86–92.

Habelitz, S., Höche, T., Hergt, R., Carl, G., Rüssel, C., 1999. Microstuctural design through epitaxial growth in extruded mica glass ceramics. Acta Mater. 47 (9), 2831–2840.

Hamedani, G.N., Yekta, B.E., Marghussian, V.K., 2012. Crystallization and optical properties of a transparent mullite glass ceramic. Opt. Mater. 34 (4), 596–599.

Harizanova, R., Bocker, C., Adeev, G., Rüssel, C., Gugov, I., 2014. Crystal growth and dielectric properties of $BaTiO_3$ obtained in aluminoborosilicate glasses. J. Non-Cryst. Solids. 401, 191–196.

Hase, H., Nasu, H., Mito, A., Hashimoto, T., Matsuoka, J., Kamiya, K., 1996. Second harmonic generation from surface crystallized $Li_2O-Ta_2O_5-SiO_2$ glass. Jpn. J. Appl. Phys. 35 (10), 5355–5356.

Hayashi, H., Tanabe, S., Hanada, T., 2001. 1.4 μm band emission properties of Tm^{3+} ions in transparent glass ceramics containing PbF_2 nano-crystals for S-band amplifier. J. Appl. Phys. 89 (2), 1041–1045.

Headley, T.J., Loehman, R.E., 1984. Crystallization of a glass-ceramic by epitaxial growth. J. Am. Ceram. Soc. 67 (9), 620–625.

Hench, L.L., 1998. Bioactive materials: the potential for tissue regeneration. J. Biomed. Mater. Res. 41 (4), 511–518.

Hench, L.L., Splinter, R.J., Allen, W.C., Greenlee, T.K., 1971. Bonding mechanism at interface of ceramic prosthetic materials. J. Biomed. Mater. Res. 5 (6), 117–141.

Herczog, A., 1964. Microcrystalline $BaTiO_3$ by crystallization from glass. J. Am. Ceram. Soc. 47 (3), 107–115.

Herczog, A., 1984. Barrier layers in semiconducting barium titanate glass ceramics. J. Am. Ceram. Soc. 67 (7), 484–490.

Hill, R., Calver, A., Stamboulis, A., Bubb, N., 2007. Real-time nucleation and crystallization studies of a fluorapatite glass–ceramic using small-angle neutron scattering and neutron diffraction. J. Am. Ceram. Soc. 90 (3), 763–768.

Hill, R.G., O'Donnell, M.D., Law, R.V., Karpukhina, N., Cochrane, B., Tulyaganov, D.W., 2010. The early stages of nucleation and crystallization of an apatite glass-ceramic: evidence for nano-scale crystallization. J. Non-Cryst. Solids 356 (52–54), 2935–2941.

Hirao, K., Tanaka, K., Makita, M., Soga, N., 1995. Preparation and optical properties of transparent glass ceramics containing β-PbF_2:Tm^{+3}. J. Appl. Phys. 78 (5), 3445–3450.

Höland, W., Beall, G.H., 2002. Glass-Ceramic Technology. American Ceramic Society, Westerville, OH.

Höland, W., Rheinberger, V., Frank, M., 1999. Mechanisms of nucleation and controlled crystallization of needle-like apatite in glass ceramics of the $SiO_2-Al_2O_3-K_2O-CaO-P_2O_5$ system. J. Non-Cryst. Solids 253 (1–3), 170–177.

Höland, W., Rheinberger, V., Schweiger, M., 2003. Control of nucleation in glass ceramics. Philos. Trans. R. Soc. Lond. A. 361 (1804), 575–589.

Höland, W., Apel, E., Van't Hoen, C., Rheinberger, V., 2006. Studies of crystal phase formations in high-strength lithium disilicate glass–ceramics. J. Non-Cryst. Solids 352 (38–39), 4041–4050.

Hopper, R.W., 1985. Stochastic theory of scattering from idealized spinodal structures: II. Scattering in general and for the basic late stage model. J. Non-Cryst. Solids 70 (1), 111–142.

Hosono, H., Abe, Y., 1995. Porous glass ceramics composed of a titanium phosphate crystal skeleton: a review. J. Non-Cryst. Solids 190 (3), 185–197.

Hsi, C.S., Cheng, H.Z., Hsu, H.J., Chen, Y.S., Wang, M.C., 2007. Crystallization kinetics and magnetic properties of iron oxide contained $25Li_2O–8MnO_2–20CaO–2P_2O_5–45SiO_2$ glasses. J. Eur. Ceram. Soc. 27 (10), 3171–3176.

Hsu, J.-Y., Speyer, R.F., 1991. Crystallization of $Li_2O \cdot Al_2O_3 \cdot 6SiO_2$ glasses containing niobium pentoxide as nucleating dopant. J. Am. Ceram. Soc. 74 (2), 395–399.

Ikeda, K., Schneider, H., Akasaka, M., Rager, H., 1992. Crystal-field spectroscopic study of Cr-doped mullite. Am. Miner. 77 (3–4), 251–257.

Imanieh, M.H., Martín, I.R., Yekta, B.E., Marghussian, V.K., Shakhesi, S., 2012. Improved cooperative emission in ytterbium-doped oxyfluoride glass ceramics containing CaF_2 nano-crystals. J. Am. Ceram. Soc. 95 (12), 3827–3833.

Imanieh, M.H., Yekta, B.E., Marghussian, V., Shakhesi, S., Martín, I.R., 2013. Crystallization of nano calcium fluoride in $CaF_2–Al_2O_3–SiO_2$ system. Solid State Sci. 17, 76–82.

Ito, S., Kokubo, T., Tashiro, M., 1978. Transparency of $LiTaO_3–SiO_2–Al_2O_3$ glass-ceramics in relation to their microstructure. J. Mater. Sci. 13 (5), 930–938.

Jackson, K.A., 1958a. Nature of solid liquid interfaces. In: Doremus, R.H., Roberts, B.W., Turnbull, D. (Eds.), Growth and Perfection of Crystals. Wiley, New York, NY.

Jackson, K.A., 1958b. Mechanism of growth. Liquid Metals and Solidification. American Society for Metals, Cleveland, OH, pp. 174–186.

Jackson, K.A., 1967. Current concepts in crystal growth from the melt. In: Reiss, H. (Ed.), Progress in Solid State Chemistry, vol. 4. Pergamon Press, New York, NY, pp. 53–80.

James, P.F., 1974. Kinetics of crystal nucleation in lithium silicate glasses. J. Phys. Chem. Glasses 15 (4), 95–105.

James, P.F., 1975. Liquid phase separation in glass forming systems. J. Mater. Sci. 10 (10), 1802–1825.

James, P.F., 1982. Nucleation in glass-forming systems—a review. In: Simmons, J.H., Uhlmann, D.R., Beall, G.H. (Eds.), Nucleation and Crystallization in Glasses, Advances in Ceramics, vol. 4. American Ceramic Society, Columbus, OH, pp. 1–48.

James, P.F., 1985. Kinetics of crystal nucleation in silicate glasses. J. Non-Cryst. Solids. 73 (1–3), 517–540.

James, P.F., 1989. Volume nucleation in silicate glasses. In: Lewis, M.H. (Ed.), Glasses and Glass-Ceramic. Chapman & Hall, London, pp. 59–105.

Jeong, E.D., Borse, P.H., Lee, J.S., Ha, M.G., Pak, H.K., Komatsu, T., et al., 2006. Second harmonic generation and fabrication of transparent $K_2O–Na_2O–Nb_2O_5–TeO_2$ glass ceramics. J. Ind. Eng. Chem. 12 (5), 790–794.

Jeyadevan, B., Tohji, K., Nakatsuka, K., 1994. Structure analysis of coprecipitated $ZnFe_2O_4$ by extended x-ray-absorption fine structure. J. Appl. Phys. 76 (10), 6325–6327.

Johnson, W.A., Mehl, R.F., 1939. Reaction kinetics in processes of nucleation and growth. Trans. Am. Inst. Min. Metall. Pet. Eng. 135, 416–442.

Jordan, A., Scholz, R., Wust, P., Fahling, H., Felix, R., 1999. Magnetic fluid hyperthermia (MFH): cancer treatment with AC magnetic field induced excitation of biocompatible superparamagnetic nanoparticles. J. Magn. Magn. Mater. 201 (1–3), 413–419.

Kalinina, A.M., Fokin, V.M., Filipovich, V.N., 1976. Method of determining the parameters characterizing crystal nucleation in glass. Fiz. Khim. Stekla. 2 (4), 298–305.

Kalinina, A.M., Fokin, V.M., Filipovich, V.N., 1977. The crystal nucleation induction period in $Li_2O \cdot 2SiO_2$ glass and its temperature dependence. Fiz. Khim. Stekla. 3 (2), 122–129.

Kalita, S.J., Bhardwaj, A., Bhatt, H.A., 2007. Nanocrystalline calcium phosphate ceramics in biomedical engineering. Mater. Sci. Eng. C. 27 (3), 441–444.

Kasuga, T., Nogami, M., Abe, Y., 1999. Titanium phosphate glass ceramics with silver ion exchangeability. J. Am. Ceram. Soc. 82 (3), 765–767.

Kazin, P.E., Trusov, L.A., Zaitsev, D.D., Tret'yakov, Y.D., 2009. Glass crystallization synthesis of ultrafine hexagonal M-type ferrites: particle morphology and magnetic characteristics. Russ. J. Inorg. Chem. 54 (14), 2081–2090.

Kazin, P.E., Trusov, L.A., Kushnir, S.E., Yaroshinskaya, N. V., Petrov, N.A., Jansen, M., 2010. Hexaferrite submicron and nanoparticles with variable size and shape via glass-ceramic route. J. Phys. Conf. Ser. 200 (072048), 1–4.

Keding, R., Rüssel, C., 1996. Oriented crystallization of fresnoite in an electric field. Ber. Bunsen Ges. Phys. Chem. 100 (9), 1515–1518.

Kerker, M., 1969. The Scattering of Light. Academic Press, New York, NY.

Khedim, H., Nonnet, H., Méar, F.O., 2012. Development and characterization of glass-ceramic sealants in the ($CaO-Al_2O_3-SiO_2-B_2O_3$) system for solid oxide electrolyzer cells. J. Power Sources. 216, 227–236.

Kim, H.G., Komatsu, T., Sato, R., Matusita, K., 1993. Crystallization of $LiNbO_3$ in tellurite glasses. J. Non-Cryst. Solids 162 (1–2), 201–204.

Kim, H.G., Komatsu, T., Sato, R., Matusita, K., 1996. Incorporation of $LiNbO_3$ crystals into tellurite glasses. J. Mater. Sci. 31 (8), 2159–2164.

Kim, H.W., Koh, Y.H., Li, L.H., Lee, S., Kim, H.E., 2004. Hydroxyapatite coating on titanium substrate with titania buffer layer processed by sol–gel method. Biomaterials 25 (13), 2533–2538.

Kim, J.E., Song, C.H., Park, H.W., Ohshimad, K., Yang, Y. S., 2007. Kinetics of crystallization in $4BaTiO_3-SiO_2$ glass. Mater. Sci. Eng. A. 449–451, 299–301.

Kishi, Y., Tanabe, S., Tochino, S., Pezzotti, G., 2005. Fabrication and efficient infrared-to-visible up-conversion in transparent glass ceramics of Er–Yb co-doped CaF_2 nano-crystals. J. Am. Ceram. Soc. 88 (12), 3423–3426.

Kissinger, H.E., 1956. Variation of peak temperature with heating rate in differential thermal analysis. J. Res. Natl. Bur. Stand. 57 (4), 217–221.

Klein, L., Uhlmann, D.R., 1974. Crystallization behavior of anorthite. J. Geophys. Res. 79 (32), 4869–4874.

Klupsch, T., Steinbeiss, E., Müller, R., Ulbrich, C., Schüppel, W., Steinmetz, H., et al., 1999. Magnetic glass ceramics—preparation and properties. J. Magn. Magn. Mater. 196–197, 264–265.

Knutson, R., Liu, H., Yen, W.M., Morgan, T.V., 1989. Spectroscopy of disordered low-field sites in Cr^{3+}: mullite glass ceramic. Phys. Rev. B. 40 (7), 4264–4270.

Kokubo, T., Tashiro, M., 1974. Dielectric properties of fine-grained $PbTiO_3$ crystals precipitated in a glass. J. Non-Cryst. Solids 13 (2), 328–340.

Kolmogorov, A.N., 1937. IZV. Akad. Nauk SSR. Ser. Fiz. 3, 355.

Komatsu, T., Tawarayama, H., Mohri, H., Matusita, K., 1991. Properties and crystallization behaviors of $TeO_2-LiNbO_3$ glasses. J. Non-Cryst. Solids 135 (2–3), 105–113.

Kord, M., Marghussian, V.K., Yekta, B.E., Bahrami, A., 2009a. Phase separation, crystallization and leaching of microporous glass ceramics in the $CaO-TiO_2-P_2O_5$ system. J. Non-Cryst. Solids 355 (2), 141–147.

Kord, M., Marghussian, V.K., Yekta, B.E., Bahrami, A., 2009b. Effect of ZrO_2 addition on crystallization behavior, porosity and chemical–mechanical properties of a $CaO-TiO_2-P_2O_5$ microporous glass ceramic. Mater. Res. Bull. 44 (8), 1670–1675.

Kržmanc, M.M., Došler, U., Suvorov, D., 2011. The nucleation and crystallization of $MgO-B_2O_3-SiO_2$ glass. J. Eur. Ceram. Soc. 31 (13), 2211–2219.

Kukkonen, L.L., Reaney, I.M., Furniss, D., Pellatt, M.G., Seddon, A.B., 2001. Nucleation and crystallization of transparent erbium III-doped oxyfluoride glass ceramics. J. Non-Cryst. Solids. 290 (1), 25–31.

Layton, M.M., Herzcog, A., 1967. Nucleation and crystallization of $NaNbO_3$ from glasses in the $Na_2O-Nb_2O_5-SiO_2$ system. J. Am. Ceram. Soc. 59 (7), 369–375.

Lee, C.-K., Speyer, R.F., 1994. Glass formation and crystallization of barium ferrite in the $Na_2O-BaO-Fe_2O_3-SiO_2$ system. J. Mater. Sci. 29 (5), 1348–1351.

Lee, Y.K., Choi, S.Y., 1996. Crystallization and properties of $Fe_2O_3-CaO-SiO_2$ glasses. J. Am. Ceram. Soc. 79 (4), 992–996.

Leonelli, C., Manfredini, T., Paganelli, M., Pozzi, P., 1991. Crystallization of some anorthite diopside glass precursors. J. Mater. Sci. 26 (18), 5041–5046.

Li, C., Song, Z., Li, Y., Lou, K., Qiu, J., Yang, Z., et al., 2013. Enhanced NIR down-conversion luminescence by precipitating nano $Ca_5(PO_4)_3F$ crystals in $Eu^{2+}-Yb^{3+}$ co-doped glass. Spectrochim. Acta A Mol. Biomol. Spectrosc. 114, 575–578.

Li, Y., Liang, K., Xu, B., Cao, J.-W., 2010. Crystallization mechanism and microstructure evolution of $Li_2O-Al_2O_3-SiO_2$ glass ceramics with Ta_2O_5 as nucleating agent. J. Therm. Anal. Calorim. 101 (3), 941–948.

Liu, C., Heo, J., 2006. Up-conversion and photon avalanche in oxy-fluoride nanostructured glasses doped with Ho^{3+}. J. Non-Cryst. Solids 352 (50–51), 5325–5330.

Liu, F., Ma, E., Chen, D., Yu, Y., Wang, Y., 2006. Tunable red-green up-conversion luminescence in novel transparent glass ceramics containing Er: $NaYF_4$ nanocrystals. J. Phys. Chem. B. 110 (42), 20843–20846.

Liu, J., Zhang, M., Yang, X., Liu, S., 2012. Preparation and characterization of $SrO-Al_2O_3-Fe_2O_3-B_2O_3$ system nano-ferrimagnetic glass-ceramics. Adv. Mater. Res. 415–417, 384–389.

Loiko, P.A., Dymshits, O.S., Zhilin, A.A., Alekseeva, I.P., Yumashev, K.V., 2013. Influence of NiO on phase transformations and optical properties of $ZnO-Al_2O_3-SiO_2$ glass–ceramics nucleated by TiO_2 and ZrO_2. Part II. Optical absorption and luminescence. J. Non-Cryst. Solids 376, 99–105.

Lopez-Esteban, S., Gutierrez-Gonzalez, C.F., Gremillard, L., Saiz, E., Tomsia, A.P., 2009. Interfaces in graded coatings on titanium-based implants. J. Biomed. Mater. Res. Part A. 88A (4), 1010–1021.

Lynch, S.M., Shelby, J.E., 1984. Crystal clamping in lead titanate glass ceramics. J. Am. Ceram. Soc. 67 (6), 424–427.

Mady, H.A., 2011. Crystallization, optical and electrical properties of lead bariumtitanate based sodium borate glass. J. Appl. Sci. Res. 7 (11), 1536–1543.

Magallanes-Perdomo, M., Luklinska, Z.B., De Aza, A.H., Carrodeguas, R.G., De Aza, S., Pena, P., 2011. Bone-like forming ability of apatite–wollastonite glass ceramic. J. Eur. Ceram. Soc. 31 (9), 1549–1561.

Maier, V., Müller, G., 1987. Mechanism of oxide nucleation in lithium aluminosilicate glass-ceramic. J. Am. Ceram. Soc. 70 (8), C176–C178.

Malakho, A., Fargin, E., Lahaye, M., Lazoryak, B., Morozov, V., Van Tendeloo, G., et al., 2006. Enhancement of second harmonic generation signal in thermally poled glass ceramic with $NaNbO_3$ nanocrystals. J. Appl. Phys. 100 (063103), 1–5.

Malyarevich, A.M., Denisov, I.A., Yumashev, K.V., Dymshits, O.S., Zhilin, A.A., 2002. Optical absorption and luminescence study of cobalt-doped magnesium aluminosilicate glass ceramics. J. Opt. Soc. Am. B. 19 (8), 1815–1821.

Marghussian, V.K., Balazadegan, U., Yekta, B.E., 2009. The effect of BaO and Al_2O_3 addition on the crystallization behavior of cordierite glass ceramics in the presence of V_2O_5 nucleant. J. Eur. Ceram. Soc. 29 (1), 39–46.

Marghussian, V.K., Beitollahi, A., Haghi, M., 2003. The effect of SiO_2 and Cr_2O_3 additions on the crystallization behavior and magnetic properties of a B_2O_3–BaO–Fe_2O_3 glass. Ceram. Int. 29 (4), 455–462.

Marotta, A., Buri, A., Branda, F., 1981. Nucleation in glass and differential thermal analysis. J. Mater. Sci. 16 (2), 341–344.

Marotta, A., Saiello, S., Branda, F., Buri, A., 1982. Activation energy for the crystallization of glass from DDTA curves. J. Mater. Sci. 17 (1), 105–108.

Matusita, K., Komatsu, T., Yokota, R., 1984. Kinetics of non-isothermal crystallization process and activation energy for crystal growth in amorphous materials. J. Mater. Sci. 19 (1), 291–296.

McMillan, P.W., 1979. Glass Ceramics. second ed. Academic Press, London.

Meinhardt, K.D., Kim, D.S., Chou, Y.S., Weil, K.S., 2008. Synthesis and properties of a barium aluminosilicate solid oxide fuel cell glass–ceramic sealant. J. Power Sources. 182 (1), 188–196.

Milhans, J., Li, D.S., Khaleel, M., Sun, X., Al-Haik, M.S., Harris, A., et al., 2011. Mechanical properties of solid oxide fuel cell glass-ceramic seal at high temperatures. J. Power Sources. 196 (13), 5599–5603.

Mirkazemi, M., Marghussian, V.K., Beitollahi, A., Dou, S. X., Wexler, D., Konstantinov, K., 2007. Effect of ZrO_2 nucleant on crystallization behavior, microstructure and magnetic properties of BaO–Fe_2O_3–B_2O_3–SiO_2 glass ceramics. Ceram. Int. 33 (3), 463–469.

Mita, Y., 1972. Luminescence processes in Yb^{3+} sensitized rare earth phosphors. J. Appl. Phys. 43 (4), 1772–1778.

Mita, Y., Hirama, K., Ando, N., Yamamoto, H., Shionoya, S., 1993. Luminescence processes in Tm^{3+}- and Er^{3+}-ion-activated, Yb^{3+}-ion-sensitized infrared up-conversion devices. J. Appl. Phys. 74 (7), 4703–4709.

Miyakawa, T., Dexter, D.L., 1970. Phonon side bands, multiphonon relaxation of excited states, and phonon-assisted energy transfer between ions in solids. Phys. Rev. B. 1 (7), 2961–2969.

Molla, A.R., Tarafder, A., Karmakar, B., 2011. Synthesis and properties of glasses in the K_2O–SiO_2–Bi_2O_3–TiO_2 system and bismuth titanate ($Bi_4Ti_3O_{12}$) nano glass–ceramics thereof. J. Mater. Sci. 46 (9), 2967–2976.

Molla, A.R., Tarafder, A., Mukherjee, S., Mohanty, S.K., Karmakar, B., 2012. Transparent Eu^{3+}-doped ferroelectric bismuth titanate glass-ceramic nanocomposites: fabrication and properties. J. Am. Ceram. Soc. 95 (10), 3056–3063.

Molla, A.R., Tarafder, A., Mukherjee, S., Mohanty, S.K., Karmakar, B., 2013. Processing and properties of Eu^{3+}-doped barium bismuth titanate ($BaBi_4Ti_4O_{15}$) glass–ceramic nanocomposites. J. Am. Ceram. Soc. 96 (8), 2387–2395.

Moskalewicz, T., Smeacetto, F., Czyrska-Filemonowicz, A., 2009. Microstructure, properties and oxidation behavior of the glass–ceramic based coating on near-α titanium alloy. Surf. Coat. Technol. 203 (16), 2249–2253.

Moulson, A.J., Herbert, J.M., 2003. Electroceramics: Materials, Properties, Applications. second ed. Wiley, Chichester.

Müller, R., Hiergeist, R., Steinmetz, H., Ayoub, N., Fujisaki, M., Schüppel, W., 1999a. Barium hexaferrite ferrofluids—preparation and physical properties. J. Magn. Magn. Mater. 201 (1–3), 34–37.

Müller, R., Hiergeist, R., Gawalek, W., Hoell, A., Wiedenmann, A., 2002. Magnetic and structural investigations on barium hexaferrite ferrofluids. J. Magn. Magn. Mater. 252, 43–45.

Müller, R., Ulbrich, C., Schüppel, W., Steinmetz, H., Steinbeiβ, E., 1999b. Preparation and properties of barium–ferrite containing glass ceramics. J. Eur. Ceram. Soc. 19 (6–7), 1547–1550.

Nass, R., Tkalcec, E., Ivankovic, H., 1995. Single phase mullite gels doped with chromium. J. Am. Ceram. Soc. 78 (11), 3097–3106.

Nasu, H., Ibar, Y., Kubodera, K.J., 1989. Optical third-harmonic generation from some high-index glasses. J. Non-Cryst. Solids 110 (2–3), 229–234.

Nasu, H., Matsushita, O., Kamiya, K., Kobayashi, H., Kubodera, K.J., 1990. Third harmonic generation from $Li_2O–TiO_2–TeO_2$ glasses. J. Non-Cryst. Solids 124 (2–3), 275–277.

O'Flynn, P.K., Stanton, T.K., 2010. Nucleation and early stage crystallization of fluorapatite in apatite–mullite glass ceramics. Cryst. Growth Des. 10 (3), 1111–1117.

Oh, S.H., Choi, S.Y., Lee, Y.K., Kim, K.N., 2001. Research on annihilation of cancer cells by glass ceramics for cancer treatment with external magnetic field. I. Preparation and cytotoxicity. J. Biomed. Mater. Res. 54 (3), 360–365.

Ozawa, T., 1971. Kinetics of non-isothermal crystallization. Polymer. 12 (3), 150–158.

Pablos-Martín, A., Hémono, N., Mather, G.C., Bhattacharyya, S., Höche, T., Bornhöft, H., et al., 2011. Crystallization kinetics of LaF_3 nano-crystals in an oxyfluoride glass. J. Am. Ceram. Soc. 94 (8), 2420–2428.

Pal, M., Brahma, P., Chakravorty, D., Bhattacharyya, D., Maiti, H.S., 1996. Nanocrystalline nickel-zinc ferrite prepared by the glass-ceramic route. J. Magn. Magn. Mater. 16, 256–260.

Pal, M., Brahma, P., Chakravorty, D., Bhattacharyya, D., Maiti, H.S., 1997. Preparation of nanocrystalline barium hexaferrite in a glass medium. Nanostruct. Mater. 8 (6), 731–738.

Pankhurst, Q.A., Connolly, J., Jones, S.K., Dobson, J., 2003. Applications of magnetic nanoparticles in biomedicines. J. Phys. D Appl. Phys. 36 (13), R167–R181.

Pardoe, H., Chua-anusorn, W., St Pierre, T.G., Dobson, J., 2001. Structural and magnetic properties of nanoscale iron oxide particles synthesized in the presence of dextran orpolyvinyl alcohol. J. Magn. Magn. Mater. 225 (1–2), 41–46.

Parkash, O., Kumar, D., Pandey, L., 1986. Glass ceramics containing ferroelectric phases. Bull. Mater. Sci. 8 (5), 557–565.

Pascuta, P., Vladescu, A., Borodi, G., Culea, E., Tetean, R., 2011. Structural and magnetic properties of zinc ferrite incorporated in amorphous matrix. Ceram. Int. 37 (8), 3343–3349.

Pascuta, P., Vladescu, A., Borodi, G., Culea, E., Tetean, R., 2012. Synthesis, structural and magnetic characterization of iron–zinc–borate glass ceramics containing nanocrystalline zinc ferrite. J. Mater. Sci. Mater. Electron. 23 (2), 582–588.

Pernice, P., Aronne, A., Sigaev, V.N., Sarkisov, P.D., Molev, V.I., Stefanovich, S.Y., 1999. Crystallization behavior of potassium niobium silicate glasses. J. Am. Ceram. Soc. 82 (12), 3447–3452.

Pinckney, L.R., Beall, G.H., 1997. Nanocrystalline non-alkali glass ceramics. J. Non-Cryst. Solids 219, 219–227.

Piriou, B., Rager, H., Schneider, H., 1996. Time-resolved fluorescence spectroscopy of Cr^{3+} in mullite. J. Eur. Ceram. Soc. 16 (2), 195–201.

Por, P.T., Golshayan, A.K., Müller, R., Rekveldt, M.T., Görnert, P., 1996. Magnetic correlations in the remanent state of nano- and micro-crystalline barium hexaferrite particles studied by neutron depolarisation. J. Magn. Magn. Mater. 161, 367–375.

Prapitpongwanich, P., Harizanova, R., Pengpat, K., Rüssel, C., 2009. Nanocrystallization of ferroelectric lithium niobate in $LiNbO_3–SiO_2$ glasses. Mater. Lett. 63 (12), 1027–1029.

Prasad, N.S., Varma, K.B.R., 2005. Evolution of ferroelectric $LiNbO_3$ phase in a reactive glass matrix ($LiBO_2–Nb_2O_5$). J. Non-Cryst. Solids 351 (16–17), 1455–1465.

Puig, J., Ansart, F., Lenorman, P., Antoine, L., Dailly, J., 2011. Sol–gel synthesis and characterization of barium (magnesium) aluminosilicate glass sealants for solid oxide fuel cells. J. Non-Cryst. Solids 357 (19–20), 3490–3494.

Qi, X., Zhou, J., Yue, Z., Gui, Z., Li, L., 2003. Permeability and microstructure of manganese modified lithium ferrite prepared by sol-gel auto-combustion method. Mater. Sci. Eng. B. 99 (1–3), 278–281.

Radonjić, L., Todorović, M., Miladinović, J., 2004. Nanostructured sodium niobate obtained by glass crystallization. Mater. Chem. Phys. 88 (2–3), 427–432.

Radonjić, L., Todorović, M., Miladinović, J., 2005. Nanostructured sodium lithium niobate and lithium niobium tantalite solid solutions obtained by controlled crystallization of glass. Mater. Sci. Eng. B. 121 (1–2), 64–70.

Rager, H., Schneider, H., Graetsch, H., 1990. Chromium incorporation in mullite. Am. Miner. 75 (3–4), 392–397.

Ranasingheh, K.S., Ray, C.S., Day, D.E., 2002. A generalized method for determining the crystal nucleation and growth rates in glasses by differential thermal analysis. J. Mater. Sci. 37 (3), 547–555.

Ray, C.S., Day, D.E., 1990. Determining the nucleation rate curve for lithium disilicate glass by differential thermal analysis. J. Am. Ceram. Soc. 73 (2), 439–442.

Ray, C.S., Fang, X., Day, D.E., 2000. New method for determining the nucleation and crystal-growth rates in glasses. J. Am. Ceram. Soc. 83 (4), 865–872.

Reinsch, S., Nascimento, M.L.F., Müller, R., Zanotto, E. D., 2008. Crystal growth kinetics in cordierite and diopside glasses in wide temperature ranges. J. Non-Cryst. Solids. 354 (52–54), 5386–5394.

Reis, S.T., Brow, R.K., 2006. Designing sealing glasses for solid oxide fuel cells. J. Mater. Eng. Perf. 15 (4), 410–413.

Reisfeld, R., Kisilev, A., Buch, A., Ish-Shalom, M., 1987. Transparent glass-ceramics doped by chromium (III): Spectroscopic properties and characterization of crystalline phases. J. Non-Cryst. Solids. 91, 333–350.

Rezlescu, L., Rezlescu, E., Popa, P.D., Rezlescu, N., 1999. Fine barium hexaferrite powder prepared by the crystallization of glass. J. Magn. Magn. Mater. 193 (1–3), 288–290.

Rezlescu, N., Rezlescu, E., Ciobotaru, I., Craus, M.L., Popa, P.D., 1998. Glass-ceramic containing lithium ferrite: magnetic and structural properties. Ceram. Int. 24 (1), 31–34.

Rezlescu, N., Rezlescu, L., Craus, M.L., Rezlescu, E., 1999. Magnetic glass-ceramic containing fine ferrite crystals. J. Magn. Magn. Mater. 196–197, 463–464.

Rezvani, M., Marghussian, V.K., Yekta, B.E., 2011. Crystal nucleation and growth rates, time–temperature transformation diagram, and mechanical properties of a SiO_2–Al_2O_3–CaO–MgO–(R_2O) glass in the presence of Cr_2O_3, Fe_2O_3 and TiO_2 nucleants. Int. J. Appl. Ceram. Technol. 8 (1), 152–162.

Rezvani, M., Yekta, B.E., Solati-Hashjin, M., Marghussian, V.K., 2005a. Effect of Cr_2O_3, Fe_2O_3 and TiO_2 nucleants on the crystallization behavior of SiO_2–Al_2O_3–CaO–MgO–(R_2O) glass ceramics. Ceram. Int. 31 (1), 75–80.

Rezvani, M., Yekta, B.E., Marghussian, V.K., 2005b. Utilization of DTA in determination of crystallization mechanism in SiO_2–Al_2O_3–CaO–MgO–(R_2O) glasses in presence of various nuclei. J. Eur. Ceram. Soc. 25 (9), 1525–1530.

Richards, B.S., 2006. Enhancing the performance of silicon solar cells via the application of passive luminescence conversion layers. Sol. Energy Mater. Sol. Cells 90 (15), 2329–2337.

Riello, P., Canton, P., Comelato, N., Polizzi, S., Verita, M., Fagherazzi, G., et al., 2001. Nucleants on the crystallization behavior of glass-ceramic materials in the Li_2O–Al_2O_3–SiO_2 system of interest for their transparency properties. J. Non-Cryst. Solids 288 (1–3), 127–139.

Rietveld, H.M., 1969. Profile refinement method for nuclear and magnetic structures. J. Appl. Crystallogr. 2 (2), 65–71.

Rodríguez, V.D., Tikhomirov, V.K., Méndez-Ramos, J., Yanes, A.C., Moshchalkov, V.V., 2010. Towards broad range and highly efficient down-conversion of solar spectrum by Er^{+3}–Yb^{3+} co-doped nanostructured glass ceramics. Sol. Energy Mater. Sol. Cells 94 (10), 1612–1617.

Rose, M.R., Lawrence, A.S., Wulff, J., 1966. The Structure and Properties of Materials, Vol. IV, Electronic Properties. Wiley, New York, NY.

Rosensweig, R.E., 2002. Heating magnetic fluid with alternating magnetic field. J. Magn. Magn. Mater. 252, 370–374.

Ruiz-Valdés, J.J., Gorokhovsky, A.V., Escalante-García, J.I., Mendoza-Suárez, G., 2004. Glass–ceramic materials with regulated dielectric properties based on the system BaO–PbO–TiO_2–B_2O_3–Al_2O_3. J. Eur. Ceram. Soc. 24 (6), 1505–1508.

Rüssel, C., 1997. Oriented crystallization of glass. A review. J. Non-Cryst. Solids. 219, 212–218.

Rüssel, C., 2005. Nanocrystallization of CaF_2 from $Na_2O/K_2O/CaO/CaF_2/Al_2O_3/SiO_2$ glasses. Chem. Mater. 17 (23), 5843–5847.

Russell, C.K., Bergeron, C.G., 1965. Structural changes preceding growth of a crystalline phase in a lead silicate glass. J. Am. Ceram. Soc. 48 (3), 162–163.

Sadhukhan, M., Modak, D.K., Chaudhuri, B.K., 1999. Study of microstructural behavior and nonadiabatic small polaron hopping conduction in $BaTiO_3$ doped lead-vanadate glass and glass ceramics dispersed with ferroelectric nano-crystals. J. Appl. Phys. 85 (7), 3477–3487.

Sahu, A.K., Kumar, D., Parkash, O., 2006a. Lead-strontium titanate glass ceramics: I—crystallization and microstructure. J. Mater. Sci. 41 (7), 2075–2085.

Sahu, A.K., Kumar, D., Parkash, O., Thakur, O.P., Prakash, C., 2006b. Lead-strontium titanate glass ceramics: II—dielectric behavior. J. Mater. Sci. 41 (7), 2087–2096.

Samson, B.N., Pinckney, L.R., Wang, J., Beall, G.H., Borrelli, N.F., 2002. Nickel-doped nanocrystalline glass-ceramic fiber. Opt. Lett. 27 (15), 1309–1311.

Sandu, V., Nicolescu, M.S., Kuncser, V., Popa, S., Pasuk, I., Ghica, C., et al., 2012. Structure and magnetic properties of nanosized magnetite obtained by glass recrystallization. J. Nanosci. Nanotechnol. 12 (6), 5043–5050.

Sangregorio, C., Wiemann, J.K., O'Connor, C.J., Rosenzweig, Z., 1999. A new method for the synthesis of magnetoliposomes. J. Appl. Phys. 85 (8), 5699–5701.

Santana-Alonso, A., Yanes, A.C., Méndez-Ramos, J., del-Castillo, J., Rodriguez, V.D.V., 2010. Sol–gel transparent nano-glass–ceramics containing Eu^{3+}-doped $NaYF_4$ nanocrystals. J. Non-Cryst. Solids. 356 (18–19), 933–936.

Sarma, K.R., Shlichta, P.J., Wilcox, W.R., Lefever, R.A., 1997. Epitaxy versus oriented heterogeneous nucleation of organic crystals on ionic substrates. J. Cryst. Growth. 174 (1–4), 487–494.

Schiffner, U., Pannhorst, W., 1987. Nucleation in a precursor glass for a $Li_2O–Al_2O_3–SiO_2$ glass ceramic—Part 2. Variation of the nucleating agent concentrations. Glastech. Ber. 60, 239–247.

Schneider, H., Okada, K., Pask, J.A., 1994. Mullite Mullite Ceramics. Wiley, Chichester, p. 201.

Scholes, S., 1970. The vitreous state, general discussion. Discuss. Faraday Soc. 50, 222–223.

Sebdani, M.M., Fathi, M.H., 2012. Preparation and characterization of hydroxyapatite–forsterite–bioactive glass nanocomposite coatings for biomedical applications. Ceram. Int. 38 (2), 1325–1330.

Secu, M., Secu, C.E., Polosan, S., Aldica, G., Ghica, C., 2009. Crystallization and spectroscopic properties of Eu-doped CaF_2 nano-crystals in transparent oxyfluoride glass ceramics. J. Non-Cryst. Solids 355 (37–42), 1869–1872.

Shah, S.A., Hashmi, M.U., Alam, S., Shamim, A., 2010. Magnetic and bioactivity evaluation of ferromagnetic $ZnFe_2O_4$ containing glass ceramics for the hyperthermia treatment of cancer. J. Magn. Magn. Mater. 322 (3), 375–381.

Shankar, J., Deshpande, V.K., 2010. Study of lead titanate based glass ceramics with addition of BaO. Integr. Ferroelectr. 119 (1), 110–121.

Shankar, J., Deshpande, V.K., 2011. Electrical and thermal properties of lead titanate glass ceramics. Physica B. 406 (3), 588–592.

Shankar, J., Deshpande, V.K., 2012. Study of $PbO–SrO–TiO_2–B_2O_3$ glass ceramics. Physica B. 407 (12), 2160–2163.

Shankar, M.V., Varma, K.B.R., 1999. Dielectric and optical properties of surface crystallized $TeO_2–LiNbO_3$ glasses. J. Non-Cryst. Solids. 243 (2–3), 192–203.

Sharma, K., Singh, S., Prajapat, C.L., Bhattacharya, S., Jagannath, S.M.R., Yusuf, S.M., et al., 2009. Preparation and study of magnetic properties of silico phosphate glass and glass ceramics having iron and zinc oxide. J. Magn. Magn. Mater. 321 (22), 3821–3828.

Shepilov, M.P., Baik, D.S., 1994. Computer simulation of crystallization kinetics for the model with simultaneous nucleation of randomly-oriented ellipsoidal crystals. J. Non-Cryst. Solids 171 (2), 141–156.

Shioya, K., Komatsu, T., Kim, H., Sato, R., Matusita, K., 1995. Optical properties of transparent glass ceramics in $K_2O–Nb_2O_5–TeO_2$ glasses. J. Non-Cryst. Solids 189 (1–2), 16–24.

Shyu, J.J., Yang, Y.S., 1995. Crystallization of a $PbO–BaO–TiO_2–Al_2O_3–SiO_2$ glass. J. Am. Ceram. Soc. 78 (6), 1463–1468.

Singh, R.K., Srinivasan, A., 2010. Apatite-forming ability and magnetic properties of glass ceramics containing zinc ferrite and calcium sodium phosphate phases. Mater. Sci. Eng. C. 30 (8), 1100–1106.

Singh, R.K., Srinivasan, A., 2011. Magnetic properties of bioactive glass ceramics containing nanocrystalline zinc ferrite. J. Magn. Magn. Mater. 323 (3–4), 330–333.

Singh, R.K., Kothiyal, G.P., Srinivasan, A., 2008. Magnetic and structural properties of $CaO–SiO_2–P_2O_5–Na_2O–Fe_2O_3$ glass ceramics. J. Magn. Magn. Mater. 320 (7), 1352–1356.

Song, S., Wen, Z., Liu, Y., Lin, J., Xu, X.X., Zhang, Q., 2010. New glass-ceramic sealants for Na/S battery. J. Solid State Electrochem. 14 (9), 1735–1740.

Stephens, E.V., Vetrano, J.S., Koeppel, B.J., Chou, Y., Sun, X., Khaleel, M.A., 2009. Experimental characterization of glass–ceramic seal properties and their constitutive implementation in solid oxide fuel cell stack models. J. Power Sources 193 (2), 625–631.

Stoch, L., 2003. Nanocrystalline glass ceramics formation. Opt. Appl. 33 (1), 115–123.

Stoch, L., 2005. Real glass crystallization high resolution electron microscopy (HREM) study and classic nucleation theory concept. Opt. Appl. 35 (4), 819–827.

Stoneham, A.M., Harding, J.H., 1998. Computer simulation of interfaces: what do we need to know? Acta Mater. 46 (7), 2255–2261.

Strnad, Z., 1986. Glass-Ceramic Materials. Elsevier, Amsterdam, the Netherlands.

Strümpel, C., McCann, M., Beaucarne, G., Arkhipov, V., Slaoui, A., Švrček, V., et al., 2007. Modifying the solar spectrum to enhance silicon solar cell efficiency. An overview of available materials. Sol. Energy Mater. Sol. Cells 91 (4), 238–249.

Stryjak, A.J., McMillan, P.W., 1978. Microstructure and properties of transparent glass ceramics. Part 1 The microstructure of spinel transparent glass ceramics. J. Mater. Sci. 13 (6), 1275–1281.

Suzuki, T., Horibuchi, K., Ohishi, Y., 2005. Structural and optical properties of $ZnO–Al_2O_3–SiO_2$ system glass–ceramics containing Ni^{2+}-doped nano-crystals. J. Non-Cryst. Solids 351 (27–29), 2304–2309.

Suzuki, T., Ohishi, Y., 2004. Broadband 1400 nm emission from Ni^{2+} in zinc-alumino-silicate glass. Appl. Phys. Lett. 84 (19), 3804–3806.

Švadlák, D., Pustková, P., Koštál, P., Málek, J., 2006. Crystal growth kinetics in $(GeS_2)_{0.2}$ $(Sb_2S_3)_{0.8}$ glass. Thermochim. Acta 446 (1–2), 121–127.

Swartz, S.L., Breval, E., Randall, C.A., Fox, B.H., 1988a. $SrTiO_3$ glass ceramics, Part I, Crystallization and microstructure. J. Mater. Sci. 23 (11), 3997–4003.

Swartz, S.L., Bhalla, A.S., Cross, L.E., Lawless, W.N., 1988b. $SrTiO_3$ glass ceramics, Part II, Dielectric properties. J. Mater. Sci. 23 (11), 4004–4012.

Szreder, N.A., Kosiorek, P., Kasiak, K., Karczewski, J., Gazda, M., Barczyński, R.J., 2014. Nanostructure and dielectric behavior of vanadate glasses containing $BaTiO_3$. J. Non-Cryst. <http://dx.doi.org/10.1016/j.jnoncrysol.2014.01.008>.

Tadic, M., Markovic, D., Spasojevic, V., Kusigerski, V., Remskar, M., Pirnat, J., et al., 2007. Synthesis and magnetic properties of concentrated α-Fe_2O_3 nanoparticles in a silica matrix. J. Alloys Compd. 441 (1–2), 291–296.

Tammann, G., 1925. The State of Aggregation. D. Van Nostrand and Co, New York, NY.

Tanabe, S., Hayashi, H., Hanada, T., Onodera, N., 2002. Fluorescence properties of Er^{3+} ions in glass ceramics containing LaF_3 nano-crystals. Opt. Mater. 19 (3), 343–349.

Tanaka, H., Yamamoto, M., Takahashi, Y., Benino, Y., Fujiwara, T., Komatsu, T., 2003. Crystalline phases and second harmonic intensities in potassium niobium silicate crystallized glasses. Opt. Mater. 22 (1), 71–79.

Tanaka, K., Kashima, K., Hirao, K., Soga, N., Mito, A., Nasu, H., 1995. Second harmonic generation in electrically poled Li_2O–Nb_2O_5–TeO_2 glasses. J. Non-Cryst. Solids 185 (1–2), 123–126.

Tanaka, K., Makita, M., Shimizugawa, Y., Hirao, K., Soga, N., 1998. Structure and high magnetization of rapidly quenched zinc ferrite. J. Phys. Chem. Solids 59 (9), 1611–1618.

Tarafder, A., Molla, A.R., Karmakar, B., 2010a. Processing and properties of Eu^{3+}-doped transparent YAG ($Y_3Al_5O_{12}$) nanoglass–ceramics. J. Am. Ceram. Soc. 93 (10), 3244–3251.

Tarafder, A., Annapurna, K., Chaliha, R.S., Tiwari, V.S., Gupta, P.K., Karmakar, B., 2010b. Structure, dielectric and optical properties of Nd^{3+}-doped $LiTaO_3$ transparent ferroelectric glass–ceramic nanocomposites. J. Alloys Compd. 489 (1), 281–288.

Tarafder, A., Molla, A.R., Dey, C., Karmakar, B., 2013a. Thermal, structural, and enhanced photoluminescence properties of Eu^{3+}-doped transparent willemite glass–ceramic nanocomposites. J. Am. Ceram. Soc. 96 (8), 2424–2431.

Tarafder, A., Molla, A.R., Karmakar, B., 2013b. Enhanced photoluminescence and structure of Dy^{3+}-doped $SrBi_2Ta_2O_9$-containing transparent glass-ceramics. Opt. Mater. 35 (8), 1549–1556.

Tartaj, P., Morales, M.P., Veintemillas-Verdaguer, S., Gonźalez-Carreno, T., Serna, C.J., 2003. The preparation of magnetic nanoparticles for applications in biomedicine. J. Phys. D Appl. Phys. 36 (13), R182–R197.

Tartaj, P., Morales, M.P., Gonźalez-Carreno, T., Veintemillas-Verdaguer, S., Serna, C.J., 2005. Advances in magnetic nanoparticles for biotechnology applications. J. Magn. Magn. Mater. 290–291 (Part 1), 28–34.

Taubert, J., Hergt, R., Müller, R., Ulbrich, C., Schüppel, W., Schmidt, H.G., et al., 1997. Phase separation in Ba-ferrite glass ceramics investigated by Faraday microscopy. J. Magn. Magn. Mater. 168 (1–2), 187–195.

Thakur, O.P., Kumar, D., Parkash, O., Pandey, L., 1995. Dielectric and microstructural behavior of strontium titanate borosilicate glass ceramic. Bull. Mater. Sci. 18 (5), 577–585.

Thakur, O.P., Kumar, D., Parkash, O., Pandey, L., 1996. Effect of 1 wt% CoO addition on dielectric and microstructural behavior of $(SrO \cdot TiO_2)$–$(2SiO_2 \cdot B_2O_3)$ glass and glass ceramic. Bull. Mater. Sci. 19 (2), 393–404.

Thakur, O.P., Kumar, D., Parkash, O., Pandey, L., 1997. Crystallization and microstructural behavior of strontium titanate borosilicate glass ceramics with Bi_2O_3 addition. Bull. Mater. Sci. 20 (1), 67–77.

Thakur, O.P., Kumar, D., Parkash, O., Pandey, L., 2002. Crystallization, microstructure development and dielectric behavior of glass ceramics in the system $[SrO \cdot TiO_2]$–$[2SiO_2 \cdot B_2O_3]$–La_2O_3. J. Mater. Sci. 37 (12), 2597–2606.

Thakur, O.P., Kumar, D., Parkash, O., Pandey, L., 2003. Electrical characterization of strontium titanate borosilicate glass ceramics system with bismuth oxide addition using impedance spectroscopy. Mater. Chem. Phys. 78 (3), 751–759.

Tick, P.A., Borrelli, N.F., Cornelius, L.K., Newhouse, M.A., 1995. Transparent glass ceramics for 1300 nm amplifier applications. J. Appl. Phys. 78 (11), 6367–6374.

Tikhomirov, V.K., Görller-Walrand, C., Driesen, K., 2008. Transparent nano-glass ceramics for efficient infrared emission. J. Alloys Compd. 451 (1–2), 542–544.

Tkalčec, E., Gržeta, B., Popović, J., Ivanković, H., Rakvin, B., 2006. Structural studies of Cr-doped mullite derived from single-phase precursors. J. Phys. Chem. Solids 67 (4), 828–835.

Todorovic, M., Radonjic, L., 1997. Lithium-niobate ferroelectric material obtained by glass crystallization. Ceram. Int. 23 (1), 55–60.

Tomozawa, M., 1972. Liquid phase separation and crystal nucleation in Li_2O-SiO_2 glasses. Phys. Chem. Glasses 13 (6), 161–166.

Uhlmann, D.R., 1970. The vitreous state, general discussion. Discuss. Faraday Soc. 50, 233–234.

Uhlmann, D.R., 1982. Crystal growth in glass-forming systems: a ten year perspective. In: Simmons, J.H., Uhlmann, D.R., Beall, G.H. (Eds.), Nucleation and Crystallization in Glasses, Advances in Ceramics, vol. 4. American Ceramic Society, Columbus, OH, pp. 80–124.

Uhlmann, D.R., Uhlmann, E.V., 1993. In: Weinberg, M.C. (Ed.), Ceramic Transactions, Nucleation and Crystallization in Liquids and Glasses, vol. 30. The American Ceramic Society, Westerville, OH, p. 109.

Varshneya, A.K., 1994. Fundamentals of Inorganic Glasses. Academic Press, San Diego, CA.

Velázquez, J.J., Rodríguez, V.D., Yanes, A.C., del Castillo, J., Méndez-Ramo, J., 2012. Photon down-shifting by energy transfer from Sm^{3+} to Eu^{3+} ions in sol-gel SiO_2-LaF_3 nano-glass ceramics for photovoltaics. Appl. Phys. B. 108 (3), 577–583.

Verma, S., Joy, P.A., 2005. Magnetic properties of superparamagnetic lithium ferrite nanoparticles. J. Appl. Phys. 98 (12), 124312 (1–9).

Wang, G., Lu, Z., Liu, X., Zhou, X., Ding, C., Zreiqat, H., 2011. Nanostructured glass–ceramic coatings for orthopaedic applications. J. R. Soc. Interface 8 (61), 1192–1203.

Wang, L., Mei, L., He, G., Li, J., Xu, L., 2011. Preparation of Ce:YAG glass-ceramics with low SiO_2. J. Am. Ceram. Soc. 94 (11), 3800–3803.

Wang, Y., Ohwaki, J., 1993. New transparent vitroceramics codoped with Er^{3+} and Yb^{3+} for efficient frequency up-conversion. Appl. Phys. Lett. 63 (24), 3268–3270.

Weinberg, M.C., 1991. Surface nucleated transformation kinetics in 2- and 3-dimensional finite systems. J. Non-Cryst. Solids 134 (1–2), 116–122.

Weinberg, M.C., 1992. Transformation kinetics of particles with surface and bulk nucleation. J. Non-Cryst. Solids 142, 126–132.

Weinberg, M.C., 1999. A few topics concerning nucleation and crystallization in glasses. J. Non-Cryst. Solids 255 (1), 1–14.

Weinberg, M.C., Birnie III, D.P., 1996. Avrami exponents for transformations producing anisotropic particles. J. Non-Cryst. Solids 202 (3), 290–296.

Wen, Z., Gu, Z., Xu, X., Cao, J., Zhang, F., Lin, Z., 2008. Research activities in Shanghai Institute of Ceramics, Chinese Academy of Sciences on the solid electrolytes for sodium sulfur batteries. J. Power Sources 184 (2), 641–645.

Widatallah, H.M., Johnson, C., Berry, F., Pekala, M., 2001. Solid State Commun. 120 (4), 171–175.

Woltz, S., Hiergeist, R., Görnert, P., Rüssel, C., 2006. Magnetite nanoparticles prepared by the glass crystallization method and their physical properties. J. Magn. Magn. Mater. 298 (1), 7–13.

Wondraczekw, L., Pradeau, P., 2008. Transparent hafnia-containing β-quartz glass ceramics: nucleation and crystallization behavior. J. Am. Ceram. Soc. 91 (6), 1945–1951.

Wu, B., Zhou, S., Ruan, J., Qiao, Y., Chen, D., Zhu, C., et al., 2008a. Energy transfer between Cr^{3+} and Ni^{2+} in transparent silicate glass ceramics containing Cr^{3+}/Ni^{2+} co-doped $ZnAl_2O_4$ nano-crystals. Opt. Express 16 (4), 2508–2513.

Wu, B., Zhou, S., Ren, J., Qiao, Y., Chen, D., Zhu, C., et al., 2008b. Enhanced luminescence from transparent Ni^{2+}-doped $MgO-Al_2O_3-SiO_2$ glass ceramics by Ga_2O_3 addition. J. Phys. Chem. Solids. 69 (4), 891–894.

Wu, B., Ruan, J., Qiu, J., Zeng, H., 2009a. Enhanced broadband near-infrared luminescence from Ni in Bi/Ni-doped transparent glass ceramics. J. Phys. D Appl. Phys. 42 (135110), 5.

Wu, C., Ramaswamy, Y., Liu, X., Wang, G., Zreiqat, H., 2009b. Plasma-sprayed $CaTiSiO_5$ ceramic coating on Ti–6Al–4V with excellent bonding strength, stability and cellular bioactivity. J. R. Soc. Interface. 6, 159–168.

Wu, Y.-H., Hsub, K.-C., Lee, C.-H., 2012. Effects of B_2O_3 and P_2O_5 doping on the microstructure evolution and mechanical strength in a lithium aluminosilicate glass–ceramic material with TiO_2 and ZrO_2. Ceram. Int. 38 (5), 4111–4121.

Xiao, Y., Song, L., Liu, X., Huang, Y., Huang, T., Wu, Y., et al., 2011. Nanostructured bioactive glass–ceramic coatings deposited by the liquid precursor plasma spraying process. Appl. Surf. Sci. 257 (6), 1898–1905.

Xu, J.L., Khor, K.A., Dong, Z.L., Gu, Y.W., Kumar, R., Cheang, P., 2004. Preparation and characterization of nano-sized hydroxyapatite powders produced in a radio frequency (rf) thermal plasma. Mater. Sci. Eng. A. 374 (1–2), 101–108.

Yamamoto, K., Abe, Y., 1998. Microporous glass ceramics in NASICON-type copper(II) titanium phosphate. J. Am. Ceram. Soc. 81 (8), 2201–2204.

Yamamoto, K., Abe, Y., 2000. Enhanced catalytic activity of microporous glass ceramics with a skeleton of NASICON-type copper (I) titanium phosphate crystal. Mater. Res. Bull. 35 (2), 211–216.

Yamamoto, K., Kasuga, T., Abe, Y., 1997. Preparation of porous glass ceramics with a skeleton of NASICON-type crystal $CuTi_2(PO_4)_3$. J. Am. Ceram. Soc. 80 (3), 822–824.

Yamane, M., Asahara, Y., 2002. Glasses for Photonics. Cambridge University Press, Cambridge.

Yamazawa, T., Honma, T., Suematsu, H., Komatsu, T., 2009. Synthesis, Ferroelectric and Electrooptic Properties of Transparent Crystallized Glasses with $Sr_xBa_{1-x}Nb_2O_6$ Nanocrystals. J. Am. Ceram. Soc. 92 (12), 2924–2930.

Yanes, A.C., Santana-Alonso, A., Méndez-Ramos, J., del-Castillo, J., 2013. Structure and intense UV up-conversion emissions in RE^{3+}-doped sol–gel glass-ceramics containing KYF_4 nanocrystals. Appl. Phys. B. 113 (4), 589–596.

Yao, K., Zhu, W., 2002. Barium titanate glass–ceramic thin films for integrated high-dielectric media. Thin Solid Films. 408 (1–2), 11–14.

Yao, K., Zhang, L., Yao, X., Zhu, W., 1996. Growth, structure, and morphology of lead titanate crystallites in the sol-gel derived glass-ceramics. Mater. Sci. Eng. B. 41 (3), 322–328.

Yekta, B.E., Marghussian, V.K., 2001. Effect of P_2O_5, B_2O_3 and PbO on the sinterability of β-quartz solid solution and gahnite glass ceramics. J. Mater. Sci. 36 (2), 477–483.

Zaitsev, D.D., Gravchikova, E.A., Kazin, P.E., Tretyakov, Y. D., Jansen, M., 2006a. Glass–ceramic composites with strontium hexaferrite ultrafine particles in the $SrO-Fe_2O_3-B_2O_3-SiO_2$ system. Int. J. Appl. Ceram. Technol. 3 (4), 279–283.

Zaitsev, D.D., Kushnir, E.S., Kazin, P.E., Tretyakov, Y.D., Jansen, M., 2006b. Preparation of the $SrFe_{12}O_{19}$-based magnetic composites via boron oxide glass devitrification. J. Magn. Magn. Mater. 301 (2), 489–494.

Zaitsev, D.D., Kazin, P.E., Tretyakov, Y.D., Jansen, M., 2005. Synthesis and magnetic properties of glass-ceramic composites $SrFe_{12}O_{19}-SrSiO_3$. J. Magn. Magn. Mater. 292, 59–64.

Zanotto, E.D., 1987. Isothermal and adiabatic nucleation in glass. J. Non-Cryst. Solids 89 (3), 361–370.

Zanotto, E.D., 1994. Crystallization of glass: a ten year perspective. "1993 Vittorio Gottardi Prize Lecture". Chim. Chron. New Ser. 23, 3–17.

Zanotto, E.D., Weinberg, M.C., 1989. Trends in homogeneous crystal nucleation in oxide glasses. Phys. Chem. Glasses 30 (5), 186–192.

Zysler, R.D., Fiorani, D., Testa, A.M., 2001. Investigation of magnetic properties of interacting Fe_2O_3 nanoparticles. J. Magn. Magn. Mater. 224 (1), 5–11.

Zysler, R.D., Mansilla, M.V., Fiorani, D., 2004. Surface effects in α-Fe_2O_3 nanoparticles. Eur. Phys. J. B. 41 (2), 171–175.

INDEX

Note: Page numbers followed by "*f*" and "*t*" refer to figures and tables, respectively.

E

F

G

FSC
www.fsc.org
MIX
Papier | Fördert gute Waldnutzung
FSC® C083411

Kolibri 360 GmbH
Ferdinand-Jühlke-Straße 7
99095 Erfurt, Deutschland
produktsicherheit@kolibri360.de